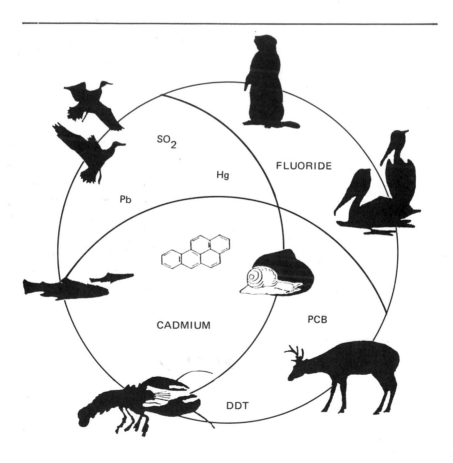

SO₂

FLUORIDE

Hg

Pb

CADMIUM

PCB

DDT

Animals as Monitors of Environmental Pollutants

Sponsored by

Northeastern Research Center for Wildlife Diseases
University of Connecticut, Storrs, Connecticut

Registry of Comparative Pathology
Armed Forces Institute of Pathology
Washington, D.C.

Institute of Laboratory Animal Resources
National Academy of Sciences, Washington, D.C.

NATIONAL ACADEMY OF SCIENCES
WASHINGTON, D.C.   1979

Support for this symposium was provided by the Northeastern Research Center for Wildlife Diseases, the Registry of Comparative Pathology, and the Institute of Laboratory Animal Resources (ILAR). This project was approved by the steering committee of the Northeastern Research Center for Wildlife Diseases: Herbert E. Doig, Director, Fish and Wildlife Division of the New York State Department of Environmental Conservation, Albany, N.Y. 12226; Theodore B. Bampton, Deputy Commissioner, State of Connecticut Department of Environmental Protection, Hartford, Conn. 06115; Edward F. Kehoe, Director, State of Vermont Fish and Game Commissioner, Montpelier, Vt. 05602; Howard N. Larsen, Director, Bureau of Sport Fisheries and Wildlife, Newton, Mass. 02159; Edwin J. Kersting, Dean, College of Agriculture and Natural Resources, University of Connecticut, Storrs, Conn. 06268.

The Registry of Comparative Pathology is supported in part by Public Health Service Grant No. RR00301-12 from the Division of Research Resources, National Institutes of Health, U.S. Department of Health, Education, and Welfare, under the auspices of Universities Associated for Research and Education in Pathology, Inc. Primary support to ILAR was provided by Contract No. N01-CM057013 from the National Cancer Institute. Additional support was provided to ILAR by Contract No. EY-76-C-02-2708-011, Department of Energy; Contract No. DNA001-78-C-0014, Defense Nuclear Agency; Contract No. 12-16-140-155-91, Department of Agriculture; Contract No. N00014-76-C-0242, U.S. Army Medical Research and Development Command, Office of Naval Research, U.S. Air Force; Contract No. NSF C310, Task Order No. 173, National Science Foundation; Contract No. N01-RR-5-2128, National Institutes of Health, Department of Health, Education, and Welfare; Grant No. RC-1, American Cancer Society, Inc; and contributions from pharmaceutical companies and other industries.

Available from:
Office of Publications
National Academy of Sciences
2101 Constitution Avenue, N.W.
Washington, D.C. 20418

Printed in the United States of America

Preface

The deleterious effect of pollution on humans and other animals living in highly industrialized urban areas is well recognized and documented, whereas there have been fewer studies on the impact of pollution on wildlife. These latter have focused largely on aquatic species.

The Symposium on Pathobiology of Environmental Pollutants: Animal Models and Wildlife as Monitors was organized in an effort to correct this deficiency and had the following goals: (1) to determine the extent to which environmental pollution can be documented in wildlife; (2) to provide a common forum for pathologists, wildlife biologists, and ecologists, who are studying environmental pollution from different perspectives; (3) to develop an informational and bibliographic resource on this important aspect of animal biology; and (4) to generate information that may be useful for legislators involved in fashioning contemporary policy decisions concerning the control of environmental pollution.

It is to be hoped that these proceedings will serve as a stimulus for further laboratory and field investigations in the environmental biology and pathology of wildlife.

The Organizing Group acknowledges the conscientious work of Mrs. Frances Peter, ALS editor, and the assistance of Ms Patricia Timmins, University of Connecticut, and Mrs. Dianne Perry, ILAR.

v

Contents

Contents

OVERVIEW
METHODOLOGY

Control Factors on Uptake and Clearance of Xenobiotic Chemicals by Fish[1]

J. RUSSELL ROBERTS, ANTHONY S. W. DEFREITAS, and MARGARET ANNE J. GIDNEY

While studying the phenomenon of pollutant bioaccumulation, we examined the utility of a model based on the specific energy requirement of a fish in a given ecosystem (Norstrom *et al.*, 1976; Roberts *et al.*, 1977). In this model, the rate of pollutant uptake from water and food is assumed to be directly linked to the food and oxygen requirements of the fish. An enormous amount of information is available on the species- and ecosystem-specific factors controlling these parameters. Thus, the available data on the bioenergetics and growth of fish should provide a particularly appealing base on which to develop models having general applicability, not only to results obtained in the laboratory, but also to those observed in the field. The overall bioaccumulation equation in terms of the concentration of pollutant in tissues (C_t), water (C_w), and food (C_f) can be written as

$$\frac{dC_t}{dt} = \underbrace{\frac{A_w C_w V_R}{w}}_{A} + \underbrace{\frac{A_f C_f V_f}{w}}_{B} - \underbrace{\left[k_{cl} w^\delta F(x) + \frac{dw\, w^{-1}}{dt} C_t \right]}_{C} \tag{1}$$

where A and B describe the uptake of pollutant from water and food, respectively, and C describes decreases in the concentration of pollutant in tissues due to clearance, $k_{cl} w^\delta C_t F(x)$, and growth dilution, $dw\, w^{-1}/dt \cdot C_t$. Here w refers to the weight of the fish in grams and k_{cl} refers

[1] National Research Council of Canada Publication No. 16789.

to the fractional clearance rate of pollutant in units of $d^{-1}w^{-\delta}F(x)^{-1}$. The function $F(x)$ refers to any other control factor that influences clearance.

The utility of this approach to pollutant uptake is appreciated when it is recognized that estimates of the daily food requirements (V_f [g/day of food]) and the quantity of water crossing the gill in one day, which is called the respiration volume (V_R [g/day of water]), can be obtained directly from the specific food and oxygen requirements of the fish. These can be estimated directly from the energy requirements. Since energy requirements are temperature-dependent, V_f and V_R are specific to a given temperature. In developing estimates of these key energy-linked parameters, consideration is given not only to the basic metabolic requirements of the fish, but also to the energy required to account for growth. Detailed discussions of the factors that must be considered when estimating such parameters are contained in Norstrom et al. (1976) and Winberg (1956).

The parameters A_w and A_f refer respectively to the fraction of pollutant in the water flowing across the gills or the food in the gastrointestinal tract that is absorbed and deposited in the tissues of the fish. These ratios, called "net assimilation coefficients," account for the competing processes of absorption and elimination of unabsorbed pollutant from the gastrointestinal tract and the gills. These parameters have been measured by relatively direct experimental techniques.

Equation 1 can be converted to an expression for bioaccumulation in terms of the body burden (P) in grams, since

$$P = C_t w \text{ and } \frac{dP}{dt} = \frac{d(C_t w)}{dt} = \frac{w dC_t}{dt} + C_t \frac{dw}{dt} \tag{2}$$

hence

$$\frac{dP}{dt} = A_w C_w V_R + A_f C_f V_f - k_{cl} F(x) P \, w^{(\delta + 1)} \tag{3}$$

Of the terms in Equation 1 only the clearance term is not inherently linked to the metabolic rate as a control factor. While clearance might be expected to be controlled by the metabolic rate, studies in this laboratory with methyl mercury have failed to demonstrate such a relation in goldfish (Carassius auratus). The fractional clearance rate of methyl mercury (Sharpe et al., 1977) was not significantly different at 5°C, 10°C, and 20°C. Furthermore, simulations of field residues of polychlorinated biphenyls (PCB's) and methyl mercury required the use of a body weight dependency for clearance of $w^{-0.58}$ (Norstrom *et al.*, 1976) not $w^{-0.2}$, as would be expected if metabolic rate was the sole control factor. Similarly, in experiments with shorthead redhorse (*Moxostoma macrolepidotum*) weighing from 24 to 398 g, we did not find that the clearance rates of *cis*- and *trans*-chlordane were significantly influenced by body size. Admittedly, any influence of metabolic rate as reflected in the body weight dependency of the clearance rate would be small and is perhaps masked by other factors. Comparison of a 50-g to a 300-g fish would probably reveal only a 30% decrease in the clearance rate (dC_t/dt) if δ equals -0.2.

Among the other factors that could influence clearance (Table 1), adiposity has been of particular interest because of its profound effect on the clearance of organochlorines by homeotherms (Pocock and Vost, 1974; deFreitas and Norstrom, 1974). We have found that the clearance of *trans*- and *cis*-chlordane is inversely related to the adiposity of the individual fish. This observation and a recent report (Addison and Zinck, 1977) support suggestions by Hamelink *et al.* (1971) and Harvey *et al.* (1971) that field residue patterns of chlorinated hydrocarbons in fish may be as much a function of lipid content of the fish as of its specific position in the food web.

DISCUSSION

It is not possible to review all control factors influencing clearance. Of particular importance are recent discussions (Moriarty, 1975a; Chambers and Yarbrough, 1976) indicating that careful work may be required to delineate the influence of body burden and length or pattern of exposure on the fractional clearance rate. Many bioaccumulation models have been successful when it is assumed that first-order processes describe the clearance of methyl mercury and chlorinated hydrocarbons by fish; however, this assumption may need to be reevaluated as models are refined to fit more sophisticated data bases.

The availability of a pollutant is affected by numerous factors both extrinsic and intrinsic to the fish (Table 1). Since the rate of pollutant uptake from both food and water is directly linked to metabolic rate, these rates are expected, and in the case of methyl mercury were found, to be related to body size (deFreitas and Hart, 1975; deFreitas *et al.*, 1977), growth rate, and temperature (Reinert *et al.*, 1974; Murphy and Murphy, 1971), as well as to C_w (deFreitas and Hart, 1975) and C_f (Roberts *et al.*, 1977). In few studies have investigators explicitly

TABLE 1   Some Factors Influencing Pollutant Bioaccumulation

Type of Factor	Factors Affecting Amount of Bioaccumulation		
	Dietary Uptake	Uptake from Water	Clearance
Extrinsic	Nature of chemical Water temperature Dietary pollutant concentration Nature of diet	Nature of chemical Water temperature Water quality parameters Water concentration of pollutant Oxygen concentration in water	Nature of chemical Water temperature Water quality parameters Presence of other pollutants in the water Concentration in tissues
Intrinsic	Body size Metabolic rate Growth rate Assimilation coefficient of pollutant Degrading capacity Satiation volume G.I. tract	Body size Metabolic rate Growth rate Assimilation coefficient of pollutant Assimilation efficiency of oxygen	Body size Metabolic rate Adiposity Degrading capacity of organism

examined the control factors that influence the net assimilation coefficients.

Table 2 demonstrates that the dietary assimilation coefficient (A_f) of methyl mercury is relatively insensitive to both extrinsic and intrinsic control factors such as body size, temperature, initial body burden, and the nature of the diet. In these studies, A_f ranged from 0.63 to 0.95, and no consistent pattern emerged. The dietary net assimilation coefficient (A_f) of *cis*- and *trans*-chlordane by shorthead redhorse suckers increased with increasing adiposity of the fish (Roberts *et al.*, 1977). Generally, it appears that persistent, highly lipophilic compounds like methyl mercury, polychlorinated biphenyls (PCB's), and chlorophenothane (DDT) have high net dietary assimilation coefficients. On the other hand, hydrophilic compounds such as inorganic mercury are assimilated less efficiently (Table 2).

One approach to determining the efficiency of pollutant uptake from water (A_w) is to relate the rate of pollutant uptake to the rate of oxygen uptake. This ratio, when multiplied by the assimilation coefficient of oxygen (A_{ox}), provides a measure of A_w. When goldfish were exposed to methylmercuric chloride in dechlorinated tap water, this ratio was approximately 0.4 (deFreitas, unpublished data, 1977). Based on a value of 0.75 for A_{ox} (Lloyd, 1961), the value of A_w is 0.3.

DeFreitas and Hart (1975) estimated that A_w is 0.2 for golden shiners (*Notemigonus crysoleucus*) maintained in dechlorinated tap water using an assumed oxygen consumption at low active metabolism. The use of these estimates for A_w allows for the possibility that uptake is not solely associated with the respiration volume.

Water quality parameters can significantly affect the net assimilation coefficients. For example, suspended solids appear to decrease the availability of methyl mercury in solution; the mercuric ion and presumably other divalent metal ions can significantly increase the uptake rate of methyl mercury; and pH shifts in the range of 5.5 to 8 have also been observed to cause a twofold shift. The effect of the chemical species on uptake is dramatically illustrated by the slower uptake of inorganic mercury from water compared to methyl mercury (deFreitas, 1977).

In the literature there are no similar studies on chlorinated hydrocarbons. After analyzing the available information, Norstrom *et al.* (1976) concluded that the A_w of persistent chlorinated hydrocarbons is generally greater than 0.75.

Equation 1 demonstrates that the fraction of the pollutant body burden acquired through the food vector, F_f, is approximated by the following equation when energy requirements associated with growth

TABLE 2 Effect of Control Factors on the Dietary Net Assimilation Coefficients of Organic Mercury, Inorganic Mercury, and Organochlorine Compounds

Control Factors	Compound	Range in Observed Net Assimilation Coefficient of Pollutant in Food, Af	Pattern of Effect	References
Species				
Brown bullhead (Ictalurus nebulosus) Yellow perch (Perca flavescens) Northern pike (Esox lucius) Walleye (Stizostedion vitreum) Ling (Lota lota) Goldfish (Carassius auratus)	Methyl mercury	0.70 → 0.95	No observed pattern	deFreitas, 1975
Brown bullhead (I. nebulosus) Yellow perch (P. flavescens) Northern pike (E. lucius) Goldfish (C. auratus)	Inorganic mercury	0.12 → 0.18	No observed pattern	deFreitas, 1975
Body size				
(1.0 g → 390 g)	Methyl mercury	0.63 → 0.95	No observed pattern	deFreitas, 1975; Sharpe et al., 1977; Roberts et al., 1977
(28 g → 218 g)	cis-Chlordane	0.24 → 0.52	No observed pattern	Roberts et al., 1977
(90 g → 288 g)	DDT	0.89	No observed pattern	Roberts, work in progress
(8 g → 320 g)	Inorganic mercury	0.12 → 0.18	No observed pattern	deFreitas, 1975

Parameter	Compound	Value	Observation	Reference
Temperature (5°C → 25°C)	Methyl mercury	0.70 → 0.95	No observed pattern	deFreitas, 1975; Sharpe et al., 1977
(8°C → 22°C)	Inorganic mercury	0.12 → 0.18	No observed pattern	deFreitas, 1975
Adiposity (0.3% → 5.2%)[a]	cis-Chlordane	0.24 → 0.52	Assimilation increased with increase in lipid content	Roberts et al., 1977
(0.3% → 5.2%)[a]	trans-Chlordane	0.19 → 0.42	Assimilation increased with increase in lipid content	Roberts et al., 1977
Initial body burden	Methyl mercury	0.66 → 0.77	No observed pattern	Roberts et al., 1977
Diet (food)	Methyl mercury	0.71 → 0.85	No observed pattern	Sharpe et al., 1977
(corn oil)		0.70 → 0.95		Roberts et al., 1977
(food)	cis-Chlordane	0.24 → 0.52	No observed pattern	Roberts et al., 1977
(corn oil)		0.35 → 0.57		Roberts et al., 1977

[a] % whole body, wet weight.

9

do not significantly affect the relative oxygen and food requirements of the fish, i.e., when the growth rate is low:

$$F_f = \frac{A_f C_f V_f}{V_R A_w C_w + A_f C_f V_f} = \left[1 + \frac{A_w C_w V_R}{A_f C_f V_f} \right]^{-1}. \qquad (4)$$

Setting the parameters A_w and A_f equal to 0.2 and 0.8 and using a value of 2×10^4 for the ratio of $V_R{:}V_f$ as suggested by the analysis of Norstrom *et al.* (1976), one finds that for the food vector to account for more than 10% or 90% of the body burden of methyl mercury, the ratio of the residues of methyl mercury in the food to those in the water would need to be approximately 5×10^2 and 5×10^4, respectively. If the water and food assimilation coefficients of a pollutant are similar, the relative concentration of pollutant in the food would need to be even higher before the food uptake vector would significantly compete with the water vector. As indicated by data in this paper, this may at times apply to some organochlorines and methyl mercury. In this instance, if the food vector accounts for 50% to 90% of the body burden, dietary concentrations of the pollutant would need to be approximately 10^4 to 10^5 times higher than those in the water. Given that the maximum concentration (tissue to water) factors in fish and other aquatic organisms for organochlorine compounds are reported to be in the 10^4 to 10^5 range (Moriarty, 1975a), the water vector could apparently compete with the food vector or even be dominant in many field situations.

Another approach to the same question is to examine the predicted steady-state dietary concentration factor (CF_{ss}) assuming that the growth rate is low. Under these conditions:

$$CF_{ss} = \frac{A_f V_f}{k_{cl} w^\delta} \qquad (5)$$

Given this constraint and assuming routine metabolism, V_f at 10°C to 15°C equals about $0.026 \, w^{0.8}$ according to the analysis of Norstrom *et al.* (1976). Assuming that the dietary assimilation coefficient is equal to 1.0, the theoretical maximum, the clearance half-life would need to be in excess of approximately 25 to 35 days before the pollutant levels in the whole body of a 1-g fish would equal pollutant levels in the diet. Reported half-lives (Moriarty, 1975a) of the persistent chlorinated hydrocarbons generally range from 30 to 75 days.

Thus, consideration of the water vector as well as trophic feeding patterns is indicated. The half-life of many organic chemicals, e.g., the commonly used "second generation" pesticides, will be less than 30 days. Particular caution should be exercised before citing food chains to explain residues of these compounds in fish.

This analysis of the energy-driven model for bioaccumulation provides quantitative support to suggestions by Moriarty (1975b) and Johnson (1973) that the concept of food chain bioaccumulation may often turn out to be a gross oversimplification.

SUMMARY

A model based on the caloric requirements of fish has been developed to describe the accumulation of polluting xenobiotics from both food and water. In this energy-driven model, pollutant uptake is determined in terms of the fraction of the assimilated dose associated with ingested food and with the flow of water across the gills. The influence of control factors such as adiposity on uptake and clearance has been analyzed. This model predicts the accumulation patterns of chlorinated hydrocarbons and methyl mercury observed in the laboratory and helps to explain the wide variation of residue levels reported in field studies.

REFERENCES

Addison, R. F., and M. E. Zinck. 1977. Rate of conversion of ^{14}C-p,p'-DDT to p,p'-DDE by brook trout (*Savelinus fontinalis*): absence of effect of pretreatment of fish with compounds related to p,p'-DDT. J. Fish. Res. Board Can. 34:119–122.

Chambers, J. E., and J. D. Yarbrough. 1976. Xenobiotic biotransformation systems in fishes. Comp. Biochem. Physiol. 55:77–84.

deFreitas, A. S. W. 1975. Mercury uptake and retention by fish. Pp. 18.1–18.29 in D. R. Miller, ed. Distribution and Transport of Persistent Chemicals in Flowing Water Ecosystems. Ottawa River Project Report 2. Division of Biological Sciences, National Research Council of Canada, Ottawa, Ontario.

deFreitas, A. S. W. 1977. Mercury uptake and retention by fish. Pp. 30.2–30.62 in D. R. Miller, ed. Distribution and Transport of Persistent Chemicals in Flowing Water Ecosystems. Ottawa River Project Final Report. Division of Biological Sciences, National Research Council of Canada, Ottawa, Ontario.

deFreitas, A. S. W., and J. S. Hart. 1975. Effect of body weight on uptake of methylmercury by fish. Pp. 356–363 in Water Quality Parameters. ASTMSTP 573. American Society for Testing and Materials, Philadelphia, Pa.

deFreitas, A. S. W., and R. J. Norstrom. 1974. Turnover and metabolism of polychlorinated biphenyls in relation to their chemical structure and the movement of lipids in the pigeon. Can. J. Physiol. Pharmacol. 52:1080–1094.

deFreitas, A. S. W., M. A. J. Gidney, A. E. McKinnon, and R. J. Norstrom. 1977. Factors affecting whole body retention of methylmercury in fish. Pp. 441–451 in

Proceedings of the 15th Hanford Life Sciences Symposium, Biological Implications of Metals in the Environment, Richland, Washington, September 29–October 1, 1975. Energy Research and Development Administration Symposium Series 42, NTIS No. CONF-750929. National Technical Information Service, Springfield, Va.

Hamelink, J. L., R. C. Waybrant, and R. C. Ball. 1971. A proposal: exchange equilibria control the degree chlorinated hydrocarbons are biologically magnified in lentic environment. Trans. Am. Fish. Soc. 100:207–214.

Harvey, G. R., V. T. Bower, R. H. Backus, and G. D. Grice. 1971. Chlorinated hydrocarbons in open-ocean Atlantic organisms. Pp. 177–186 in D. Dyrssen and D. Jagner, ed. The Changing Chemistry of the Oceans. Proceedings of the 20th Nobel Symposium, Göteborg. John Wiley & Sons, New York.

Johnson, D. W. 1973. Pesticide residues in fish. Pp. 181–212 in C. A. Edwards, ed. Environmental Pollution by Pesticides. Plenum Press, London and New York.

Lloyd, R. 1961. Effect of dissolved oxygen concentration on the toxicity of several poisons to rainbow trout (*Salmo gairdneri* Richardson). J. Exp. Biol. 38:447–455.

Moriarty, F. 1975a. Exposure and residues. Pp. 29–40 in F. Moriarty, ed. Organochlorine Insecticides: Persistent Organic Pollutants. Academic Press, London and New York.

Moriarty, F. 1975b. Pollutants and Animals. George Allen and Unwin Ltd., London. 140 pp.

Murphy, P. G., and J. V. Murphy. 1971. Correlation between respiration and direct uptake of DDT in the mosquito fish (*Gambusia affinis*). Bull. Environ. Contam. Toxicol. 6:581–588.

Norstrom, R. J., A. E. McKinnon, and A. S. W. deFreitas. 1976. A bioenergetics-based model for pollutant accumulation by fish. Simulation of PCB and methyl-mercury residue levels in Ottawa River yellow perch (*Perca flavescens*). J. Fish. Res. Board Can. 33:248–267.

Pocock, D. M. E., and A. Vost. 1974. DDT absorption and chylomicron transport in rat. Lipids 9:374–381.

Reinert, R. E., L. J. Stone, and W. A. Willford. 1974. Effect of temperature on accumulation of methyl mercuric chloride and p,p′-DDT by rainbow trout (*Salmo gairdineri*). J. Fish. Res. Board Can. 31:1649–1652.

Roberts, J. R., A. S. W. deFreitas, and M. A. J. Gidney. 1977. Influence of lipid pool size on bioaccumulation of the insecticide chlordane by northern redhorse suckers (*Moxostoma macrolepidotum*). J. Fish. Res. Board Can. 34:89–97.

Sharpe, M. A., A. S. W. deFreitas, and A. E. McKinnon. 1977. The effect of body size on methylmercury clearance by goldfish (*Carassius auratus*). Environ. Biol. Fish. 2:37–43.

Winberg, G. G. 1956. Rate of metabolism and food requirements of fishes. Nauchn. Tr. Beloruss. Gos. Univ. V.I. Lenina, Minsk. 253 pp. (Transl. from Russian by Fisheries Research Board of Canada Translation Series No. 194, 1960.)

QUESTIONS AND ANSWERS

R. GARMAN: Was there variation in the assimilation coefficients of *cis*-chlordane?

J. ROBERTS: Yes, there was considerable variation. In spite of the overall variability in the net assimilation coefficients, there is a definite trend suggesting that in fish with less than 2% body fat, the net assimilation coefficient was consistently low and directly related to adiposity. Above a body fat content of about 2%, the net assimilation coefficient seemed to remain relatively constant. The trend was probably masked by the overall variability of the data and the relatively small changes expected in the net assimilation coefficient.

R. GARMAN: Do you induce certain body fat percentages in different fish by controlling the amount of feed or some regimen of that nature, or are these fish of different ages?

J. ROBERTS: No. These fish were from one wild population of shorthead redhorse suckers. They had been acclimated for about 90 days to a slightly greater than maintenance diet which produced no significant growth. We monitored the fish to confirm that no growth occurred throughout the study. We were quite fortunate in that the fish naturally had a range in lipid content from approximately 0.7% to about 5%. This is why we were able to demonstrate a clear relationship between adiposity and clearance rate.

G. CHOULES: How do you measure your uptakes?

J. ROBERTS: We used ^{203}Hg-methyl mercury and measured the decrease in radioactive body burden after the dose had been absorbed from the gastrointestinal tract rather than measuring it immediately after the dose was ingested. We have just published a technique for measuring the quantity of chlorinated hydrocarbons in the tissues by standard gas-liquid chromatography (Roberts *et al.*, 1977).[1] In this procedure, we used labeled ^{203}Hg-methyl mercury as a tracer in the diet, which also contained chlorinated hydrocarbons, so we could define the actual quantity of feed ingested. Then, we back-calculated from the uptake and clearance parameters for methyl mercury to determine the amount of feed and consequently the dose of chlorinated hydrocarbons ingested.

J. WEIS: I was interested in your statement that the presence of inorganic mercury affected the uptake rate of methyl mercury. Are there other heavy metals in the water that may have an effect on the uptake of methyl mercury?

J. ROBERTS: Yes, they increased the rate of uptake. One might assume that it would be observed with other compounds similar to the mercuric ion, such as other divalent metal ions.

D. SCARPELLI: Chlorinated hydrocarbons appear to have a prolonged residency time in adipose tissue. Do you have any clues on what factors regulate the loss of such compounds from the host?

J. ROBERTS: I do not. There is a report in the literature in which the authors postulate that the last molecule of a chlorinated hydrocarbon to be deposited in the adipose tissue of rats is the first to be cleared (Baron and Walton, 1971).[2] This suggests a complicated rather than simple model.

[1]Roberts, J. R., A. S. W. deFreitas, and M. A. J. Gidney. 1977. Influence of lipid pool size on bioaccumulation of the insecticide chlordane by northern redhorse suckers (*Moxostoma macrolepidotum*). J. Fish. Res. Board Can. 34:89–97.
[2]Baron, R. L., and M. S. Walton. 1971. Dynamics of HEOD (Dieldrin) in adipose tissue of the rat. Toxicol. Appl. Pharmacol. 18:958–963.

D. SCARPELLI: Are there clues as to mechanism?

A. de FREITAS: I think there may be other clues in nonfish literature, indicating that lipid pool size and turnover of lipid should be the main factors controlling the rate of metabolism and clearance of organochlorine compounds. The induction of liver enzymes, which might cause an increase in the detoxification and clearance of such compounds, is an important additional complication. However, the actual movement of the organochlorine compounds from adipose tissue to, say, liver is definitely related to metabolic rate and adipose tissue size.

J. LINASK: I was very intrigued by the fact that you found a decreased uptake rate of inorganic mercury for river water. Is there any possibility that there are chelating agents in the river water that account for this pattern?

J. ROBERTS: Yes, we do observe a dramatic decrease in the uptake of inorganic mercury for river water compared to tap water. There are several possibilities that could explain this decrease but we have not been able to examine them.

A Discrete-Event Approach to Predicting the Effects of Atmospheric Pollutants on Wildlife Populations Using ^{14}C Exposure

KENNETH R. DIXON and BRIAN D. MURPHY

The environmental impact of atmospheric pollutants emitted from a point source depends upon both meteorological factors (speed and direction of wind, and atmospheric stability) and biological factors (differences in seasonal activity of plant and animal species, food habits, and mobility). Most environmental impact assessments use long-term averaged values of meteorological parameters and largely ignore biological factors. Described below is a "discrete-event" approach to atmospheric pollutant impact assessment that provides for recurrent exposure to short-term pulses of high concentration and can be correlated to seasonal variability in biological activity. Such an approach is particularly important in the study of certain wildlife populations that, due to their relatively low mobility, are continuously exposed to ambient pollutant concentrations. A study of the effects of ^{14}C as estimated by the discrete-event approach is compared with the standard methods to study ^{14}C exposure in the meadow vole (*Microtus pennsylvanicus*). Using wildlife as monitors should provide conservative estimates of pollutant effects on human populations.

Exposure to ground-level concentrations of atmospheric pollutants depends on the emission rate from the point source and meteorological conditions that determine the dispersion of the pollutant in the atmosphere. Exposure can be through inhalation, immersion, and ingestion, or a combination of these routes.

The standard method of estimating ground-level concentrations of gaseous effluents (χ) at a given distance (r) and in a sector angle (θ) uses

15

an annual release rate to the atmosphere (Q') and an annual average atmosphere dispersion factor (χ/Q') (UNSCEAR, 1972; USNRC, 1976a,b). The Gaussian plume straight-line trajectory is the most commonly used model (Gifford, 1968). However, emission rates vary according to the level of plant activity and the degree to which emission controls are in effect. Ground-level concentrations vary according to atmospheric turbulence. For example, measurements of off-site concentrations of radioactive gaseous effluents from a boiling water reactor (BWR) showed high variability (peak values two orders of magnitude above the average) associated with meteorological dispersion parameters (Carey et al., 1974). The importance of considering this variability in radiation dosage calculations was stated in the report of the National Research Council's Advisory Committee on Biological Effects of Ionizing Radiation (1972) and more recently by Neyman (1977).

To simulate this variability in ground-level concentrations, we developed a discrete plume event model. A "plume event" is defined as the occurrence of a plume over a given point for a discrete period, e.g., 1 h.

MATERIALS AND METHODS

One of the gaseous radioisotopes produced by various nuclear power facilities, such as power reactors and fuel reprocessing facilities, is ^{14}C (Magno et al., 1974), primarily in the form of $^{14}CO_2$. Radiological doses occur primarily through the transfer of carbon via the food chain and inhalation.

The estimation of radiological dose from ^{14}C involves three basic steps: estimation of ground-level concentrations of $^{14}CO_2$; estimation of the ^{14}C specific activity in vegetation (S_p) due to exposure to the $^{14}CO_2$ estimated in the first step; and calculation of dose from inhalation of $^{14}CO_2$ and ingestion of vegetation.

We began by using an atmospheric transport model (ATM) (Culkowski and Patterson, 1976) to estimate the ground-level concentrations of $^{14}CO_2$ for each point on a grid. Each concentration was associated with a given set of meteorological conditions, including windspeed and stability. Six Pasquill–Gifford stability classes were considered, ranging from A (very unstable) to F (very stable) (Gifford, 1961). Classes A, B, and C are essentially daytime occurrences. Classes E and F are nighttime occurrences, and Class D can be assumed to occur with equal probability during the day or night. Using meteorological data obtained from the Tennessee Valley Authority for Oak

Ridge, Tennessee, the probability of occurrence of each windspeed-stability class during each month of the year was calculated. It is important that the stability categories be separated into daytime and nighttime conditions, since the higher ground-level concentrations near a point source associated with the unstable daytime conditions are present during the photosynthetically active part of the day.

We then constructed a simple Monte Carlo algorithm incorporating the monthly probabilities of windspeed-stability conditions. The algorithm was invoked for each hour of the simulated year, giving a ground-level concentration of $^{14}CO_2$ for that hour. Predicted ground-level concentrations in the predominantly downwind sector at a point 500 m from the source are shown in Figure 1 for five days during the growing season. The source of $^{14}CO_2$ was assumed to be a 90-m stack which emitted ^{14}C as $^{14}CO_2$ at a rate of 2 μCi/s. This rate is based on a

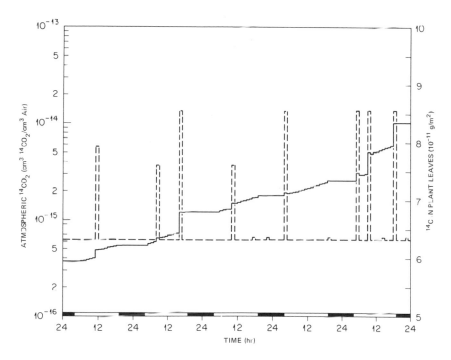

FIGURE 1 Atmospheric concentrations of $^{14}CO_2$ and ^{14}C content of leaves for 5 days during the growing season. The dashed line represents ground-level atmospheric concentrations of $^{14}CO_2$; the solid line, the ^{14}C content of leaves. Hours of dark are represented by solid bars; hours of light by clear bars.

constant annual release rate used in the standard method. Release rates undoubtedly vary, but a more accurate forecast would require data on the distribution of these rates.

Using standard methods to estimate the ^{14}C specific activity in vegetation (S_p), one must assume that the ^{14}C activity is equal to the annual average specific activity in the atmosphere ($\bar{S}_a$) surrounding the vegetation (USNRC, 1976a,b). This assumption is based on the observations of ^{14}C content of plants exposed to constant levels of natural $^{14}CO_2$ or widely dispersed atmospheric ^{14}C from the fallout of atomic tests (Broecker and Olson, 1960; Broecker and Walton, 1959; Vogel and Marais, 1971). However, there is little information on ^{14}C levels in vegetation exposed to highly variable concentrations such as those observed near a point source (Carey et al., 1974).

To examine the response of plants to highly variable concentrations, which may result from a series of plume events, and to predict values of S_p in such plants, the investigators developed a carbon assimilation model based on a model of plant biomass dynamics (Dixon et al., 1976), which was modified to include ^{14}C.

Essential features of the carbon assimilation model include a photosynthesis model based on gaseous diffusion (Gaastra, 1959; Penman and Schofield, 1951), a labile sugar transport model based on sugar gradients that transport sugar between plant parts (Thornley, 1972), and equations that fix labile sugar as insoluble structural carbon.

RESULTS

Figure 1 illustrates the response of plant leaves to the ground-level atmospheric concentration of $^{14}CO_2$. It shows daylight and nighttime hours, predicted plume events, and the simulated ^{14}C in leaves over a 5-day simulation. Note that when a plume event occurs in the daylight, the ^{14}C in leaves increases. No accumulation occurs at night because of the absence of photosynthesis.

A 1-yr simulation with a growing season from April to October is shown in Figure 2. In this hypothetical case, the meteorological conditions in the Oak Ridge area were used. A maximum specific activity in the plant leaves (S_p) of 26.8 pCi/gC was predicted. Figure 2 also includes the calculated average annual ground-level atmospheric ^{14}C specific activity ($\bar{S}_a$) based on the monthly windspeed-stability class probabilities used in the Monte Carlo algorithm. The value of S_p represents an increase of 89% over the calculated value of $\bar{S}_a$. The predicted mean and standard deviation of the maximum value of S_p

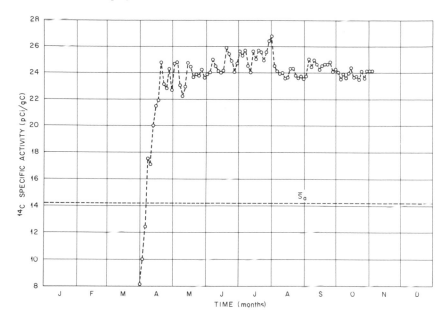

FIGURE 2 Environmental ^{14}C specific activities. The open circles represent the ^{14}C specific activity in plant leaves (S_p) sampled every 60 h. The dotted line represents the yearly average atmospheric specific activity ($\bar{S}_a$).

(sampled every 60 h) from 10 Monte Carlo simulations was 23.5 ± 3.6 pCi/gC, indicating the relative insensitivity of S_p to the random order of plume events.

DISCUSSION

The reason for the difference between S_p and $\bar{S}_a$ is that the flux of CO_2 into the leaves depends on the instantaneous partial pressure of CO_2 outside the leaf's stomata. Therefore, when a plume event occurs, the flux of $^{14}CO_2$ increases significantly (Figure 1). Once the labile photosynthate is produced, a certain fraction is fixed as nonsoluble structural material (cell walls, etc.). Each time the ^{14}C in the labile pool increases in response to a plume event, the fixed ^{14}C also increases until a maximum, determined by leaf size, is reached.

The two major exposure pathways of ^{14}C are inhalation of $^{14}CO_2$ and ingestion of food containing radiocarbon. In general, the dose rate to a body organ in rem/h due to deposition of a radionuclide in that organ can be described as

$$\frac{dD_\beta}{dt} = \frac{2.13 \cdot \epsilon \cdot q(t)}{m}, \tag{1}$$

where dD_β/dt = the dose rate of a β emitter to the reference organ (rem/h),

ϵ = the effective absorbed energy per disintegration (MeV/dis) (rem/rad),

m = mass of the reference organ (g),

$q(t)$ = activity burden of the radionuclide in the organ (μCi), = body concentration,

t = time (h).

The total dose received for a given interval is obtained by integrating Equation 1 over that interval. The activity burden $q(t)$ is determined by the differential equation:

$$\frac{dq(t)}{dt} = -(\lambda_r + \lambda_b)q(t) + f \cdot I(t), \tag{2}$$

where λ_r = radioactive decay constant (h^{-1}),

λ_b = metabolic removal rate constant (h^{-1}),

f = fraction of intake reaching the reference organ, and

$I(t)$ = rate of intake into the body of activity at time t (μCi/h).

Since ^{14}C has a radiological half-life of 5,730 yr, the radioactive decay constant can be ignored for most dose calculations. The metabolic removal rate constant will depend upon the particular carbon compound in which ^{14}C occurs and the rate of metabolism of the individual animal. Bernard (1974) presents a rate constant in mice based on a mean residence time of 6 days. He used data on ^{14}CO$_2$ inhalation from Buchanan (1951).

Ingestion dose will depend on the seasonal dynamics of ^{14}C in vegetation that is consumed by the individual voles. A detailed simulation of seasonal feeding habits is beyond the scope of this paper. However, a comparison of the discrete-event approach with the standard method can be made using the estimates of specific activity given above, assuming all other factors to be equal. The International

Commission on Radiological Protection (ICRP) (1959) gives a total body value of f for ingestion of 1.0. An estimated ingestion rate by laboratory voles is 0.25 gC/h (Golley, 1960). The rate of intake into the body, $I(t)$, then, is the ingestion rate times the specific activity of ^{14}C in the vegetation. Solving Equations 1 and 2, using the value of $\bar{S}_a$ as 14.2 pCi/gC, yields an annual dose to the voles of 9.9 mrem. Assuming a daily ingestion of leafy material with an S_p value of 23.5 pCi/gC, the calculated annual dose to the voles is 16.4 mrem, which is 65% above the dose based on the time-averaged value of specific activity.

The radiation dose from inhalation is directly related to the ambient $^{14}CO_2$ concentrations, including the plume events (Figure 1). Golley (1960) reports an inhalation rate of 82.2 ml/h for voles. The ICRP (1959) total body value of f for inhalation is 0.75. The value of $I(t)$ in Equation 2 is calculated by multiplying the inhalation rate times the activity of $^{14}CO_2$ in μCi/ml. Equations 1 and 2 now can be solved with f and $I(t)$ values for inhalation. First using a $^{14}CO_2$ activity value of 2.7×10^{-12} μCi/ml corresponding to the annual average value ($\bar{S}_a$), the annual dose from inhalation of $^{14}CO_2$ is 4.7×10^{-4} mrem. The annual dose calculated from exposure to plume event concentrations is 4.9×10^{-4} mrem or 5% above that using time-averaged concentrations of $^{14}CO_2$. With a discrete-event approach to the modeling of pollutant transport and biological effects, allowances can be made for exposure to highly variable ground-level concentrations of atmospheric pollutants. Further analyses using the discrete-event approach should examine additional sources of variability, including emission rates and animal activity. We agree with Neyman (1977) that "attempts to formulate radiation protection guides in terms of averages over areas or time per individual population member would be imprudent."

SUMMARY

The reported results suggest that the biological responses to variable concentrations of pollutants may differ from responses to chronic low-level exposures. The effects of atmospheric pollutants will depend upon the dispersion of emissions from a point-source and the resulting ground-level pollutant concentrations. The model in this paper calculates an hour-by-hour exposure of such atmospheric pollutants as gaseous radioisotopes and toxic metals using the probabilities of different atmospheric conditions. This discrete-event approach yields annual dose estimates significantly greater than those based on standard methods.

ACKNOWLEDGMENTS

The computer time for this project was supported in part through the facilities of the Computer Science Center of the University of Maryland. Special thanks go to J. O. Doyle for his assistance in developing the inhalation dose model computer program.

REFERENCES

Bernard, S. R. 1974. A human metabolic model for ^{14}c-labelled metabolites useful in dose estimation. Pp. 1400–1405 in Proceedings of the 3rd International Congress of the International Radiation Protection Association. CONF-730907-P1, usaec. Pergamon Press, Oxford. 1475 pp.

Broecker, W. S., and E. A. Olson. 1960. Radiocarbon from nuclear tests, II. Science 132:712–721.

Broecker, W. S., and A. Walton. 1959. Radiocarbon from nuclear tests. Science 130:309–314.

Buchanan, D. L. 1951. Uptake and retention of fixed carbon in adult mice. J. Gen. Physiol. 34:737–759.

Carey, W. M., L. Battist, and W. E. Keene. 1974. Measurements of the dispersion of gaseous radioactive effluents from an operating boiling water reactor. Pp. 291–295 in Proceedings of the 3rd International Congress of the International Radiation Protection Association. CONF-730907-P1, usaec. Pergamon Press, Oxford. 1475 pp.

Culkowski, W. M., and M. R. Patterson. 1976. A Comprehensive Atmospheric Transport and Diffusion Model. ORNL/NSF/EATC-17. 117 pp. Oak Ridge National Laboratory, Oak Ridge, Tenn.

Dixon, K. R., R. J. Luxmoore, and C. L. Begovich. 1976. ceres—A Model of Forest Stand Biomass Dynamics for Predicting Trace Contaminant, Nutrient, and Water Effects. ORNL/NSF/EATC-25. 102 pp. Oak Ridge National Laboratory, Oak Ridge, Tenn.

Gaastra, P. 1959. Photosynthesis of crop plants as influenced by light, carbon dioxide, temperature, and stomatal diffusion resistance. Meded. Landouwhogesch. Wageningen 59:1–68.

Gifford, F. A., Jr. 1961. Use of routine meteorological observations for estimating atmospheric dispersion. Nucl. Saf. 2:47–57.

Gifford, F. A., Jr. 1968. An outline of theories of diffusion in the lower layers of the atmosphere. Pp. 65–116 in D. H. Slade, ed. Meteorology and Atomic Energy. TID-24190. U.S. Atomic Energy Commission, Oak Ridge, Tenn. 445 pp.

Golley, F. B. 1960. Energy dynamics of a food chain of an old-field community. Ecol. Monogr. 30:187–206.

International Commission on Radiological Protection (icrp). 1959. Report of Committee II on Permissible Dose for Internal Radiation. icrp Publ. No. 2. Pergamon Press, New York. 233 pp.

Magno, P. J., C. B. Nelson, and W. H. Ellett. 1974. A consideration of the significance of carbon-14 discharges from the nuclear power industry. Pp. 1047–1055 in Proceedings of the 13th aec Air Cleaning Conference. National Technical Information Service, Springfield, Va. 1121 pp.

National Research Council. Advisory Committee on the Biological Effects of Ionizing

Radiations. 1972. The Effects on Populations of Exposure to Low Levels of Ionizing Radiation. National Academy of Sciences, Washington, D.C. 217 pp.

Neyman, J. 1977. Public health hazards from electricity-producing plants. Science 195:754–758.

Penman, H. L., and R. K. Schofield. 1951. Some physical aspects of assimilation and transpiration. Pp. 115–129 in Carbon Dioxide Fixation and Photosynthesis, Symposium of the Society for Experimental Biology No. 5. Academic Press, New York.

Thornley, J. H. M. 1972. A model to describe the partitioning of photosynthate during vegetative plant growth. Ann. Bot. 33:419–430.

United Nations Scientific Committee on the Effects of Atomic Radiation (UNSCEAR). 1972. Ionizing Radiation: Levels and Effects. United Nations, New York. 447 pp.

U.S. Nuclear Regulatory Commission (USNRC). 1976a. Calculation of Annual Doses to Man from Routine Releases of Reactor Effluents for the Purpose of Evaluating Compliance with 10 CFR Part 50, Appendix I, Regulatory Guide 1.109.

U.S. Nuclear Regulatory Commission (USNRC). 1976b. Methods for Estimating Atmospheric Transport and Dispersion of Gaseous Effluents in Routine Releases from Light-Water-Cooled Reactors, Regulatory Guide 1.111.

Vogel, J. C., and M. Marais. 1971. Pretoria radiocarbon dates 1. Radiocarbon 13:378–394.

QUESTIONS AND ANSWERS

J. NEWMAN: What findings led you to develop this model and make those predictions?

K. DIXON: The model was developed to assess the impact of a proposed high-temperature, gas-cooled reactor fuel reprocessing facility. There was a significant difference between the estimated dose of ^{14}C ($\sim$87 mrem/yr) and the dose calculated through the discrete-event approach, which is approximately twice that rate. Since ^{14}C primarily takes the form of $^{14}CO_2$, we thought that there would be a problem in the ^{14}C dose unless some system of CO_2 removal was attached to the fuel processing facility.

A. de FREITAS: You indicated that you used a very short time in which to project leaf response to $^{14}CO_2$. If you had studied the leaves over a longer period, while they were still growing, would there have been an even greater difference between the annual average approach and the discrete-event approach? Secondly, since the growing season for any given leaf is about 3 weeks, would your data be different for a leaf that begins to grow later in the season?

K. DIXON: The length of the growing season probably wouldn't make a difference because the specific activity of plants is reached early in the growing season and levels off after the leaf area has reached its maximum. A problem may be the differences in growing seasons of various species. These may be affected in different ways by seasonal meteorological conditions. We only looked at one species with this particular growing season and are presently looking at other species which grow at other times of the year. I think that most plants would show essentially the same pattern of reaching an early maximum, although different species may not reach exactly the same levels due to the meteorological differences I just mentioned.

A. de FREITAS: Do you feel that you have enough information on all of the variables to actually apply this approach and make predictions? Or do you feel it is simply a useful tool to reflect what could happen?

K. DIXON: I would like to think that it is possible to make predictions; however, at present our data are not resolute enough for us to consider the variability of all of the factors. For example, we need estimates of the variability of emission rates, meteorological conditions, and information on metabolic changes in different animal species as affected by seasonal changes. Furthermore, we do not have reliable seasonal estimates of the various plant species with reference to concentrations of pollutants. Finally, we do not know whether these plants represent significant food items in the diet of wildlife species.

H. CASEY: You mentioned that a dose of 16 mrem/yr of ^{14}C was predicted. How does that compare with exposure to ^{40}K, a naturally occurring radioisotope?

K. DIXON: I don't know the answer for ^{40}K. However, the natural ^{14}C exposure dose for man is approximately 1 mrem/yr.

G. CHOULES: I was puzzled by the example where the concentration increased in plants with an increase in pollution. How does that differ from instances where the concentration of pollutant rose to a certain level and then plateaued? Were you dealing with young leaves in the one instance and leaves that had come to maturity in the other?

K. DIXON: The examples I showed were on different time scales; one for a 5-day period, the other for an interval of 1 yr. In the 5-day study, the leaf was still growing since the growing season was just beginning.

D. SCARPELLI: Has anyone looked at the body burden that is imposed on wildlife around this particular reactor in terms of the ingestion of ^{14}C and other radionuclides? I'm reminded of the excitement that was caused a number of years ago when someone reported an osteogenic sarcoma in a muskrat trapped in the environs of an atomic fuel processing plant and found a considerable amount of strontium 90 in the bone. There was a question about whether pollution from the reactors had been responsible. Is there monitoring of wildlife that live around atomic energy plants?

K. DIXON: Wildlife is monitored at all the nuclear power plants, but I do not know how exhaustive or sustained this is. As far as I know, none of these monitoring programs looks at body concentrations of various radionuclide pollutants.

AQUATIC POLLUTANTS

Effects of Crude Oil Ingestion on Immature Pekin Ducks (*Anas platyrhynchos*) and Herring Gulls (*Larus argentatus*)

DAVID S. MILLER, WILLIAM B. KINTER, and
DAVID B. PEAKALL

The several million metric tons of petroleum hydrocarbons entering the aquatic environment annually (Travers and Luney, 1976) are a major source of oceanic pollution. Oil is clearly lethal to birds as evidenced by the large numbers of oiled sea and shore birds that wash up after a spill. These birds presumably die of exposure resulting from the alteration of the waterproofing properties of their plumage by oil. Consequently, studies on the effect of oil on sea birds have largely been concerned with direct mortality. However, the low survival rate among apparently healthy oiled birds that have been cleaned and released suggests that there are delayed toxic effects in these waterfowl (Conder, 1968). Certainly, mortality could be caused by oil that is swallowed or aspirated when birds preen or ingest contaminated food or water. In 1966, Hartung and Hunt reported that relatively high oral doses (1 to 3 ml/kg body weight [BW]) of industrial oils caused lipid pneumonia and gastrointestinal irritation in several species of waterfowl. More recently, the *in vitro* measurements by Crocker *et al.* (1974, 1975) demonstrated that crude oil ingestion by saline-adapted Pekin ducklings (*Anas platyrhynchos*) inhibits intestinal salt and water transport. Unfortunately, they did not measure plasma electrolyte levels to determine if overall osmoregulation was affected.

Pekin ducks, which are genetically mallards, possess a limited ability to osmoregulate while maintained on hypertonic drinking water (Fletcher and Holmes, 1968). In contrast, herring gulls (*Larus argentatus*) can osmoregulate in a simulated marine environment (see below). Thus, these two species are convenient laboratory models for studying the toxicity of crude oil to coastal and oceanic birds.

27

METHODS

Four- to 6-month old Pekin ducks (purchased from C & R Duck Farms, Westhampton, N.Y.) and 3- to 4-week old herring gull chicks (captured on Old Man Island, off Cutler, Maine) were housed in sheds. Ducks were maintained on dry feed and fresh water (FW) or 50% to 100% Frenchman's Bay seawater (SW), gulls on whole unsalted herring and SW. The 100% SW had the following composition: 440 meq/l sodium (Na), 9.4 meq/l potassium (K), and 940 mosM/kg. Birds were assigned to exposure or control sheds. The exposure group was given a single oral dose by intubation of either a Kuwait crude (KC) oil containing 22% aromatics or a slightly heavier South Louisiana crude (SLC) oil with 17% aromatics (analysis given in American Petroleum Institute Report, 1974). The dose generally used, 0.2 ml, is approximately equivalent to 0.06 ml/kg BW for ducks and 0.3 ml/kg BW for gulls. It is both environmentally realistic and well below the concentration that causes lipid pneumonia in waterfowl (Hartung and Hunt, 1966). After treatment, birds were weighed and 0.2-ml blood samples were taken from their wing veins daily. At the end of the experiment, birds were decapitated and their livers, salt glands, and small intestines were fixed in 6% glutaraldehyde and processed for light-microscopic study (Karnaky *et al.*, 1976). Sodium, potassium–adenosinetriphosphatase (Na,K–ATPase) activity was assayed in freeze-dried homogenates of nasal gland and intestinal mucosa by the procedure developed by Miller *et al.* (1976a,b). Protein was determined by the method of Lowry *et al.* (1951) using crystalline serum albumin as standard; plasma sodium and potassium concentrations were determined by flame photometry.

RESULTS

Initial experiments showed that neither KC nor SLC administered at the 0.2-ml dose level affected plasma electrolytes or BW in ducks given FW. When the SLC dose level was raised to 2.0 ml, however, there was a small decrease in BW. In 1968, Fletcher and Holmes demonstrated that Pekin ducks possess only a limited ability to tolerate marine conditions since they cannot osmoregulate when they drink water containing more than 60% SW. To determine if crude oil ingestion affects the ability of Pekins to tolerate salt loading, the salinity of the drinking water was increased first to 60% SW and then to 100% SW. A representative experiment with KC-dosed ducks and paired controls is shown in Figure 1.

Crude oil did not affect osmoregulation in ducks maintained on 60% SW. In contrast, exposed ducks on 100% SW exhibited greater body

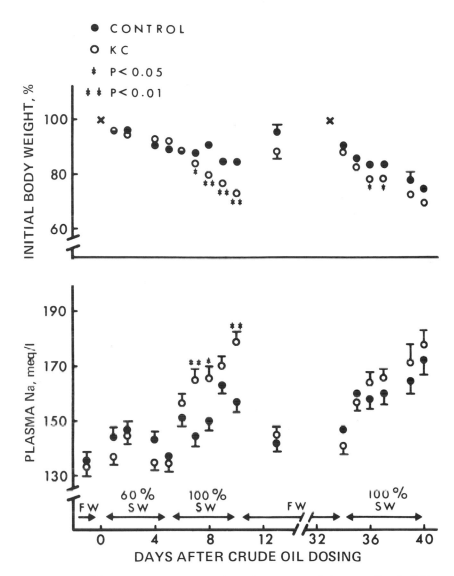

FIGURE 1 Effect of a single oral 0.2-ml dose of KC on BW and plasma sodium in Pekin ducks maintained on FW, 60% SW, or 100% SW. Each point represents the mean value derived from six birds.

weight loss and higher plasma sodium levels than controls. Plasma potassium was not affected by crude oil (not shown in the figure). Qualitatively similar results were obtained when these same ducks were again transferred to 100% sw 1 month after dosing (Figure 1). Significantly, in this and other experiments with kc, slc, and two refined oils, differences between the exposed and the control ducks were greatest during periods of heavy rain (Figure 1, days 7, 8, and 10) when ducks had some access to fw. To determine if crude oil impaired the ability of 100% sw-stressed ducks to utilize drinking water of lower salinity, we transferred 100% sw-stressed ducks (0.2-ml slc-dosed exposed ducks and paired controls) back to 60% sw. As shown in Figure 2, controls began to recover immediately after transfer, while the exposed ducks continued to exhibit elevated plasma sodium levels for an additional 3 days. Finally, when these birds were given fw to drink, plasma sodium levels in both exposed and control groups fell to fw values within 1 day (Figure 2).

In 1978, Miller et al. reported that herring gull chicks maintained on 100% sw showed minimal osmoregulatory impairment with kc ingestion and some significant impairment with slc (Figure 3). Both crude oils did, however, cause immediate cessation of growth, which for slc appeared to be related to reduced nutrient transport in the small intestine. slc-exposed gull intestines also exhibited major pathological changes in tissue morphology that may be related to impaired absorptive capacity (Figure 4).

These findings indicate that salt and water balance may be affected in oil-dosed ducks and gulls and that osmoregulatory organs are possible targets of crude oil action. Marine birds drinking hypertonic solutions osmoregulate by absorbing salt and water from the intestine and excreting excess salt via the nasal salt glands (Peaker and Linzell, 1975). Since the enzyme, Na,K–ATPase, is believed to be involved in the active ion-pumping process in both osmoregulatory organs (Ernst et al., 1967; Fletcher et al., 1967; Edmonds, 1974), Na,K–ATPase activity in homogenates of duck and gull salt glands and intestinal mucosa was assayed.

Preliminary experiments with slc indicate that effects on intestinal and salt gland ATPase activities are qualitatively similar to those found for kc-exposed ducks. Although kc did not affect salt gland ATPase activity in fw ducks, there was a 30% reduction in gland specific activity in kc-dosed ducks on 60% sw, and in both kc- and slc-dosed gulls (Table 1). Duck gland weight was not affected by crude oil, and the total gland ATPase activity was 35% lower than in controls (differ-

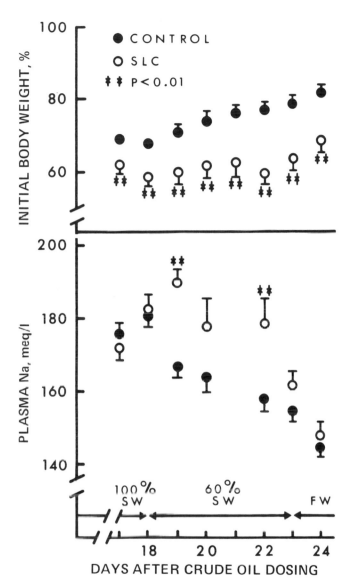

FIGURE 2 Effect of a single oral 0.2-ml dose of SLC on BW and plasma sodium in 100% SW-stressed Pekin ducks that have been transferred to 60% SW. Each point represents the mean value derived from 11 controls or six SLC birds.

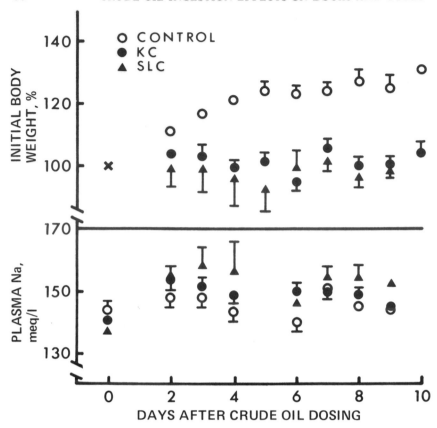

FIGURE 3 Effect of a single oral 0.2-ml dose of KC or SLC on BW and plasma sodium concentrations in herring gulls maintained on 100% SW. Each point represents the mean value derived from seven KC-dosed, six control, or six SLC-dosed birds. Body weight for dosed gulls was significantly lower (P < 0.05) than controls from day 2 on. On day 6, plasma sodium in KC-dosed birds was significantly higher than in controls. On days 6, 8, and 9, plasma sodium in SLC-dosed birds was significantly higher than in controls. Data from Miller *et al.*, 1978.

ence not significant). In gulls, the reduction in enzyme specific activity was accompanied by hypertrophy of salt gland tissue so that total gland Na,K–ATPase activity was not significantly reduced in KC-dosed gulls and reduced only 20% in SLC-dosed gulls (Table 1). Preliminary light microscopy of salt gland tissue from oil-dosed ducks and gulls indicated noticeably less complexity of basal and lateral membranes of secretory cells than in controls. Since histochemical studies have shown these membranes to be the site of the Na,K–ATPase (Ernst,

1972), our histological observations are consonant with findings of reduced enzyme activity in salt gland homogenates from oil-dosed birds. With regard to ATPase activity of the intestinal epithelium, neither the ducks nor the gulls were affected by KC (Table 1). SLC reduced epithelial Na,K–ATPase activity in the gull by 38%; however, because of the low enzyme activity in this tissue and the small sample size, SLC levels were only significantly different from controls at the 90% probability level.

In the gull, both KC and SLC caused liver hypertrophy and an approximately 100% increase in hepatic microsomal cytochrome P_{450} specific activity (Table 1; Miller *et al.*, 1978). In the duck, however, neither KC nor SLC ingestion (0.2- to 2.0-ml doses) caused liver hypertrophy (Table 1 and unpublished data of Miller, Kintcr, and Peakall). We have not yet determined whether duck hepatic microsomal enzyme activities are induced by crude oil ingestion.

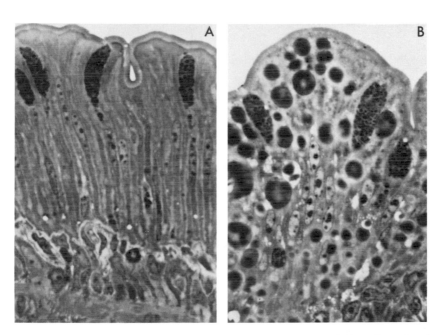

FIGURE 4 Light micrographs of small intestinal tissue from control (A) and SLC-dosed (B) gulls. Tissue from the dosed gulls exhibits proliferative edema with numerous dark bodies in the epithelial mucosa. Bodies were Sudan black-positive suggesting droplets of lipid (Lillie, 1965). Intestinal slices were fixed in 6% glutaraldehyde, exposed to osmium vapor, and embedded in Epon. Epon sections were stained with methylene blue-azure II and photographed using phase contrast optics. (×1,250.)

TABLE 1   Effects on Organ Weights and Na,K–ATPase Activities in Pekin Ducks and Herring Gulls Given Single Oral 0.2-ml Doses of Crude Oil[a]

	Duck				Gull		
	FW		60% sw (5d)		100% sw		
	Control (N = 3)	KC (N = 3)	Control (N = 6)	KC (N = 6)	Control (N = 5)	KC (N = 6)	SLC (N = 4)
Salt Gland (one gland)							
gland wt, mg	158 ± 10	155 ± 24	398 ± 60	399 ± 39	387 ± 22	436 ± 28	451 ± 32
gland wt, mg/kg BW	55 ± 3	52 ± 5	124 ± 16	138 ± 13	454 ± 17	563 ± 14[c]	575 ± 37[b]
Na,K–ATPase specific activity, μmol Pi/mg protein × h	8 ± 2	13 ± 2	40 ± 3	29 ± 3[b]	47 ± 2	36 ± 1[c]	32 ± 2[c]
Na,K–ATPase total activity, μmol Pi/mg gland × h	67 ± 20	93 ± 6	868 ± 156	561 ± 94	1,037 ± 74	908 ± 76	817 ± 50
Intestine							
Na,K–ATPase specific activity, μmol Pi/mg protein × h	6 ± 1	5 ± 1	5 ± 1	5 ± 1	2.8 ± 0.4	2.3 ± 0.6	1.8 ± 0.6
Liver							
organ wt, g	53 ± 3	50 ± 2	45 ± 3	41 ± 2	22 ± 1	30 ± 3[b]	33 ± 2[c]
organ wt, g/kg BW	19 ± 1	17 ± 1	15 ± 2	14 ± 1	26 ± 1	38 ± 2[c]	42 ± 1[c]

[a] Birds sacrificed 7 to 9 days after dosing. Data given as mean ± SE with the number of birds in parentheses. ATPase total activity calculated from specific activity, gland weight, and approximate protein content of gland. Gull data taken from Miller et al. (1978). BW = body weight.
[b] Significantly different from controls, P < 0.05.
[c] Significantly different from controls, P < 0.01.

DISCUSSION

Because of the proposed use of the Pekin duck and herring gull as laboratory models for coastal and oceanic birds, respectively, it is of particular interest to compare the effects of crude oil in these two species. Crude oil ingestion only affected 100% sw-stressed ducks, slightly decreasing their capacity to extract solute-free water from sw and significantly delaying their ability to recover when water of lower salinity was available. In 100% sw-adapted gulls, crude oil caused similar osmoregulatory impairment and substantial growth inhibition. Significantly, in preliminary experiments with the black guillemot (*Cepphus grylle*), a true oceanic species, there were slightly elevated plasma sodium levels and substantially reduced BW gains after KC dosing (Miller *et al.*, 1978).

Fletcher *et al.* (1967) and Ernst *et al.* (1967) have shown that salt gland sodium extrusion rates in the Pekin duck exhibit a significant positive correlation with gland Na,K–ATPase activity. Since ducks maintained on hypertonic saline excrete approximately 90% of the excess ingested sodium by extrarenal routes (Fletcher and Holmes, 1968), a reduction in sodium pump activity in the salt gland should be detectable as a decrease in osmoregulatory ability in birds drinking hypertonic solutions. Thus, our data suggest a cause and effect relationship between reduced Na,K–ATPase activity and impaired osmoregulation in oil-dosed ducks and gulls. In gulls, but not ducks, reduced salt gland ATPase specific activity was partially offset by hypertrophy of the gland tissue. Since Miller *et al.* (1976a) found similar compensatory hypertrophy of salt gland tissue in DDE-fed common puffins (*Fratercula arctica*), it is possible that marine birds possess an "osmoregulatory reserve" that allows them to tolerate limited reduction in nasal gland Na,K–ATPase levels without serious impairment of osmoregulatory ability.

The importance of the intestine as a target for crude oil toxicity remains to be determined. The reduced nutrient transport rates and Na,K–ATPase activities in the SLC-dosed gull intestine may be related to the observed growth inhibition and limited impairment of plasma osmoregulation (Miller *et al.*, 1978); but, growth was also inhibited in KC-dosed gulls whose intestinal transport rates were not reduced. Crocker *et al.* (1975) have demonstrated that ingestion of several crudes (including one KC) substantially inhibited salt and water transport in the intestine of saline-adapted Pekin ducklings. Our data clearly show no reduction in intestinal Na,K–ATPase activity in KC-dosed ducks (Table 1). However, the observation that saline adaptation causes increased intestinal salt and water transport in ducks (Crocker and Holmes, 1971) with no increase in Na,K–ATPase activity (Table 1 and D. Miller, unpublished observations) certainly suggests compatibility of our data with those of Crocker *et al*. Perhaps another essential component of intestinal transport, e.g., a carrier in the brush border, is also affected by crude oil.

Recently, an epiornithic (large kill) of young alcids (family, Alcidae) and shearwaters (family, Procellariidae) and severe population declines in the puffin have been reported on both sides of the Atlantic (Watson, 1970; Holdgate, 1971; Flegg, 1972; Nettleship and Lock, 1973; Lloyd *et al*., 1974). Evidence provided by Miller *et al*. (1976a) and Lincer and Peakall (1973) suggests that organochlorine pollutants, such as dichlorodiphenylethylene (DDE) or polychlorinated biphenyls (PCB's), are not responsible. The data reported in this paper indicate, however, that small amounts of ingested crude oils do produce multiple sublethal effects that could reduce a sea bird's capacity for long-term survival. Limited evidence also suggests a long-term course of crude oil action. Therefore, when considered in conjunction with stress factors which birds at sea might normally encounter, e.g., food shortages and severe storms, the potential for lethality is considerable. Before we can conclude that oil pollution is responsible for the recently reported sea bird mortalities, additional work is needed to define fully the time course and dose–response for each of the crude oil effects, to identify which component(s) of these extremely complex mixtures are responsible, and to compare the results of such laboratory toxicity studies with available data on species affected in the wild.

SUMMARY

A single small oral dose of either Kuwait or South Louisiana crude oil was administered to Pekin ducks and herring gull chicks. Crude oil dosing caused significant osmoregulatory impairment in seawater-stressed ducks. In seawater-adapted gulls osmoregulation was only slightly impaired by crude oil; however, growth was substantially inhibited. These findings suggest that crude oil causes sublethal effects which might impair a bird's ability to survive in a coastal or marine environment.

ACKNOWLEDGMENTS

This study was supported by U.S. Public Health Service grant ES-00920 and National Science Foundation grant PCM 75-03098. We thank C. Bunker, E.

Burnham, H. Church, H. Doerstling, B. Peakall, and M. Ratner for excellent technical assistance. Analyzed reference oil samples were kindly provided by the American Petroleum Institute.

REFERENCES

American Petroleum Institute. 1974. American Petroleum Institute Reference Oils. API Report No. AID.1BA.74. American Petroleum Institute, Washington, D.C. 5 pp.

Conder, P. 1968. To clean or kill. Birds (London) 2:56.

Crocker, A. D., and W. N. Holmes. 1971. Intestinal absorption in ducklings (*Anas platyrhynchos*) maintained on fresh water and hypertonic saline. Comp. Biochem. Physiol. 40A:203–211.

Crocker, A. D., J. Cronshaw, and W. N. Holmes. 1974. The effect of crude oil on intestinal absorption in ducklings (*Anas platyrhynchos*). Environ. Pollut. 7:165–177.

Crocker, A. D., J. Cronshaw, and W. N. Holmes. 1975. The effect of several crude oils and some petroleum distillation fractions on intestinal absorption in ducklings (*Anas platyrhynchos*). Environ. Physiol. Biochem. 5:92–106.

Edmonds, C. J. 1974. Salts and water. Pp. 711 in D. H. Smyth, ed. Intestinal Absorption. Plenum Press, London.

Ernst, S. A. 1972. Transport adenosine–triphosphatase cytochemistry. 2. Cytochemical localization of ouabain-sensitive, potassium-dependent phosphatase activity in the secretory epithelium of avian salt gland. J. Histochem. Cytochem. 20:23–28.

Ernst, S. A., C. C. Goertmiller, and R. A. Ellis. 1967. The effect of salt regimens on the development of (Na$^+$K$^+$)-dependent ATPase activity during the growth of salt glands in ducklings. Biochem. Biophys. Acta 135:682–692.

Flegg, J. M. 1972. The puffin on St. Kelda, 1969–1971. Bird Study 19:7–17.

Fletcher, G. L., and W. N. Holmes. 1968. Observations on the intake of water and electrolytes by the duck (*Anas platyrhynchos*) maintained on fresh water and on hypertonic saline. J. Exp. Biol. 49:325–339.

Fletcher, G. L., I. M. Stainer, and W. N. Holmes. 1967. Sequential changes in adenosine triphosphatase activity and the electrolyte excretory capacity of the nasal glands of the duck during the period of adaptation to salt-water. J. Exp. Biol. 47:375–391.

Hartung, R., and G. S. Hunt. 1966. Toxicity of some oils to waterfowl. J. Wildl. Manag. 30:564–570.

Holdgate, M. W., ed. 1971. The Seabird Wreck of 1969 in the Irish Sea. Publ. C, No. 4. Natural Environment Research Council, London. 17 pp.

Karnaky, K. J., Jr., L. B. Kinter, W. B. Kinter, and C. E. Stirling. 1976. Teleost chloride cell. II. Autoradiographic localization of gill Na,K–ATPase in killifish (*Fundulus heteroclitus*) adapted to low and high salinity environments. J. Cell Biol. 70:157–177.

Lillie, R. D. 1965. Pp. 129–132 in Histopathologic Technic and Practical Histochemistry. McGraw-Hill, New York.

Lincer, J. L., and D. B. Peakall. 1973. PCB pharmacodynamics in the ring dove and early gas chromatographic peak diminution. Environ. Pollut. 4:59–68.

Lloyd, D., J. A. Bogan, W. R. P. Bourne, P. Dawson, F. L. F. Parslow, and A. G. Stewart. 1974. Seabird mortality in the North Irish Sea and Firth of Clyde early in 1974. Mar. Pollut. Bull. 5:136–140.

Lowry, H., N. M. Rosenbrough, A. L. Farr, and R. J. Randall. 1951. Protein measurement with the Folin phenol reagent. J. Biol. Chem. 193:265–275.

Miller, D. S., W. B. Kinter, D. B. Peakall, and R. W. Risebrough. 1976a. DDE feeding and plasma osmoregulation in ducks, guillemots and puffins. Am. J. Physiol. 231:370–376.

Miller, D. S., D. B. Peakall, and W. B. Kinter. 1976b. The enzymatic basis for DDE-induced eggshell thinning in a sensitive bird. Nature 259:122–124.

Miller, D. S., D. B. Peakall, and W. B. Kinter. 1978. Crude oil ingestion: sublethal effects in herring gull chicks. Science 199:315–317.

Nettleship, D. N., and A. R. Lock. 1973. Tenth census of seabirds in the sanctuaries of the North Shore of the Gulf of St. Lawrence. Can. Field Nat. 87:395–402.

Peaker, M., and J. L. Linzell. 1975. Pp. 191–194 in Salt Glands in Birds and Reptiles. Cambridge University Press, Cambridge.

Travers, W. B., and P. R. Luney. 1976. Drilling, tankers and oil spills on the Atlantic outer continental shelf. Science 194:791–796.

Watson, G. E. 1970. A shearwater mortality on the Atlantic Ocean. Atl. Nat. 25:75–80.

QUESTIONS AND ANSWERS

G. CHOULES: We've done some work with an industrially polluted pond in Colorado (Choules et al., 1978)[1] and found that mallard ducks (Anas platyrynchos) would not drink pond water. Also, in these cases the ducks would not preen. We believe that this was an external effect due more to the detergent in the water than to any of the poisonous materials. Is it possible that crude oil is less disturbing to ducks in the wild than detergents, and that the ducks in your study ingested crude oil through preening?

D. MILLER: About 10 years ago, Hartung and Hunt[2] reported that lightly oiled waterfowl ingested a considerable amount of oil during preening. I do not know of any published reports indicating that birds avoid water or food that is contaminated by oil.

J. ZINKL: I have studied several oiled western grebes (Aechmophorus occidentalis) from a pond in Colorado. The birds had a considerable amount of oil on their feathers. Necropsy revealed large amounts of oil in the intestinal tract, proventriculus, and gizzard.

D. MILLER: Did you notice any evidence of enteritis?

J. ZINKL: Since the birds were autolyzed only a gross necropsy was performed.

J. PAYNE: I wouldn't generalize from the doses that you gave because they're very light crudes. I think most of the toxicity is from the low molecular weight aromatics, which are rapidly lost by evaporation in most spills and are not present in certain crude oils.

D. MILLER: We chose these crude oils because they are representative of crude oils that are found in oil spills. We are going to study the effects of weathering on the toxicity of these oils.

---

[1] Choules, G. L., W. C. Russell, and D. A. Gauthier. 1978. Duck mortality from detergent polluted water. J. Wildl. Manag. 42(2):410–414.

[2] Hartung, R., and G. S. Hunt. 1966. Toxicity of some oils to waterfowl. J. Wildl. Manag. 30:564–570.

J. COUCH: Has anyone conducted long-term studies on the interaction between polycyclic hydrocarbons and natural pathogens in birds? We have been studying the effect of polychlorinated biphenyls (PCB's) on certain marine invertebrates and have found that, in some cases, natural pathogens (e.g., a virus) and other diseases were enhanced by exposure to PCB's.

D. MILLER: I know of no reports attributing mortality in oil-dosed birds to specific pathogens. However, Hartung and Hunt did mention that oil-dosed birds stressed by either cold or crowding showed increased mortality over nonstressed oil-dosed birds.

J. COUCH: I think this is critical because in a natural situation, wildlife interacting over the long-term with pollutants, particularly crude oils and PCB's, may show an impaired immune competence which renders them susceptible to a wide variety of opportunistic pathogens.

J. ROBERTS: Have you considered the combined interactions of emulsifiers and solvents that may be used in programs of clean-up following oil spills? This may be an important point for consideration.

D. MILLER: In 1974, Crocker *et al.*[3] reported that detergent augments the inhibiting effect of crude oil on intestinal salt and water transport.

D. SCARPELLI: Has anyone studied the livers of these birds with special attention to the endoplasmic reticulum to see if the smooth-surfaced reticulum is hypertrophied and if drug metabolizing enzymes are increased?

D. MILLER: Not that I know of.

D. SCARPELLI: It is significant that you found changes in the sodium–potassium pump enzyme which is membrane bound. It may be that long-chain hydrocarbons of crude oil are soluble in lipid, concentrate in the lipid layer of cell membranes, and affect sodium potassium-activated ATPase and ion transport. It is also interesting that amino acid uptake appears to be impaired, since there is evidence that amino acid uptake is probably coupled to sodium–potassium-activated ATPase activity.

M. FRIEND: I have been working on salt glands with other environmental pollutants and the results differ from your results with crude oil. Did you find that salt glands increase relative to body weight with the oil-treated birds?

D. MILLER: We found that salt glands increased in gulls but not in ducks.

M. FRIEND: As the bird gets older, there is an increase in salt gland weight relative to body weight. In birds exposed to parathion, salt gland weights decreased markedly relative to body weights due to atrophy secondary to cholinesterase inhibition. Such atrophied glands also have impaired sodium transport. Simultaneous exposure to crude oils and organophosphates could exert a synergistic lowering effect on sodium transport, which in the long run could seriously affect waterfowl. In 1964, a Canadian scientist, Cooch[4],

---

[3]Crocker, A. D., J. Cronshaw, and W. N. Holmes. 1974. The effect of crude oil on intestinal absorption in ducklings (*Anas platyrhynchos*). Environ. Pollut. 7:165–177.
[4] Cooch, F. G. 1964. A preliminary study of the survival value of a functional salt gland in prairie *Anatidae*. Auk 81:380–393.

hypothesized that sublethal levels of botulinum toxin inhibit salt gland function, which may suggest part of the disease mechanism in botulism.

D. MILLER: Another possibility that comes to mind is that birds exposed simultaneously to carbamates and crude oils could show a marked impairment of osmoregulation due to the additive effect of the two substances.

M. FRIEND: Did you analyze the secretions from the salt glands in oil-treated birds? We found that after exposure to certain organophosphates, the salt gland secreted water but not salt.

D. MILLER: We have tried this and have found that after an intravenous salt load, secretion rates are highly variable. We have therefore discontinued that means of assessing salt gland function and now monitor plasma electrolyte levels in birds continuously exposed to seawater. We believe that this approach is more environmentally realistic.

Hydrocarbon Pollution and the Prevalence of Neoplasia in New England Soft-Shell Clams (*Mya arenaria*)

ROBERT S. BROWN, RICHARD E. WOLKE,
CHRIS W. BROWN, and SAUL B. SAILA

Because of the increasing presence of hazardous organic chemical pollutants in the aquatic environment and the growing number of reports of a variety of lesions in aquatic organisms, especially neoplasms in molluscs (N.Y. Academy of Sciences, 1977), it is imperative that the role of pollution in aquatic environmental pathology be investigated. Neoplastic lesions have been found in molluscs from areas previously regarded as relatively unpolluted (Christensen *et al.*, 1974; Farley, 1969a,b; Farley and Sparks, 1970; Frierman and Andrews, 1976; Goner, 1971; Mix, 1975). There has also been a high prevalence of neoplastic lesions in shellfish exposed to petroleum-derived hydrocarbon (PDH) pollution (Barry and Yevich, 1972, 1975; Brown *et al.*, 1976, 1977).

These findings provided the impetus for us to begin a 3-year study in January 1976 to determine the occurrence of these lesions in soft-shell clam populations and to study their relationship to PDH pollution.

The purpose of this paper is to present the ecological, pathological, and chemical data collected in the first year of the study, and to relate these findings to the biology, prevalence, and etiology of neoplasms in molluscs.

MATERIALS AND METHODS

The organization of this study and the ecological and histopathological methods used have been detailed by Brown *et al.* (1976, 1977). Briefly, the ecologists collected clams from selected New England sites that

41

had different types and degrees of pollution (based on data gathered from the literature or from records of state natural resource agencies). The chemists performed infrared spectrophotometric and/or gas chromatographic analyses of the clams and associated sediments. The pathologists collected size data (length, width, height, total weight, tissue weight) and examined Dietrich's fixed, hematoxylin- and eosin-stained tissue sections to determine sex and to identify lesions, especially those resembling neoplasms.

A mortality study was initiated to evaluate the lethality of the neoplastic condition. Two hundred and sixty-three clams collected from two areas where the disease was prevalent were maintained in flowing seawater systems for 9 months. Histopathological analyses were performed on animals that died during the study.

For chemical analysis, composite clam samples (~30 g wet weight) were freeze-dried to remove all water. Then, clam and sediment samples were treated identically. The dry weight of a sediment sample was determined by weighing 5 g, drying at 100°C, then reweighing. Approximately 5 g of the freeze-dried clams or ~30 g (wet weight) of sediments were added to a 100-ml mixture of 70% 0.5 N potassium hydroxide/methyl hydroxide (KOH/MeOH) and 30% benzene. Then, 10 ml of water was added. For gas chromatographic measurements, an internal standard of 5 μg of eicosane in benzene was added. The mixture was refluxed for 2 h. This released the hydrocarbons into the benzene layer and the salts of the esters into the MeOH/H$_2$O layer. The mixture was cooled and filtered and the sediment or clam sample washed with 20 ml of methanol and 20 ml of hexane. The filtrates were combined and enough water added to bring its volume up to that of the methanol. The organic phase containing most of the hydrocarbons was separated; the aqueous phase was extracted twice with 20 ml of hexane; and the organic phases were combined.

After the organic phase was separated, the relatively nonpolar hydrocarbons were separated from the other highly polar compounds using an alumina/copper column with 95/5 hexane/benzene to elute the column. The effluent was then concentrated by evaporating most of the solvent, and the sample was further separated by thin layer chromatography using silica gel plates and a mixture of 100/1 hexane/ammonium hydroxide to develop the plates. Following this, the gas chromatogram (GC) of the sample was measured. A baseline was measured on the GC prior to each sample injection. The baseline was subtracted from the sample's chromatogram, and the area of the eicosane peak (internal standard) and the total area minus the eicosane peak were measured with a planimeter. The area was then converted to weight by comparison to the eicosane internal standard.

GC's were measured on a Hewlett-Packard Model 5710H using flame ionization detection and a 3-m column of 3% SP 2250 on 100/200 Suppeloport. The column was temperature-programmed between 90°C and 300°C at 8°/min.

RESULTS

One thousand eight hundred thirteen clams were collected from 10 sites with differing types and degrees of PDH pollution (Table 1). The percent of clams having neoplastic lesions ranged from 0% (clams collected near Providence, R.I.) to 39% (clams collected near Quonset, R.I.). These lesions, described in detail by Brown *et al.* (1976, 1977), consisted of markedly anaplastic, mitotically active cells occurring either in connective tissue or gonadal follicles, and displayed both invasive and metastatic behavior. All the neoplasms were apparently of hematopoietic origin, with the exception of the majority of those in clams from Searsport, Maine, which were of gonadal origin (Figures 1 and 2).

In the mortality study, to evaluate the long-term effects of these lesions, 22 of the 33 clams that died were examined histopathologically; 13 hematopoietic and 4 gonadal neoplasms were found (Table 2).

TABLE 1 Prevalence of Neoplasia

Area of Collection	Condition of Environment	Number of Clams Examined	Clams Having Neoplasia, %
Sandy Point, R.I.	Clean	10	20
Wickford, R.I.	Clean	203	4
Providence, R.I.	Industrial/ domestic sewage	90	0
East Greenwich, R.I.	Domestic sewage	349	4
Quonochontaug, R.I.	Clean	199	2
Goose Cove, Maine	Heavy metals	192	1
Portland, Maine	Industrial/ domestic sewage	197	4
Bourne, Mass.	#2 fuel oil spill	320	31[a]
Searsport, Maine	#2 fuel/JP-5 Oil	204	13[a]
Quonset, R.I.	Dumpsite	49	39[a]
TOTAL		1,813	

[a] $P \leq 0.05$.

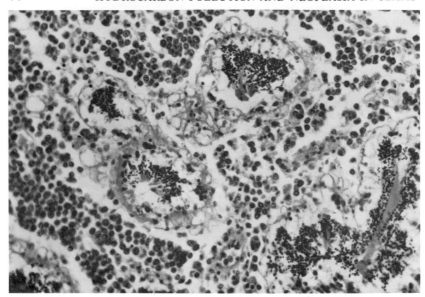

FIGURE 1 Hematopoietic neoplasm. Note individual anaplastic cells in connective tissue between gonadal follicles. (Hematoxylin and eosin. ×500.)

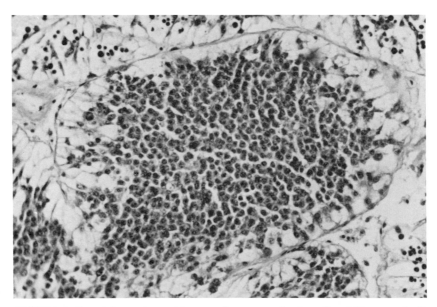

FIGURE 2 Gonadal neoplasm. Note contiguous anaplastic cells within a markedly enlarged gonadal follicle. (Hematoxylin and eosin. ×500.)

Predictive regression analyses were computed in order to compare the following subsamples: male, female, unsexed (no gametes present), and both neoplastic and nonneoplastic males, females, and unsexed clams. The slopes of the regression lines of these data were compared. There were no substantial mean size differences between neoplastic and nonneoplastic normal animals (this implied no age-related susceptibility to neoplasia); however, there was a significant difference between their tissue weights. The rate at which the tissue weight increased as a function of length was lower for the neoplastic animals. Chi-square analyses comparing frequencies of neoplasia by sex showed no significant differences in the proportions of neoplasia. Chi-square analyses of proportions of all neoplastic versus all nonneoplastic clams by sample site showed significant differences among the sample areas. Samples from Bourne, Mass., Quonset, R.I., and Searsport, Maine, had a higher observed frequency of neoplasia than all other sites examined.

Hydrocarbons in composite clam samples from seven sites and in an associated sediment sample from one site were analyzed by gas chromatography (Table 3). Clam samples from two of the sites were collected and analyzed on two different occasions. The total hydrocarbons in the first five samples were of the same order of magnitude (0.0148 to 0.0264 mg/g). The two samples from East Greenwich, R.I., contained widely varying amounts of hydrocarbons; the first sample contained the largest amount (0.25 mg/g), whereas the second sample contained the smallest detectable amount (0.0009 mg/g). These samples were collected near a sewage outfall, which may account for the variation.

High percentages of neoplasia were found at sites with significant amounts of hydrocarbons. Two exceptions to this were the Portland and East Greenwich I samples, which were high in hydrocarbons but low in neoplasia. The large amount of hydrocarbons in the East Greenwich sample was also probably coincidental.

To explain the Portland values, we performed an additional gas chromatographic analysis. The ratio of the unresolved "hump" in the chromatogram to the resolved peaks was determined. In general, the peaks were primarily due to normal paraffins (i.e., straight chain paraffins). This ratio reflects the amount of all other hydrocarbons including saturated and unsaturated cyclics to the normal paraffins. In many of the gas chromatograms, the "hump" had two maxima; one below and one above the retention time for eicosane. The ratio of the heights of these two maxima was determined by dividing the height of the maximum at the longer retention time (highest temperature) by that

TABLE 2 Mortality Study

Area of Collection	Number of Mortalities								Totals, Number Dead/Total Collected	%	Autopsies/ Total Dead	Number of Clams with Neoplastic Lesions	
	1976				1977							Hematopoietic	Gonadal
	Sept.	Oct.	Nov.	Dec.	Jan.	Feb.	Mar.	Apr.					
Bourne, Mass. I	3	9	3	1	0	0	0	1	17/88	14	} 18/29[a]		
Bourne, Mass. II	1	6	3	1	0	0	1	0	12/45	27		13	
Searsport, Maine	0	0	1	2	0	0	0	1	4/100	4	4/4		4

[a] Combined total of dead from sites I and II. Only 18 were autopsied because the autolysis in the remaining 11 was sufficient to prevent diagnosis.

TABLE 3 Comparison of the Hydrocarbon Content to the Prevalence of Neoplastic Lesions in Soft-Shell Clams (*Mya arenaria*) from Nine Regions in New England

Location	Total Hydrocarbons in Clams, mg/g wet wt	Ratio of Resolved: Unresolved Hydrocarbons	Ratio of Large: Small Cyclic Hydrocarbons	Clams with Neoplasia, %
Searsport	0.0280	4.6	1.36	13
Quonset[a]	0.0177	3.7	0.73	39
Bourne	0.0148	7.2	0.85	31
Portland	0.0264	18.0	0.38	4
East Greenwich				
I	0.2500	7.5	0.33	4
II	0.0009	—	—	0
Goose Cove	Nondetectable	—	—	1
Quonochontaug	Nondetectable	—	—	2
Deer Island	Nondetectable	—	—	0

[a] Associated sediments.

of the shorter retention time. Since the "hump" is primarily due to saturated and unsaturated cyclics, this ratio is indicative of the ratio of large to small cyclic compounds (Table 3).

The ratio of unresolved:resolved was very high for the clams from Portland compared to those of the other samples; this indicated that the amount of *n*-paraffins was low, which may be accounted for by biodegradation of aged oils. Since large quantities of heavy crudes are transferred from tankers to pipelines in the Portland Harbor, oil has probably been accumulating in the bottom sediments for many years. Similar chromatograms were obtained from extracts of clams collected in the Providence Harbor, which is the terminus for shipment of large quantities of heavy, industrial fuel oils to New England.

DISCUSSION

Neoplasia was prevalent in the New England soft-shell clam populations that were sampled. The condition was found in geographically distant sites and was epizootic in three areas. Based on cytological findings and preliminary results in which a large percentage of dying clams had disseminated neoplastic lesions, the disease appears to be malignant and chronic. The finding that neoplastic clams weighed significantly less than nonneoplastic ones suggested that they were in a state of debilitation, perhaps analogous to the cachexia seen in vertebrates suffering from malignant neoplasia. A comparison of the ecological, histopathological, and chemical data suggests a multifactorial etiology.

The current chemical data base is too small and the pathology data too equivocal to make valid statistical analyses and conclusions regarding the relationship between the prevalence of the lesions and the presence or types of hydrocarbons. There appeared to be a trend of neoplasia occurring primarily in populations retaining hydrocarbons, but this was not a clear and direct relationship. Clams from all areas historically recognized as contaminated by PDH pollution had similar concentrations of hydrocarbons, but they had both the lowest (Providence and Portland) and the highest (Bourne, Searsport, and Quonset) prevalences of neoplasia.

The hydrocarbons in clams were composed of smaller cyclics rather than larger cyclics, with the exception of those from Searsport, Maine. Since many of the large unsaturated cyclics are oncogenic in mammals, there may be a correlation between the ratio of large to small cyclic compounds and prevalance of neoplasia. However, this does not appear to be the case for the limited number of samples in our study.

Factors other than PDH pollution are probably involved in the causation and prevalence of neoplasia in molluscs. A smaller percentage of clams from other pollution-impacted sites, as well as pristine sites having no evidence or history of PDH pollution, also appeared to be susceptible to the development of neoplasia. This suggests that other factors such as an infectious agent or genetic susceptibility may play a role in oncogenesis. In fact, if the prevalence of such lesions in soft-shell clams is contrasted with other species that have been studied, the soft-shell clam appears to be the most susceptible molluscan species (Table 4). In preliminary results from a comparative study (Brown, 1977), 39% of the soft-shell clams from Quonset, R.I., a PDH-polluted site, had neoplasms, while none were found in 156 quahogs (*Mercenaria mercenaria*) examined.

Results did not identify the etiology of neoplastic lesions in soft-shell clams. The information gathered in the next year of the study may help to determine this point.

SUMMARY

Ecological, chemical, and pathological analyses were made on soft-shell clams from 10 New England sites of varying types and degrees of hydrocarbon pollution (nominal, oil spill, industrial and domestic sewage, and heavy metal). To date, over 1,300 clams have been histopathologically examined. Of these, 162 had neoplastic lesions.

TABLE 4   Occurrence of Neoplasia in Molluscan Species

Species	Highest Percentage of Molluscs with Neoplasia as Reported in Other Studies	Reference
American oyster		
(*Crassostrea virginica*)	{ 0.02	Farley, 1969b
	5.94	Frierman and Andrews, 1976
Native Pacific oyster		
(*Ostrea lurida*)	12.00	Farley and Sparks, 1970
Quahog		
(*Mercenaria mercenaria*)	3.70	Barry and Yevich, 1972
Blue mussel		
(*Mytilus edulis*)	12.00	Farley, 1969a
Irus macoma		
(*Macoma irus*)	5.00	Goner, 1971
Bent-nose macoma		
(*Macoma nasuta*)	5.00	Goner, 1971
Balthic macoma		
(*Macoma balthica*)	17.00	Christensen *et al.*, 1974
Soft-shell clam		
(*Mya arenaria*)	{ 26.60	Barry and Yevich, 1972
	31.00	Brown *et al.*, 1976, 1977

Clams from one site had predominately gonadal neoplasms, while the majority were of hematopoietic origin. Cells of both types were markedly anaplastic, invasive, and appeared to have metatasized.

The amounts of hydrocarbons found in both clams and sediments by gas chromatography and infrared spectroscopy were in agreement with the ecological histories of the sites. The amount of hydrocarbons in clams was related to the amount in the associated sediments. There was no correlation between the amount of total hydrocarbons in clams and the prevalence of neoplasms. However, there was a correlation between the amount of polynuclear aromatic hydrocarbons in clams and the prevalence of these lesions.

ACKNOWLEDGMENTS

This study was supported by American Petroleum Institute grant #98-20-7372. Robert S. Brown wishes to thank Vicky Murray for preparing tissues for histopathological analysis and Richard Appeldoorn for selecting sites and collecting clams. This paper is contribution #1791 of the Rhode Island Agricultural Experiment Station.

REFERENCES

Barry, M. M., and P. P. Yevich. 1972. Incidence of gonadal cancer in the quahaug, *Mercenaria mercenaria*. Oncology 26:87–97.

Barry, M. M., and P. P. Yevich. 1975. The ecological, chemical and histopathological evaluation of an oil spill site. Part II. Histopathological studies. Mar. Pollut. Bull. 6:171–173.

Brown, R. S. 1977. The redevelopment of Quonset/Davisville: An environmental assessment. Technical Appendix No. 4. Histopathological Findings in *Mya arenaria* and *Mercenaria mercenaria* samples from Quonset Point/Davisville, Rhode Island. Marine Technical Report No. 55 Coastal Resources Center, University of Rhode Island, Narragansett. 9 pp.

Brown, R. S., R. E. Wolke, and S. B. Saila. 1976. Preliminary report on neoplasia in feral populations of the soft-shell clam. *Mya arenaria*: prevalence, histopathology and diagnosis. Pp. 151–158 in Proceedings of the First International Colloquium on Invertebrate Pathology. Printing Department, Queen's University, Kingston, Canada.

Brown, R. S., R. E. Wolke, S. B. Saila, and C. W. Brown. 1977. Prevalence of neoplasia in ten New England populations of the soft-shell clam, *Mya arenaria*. Ann. N.Y. Acad. Sci. 298:522–534.

Christensen, D. J., C. A. Farley, and F. G. Kern. 1974. Epizootic neoplasms in the clam *Macoma balthica* (L.) from Chesapeake Bay. J. Natl. Cancer Inst. 52:1739–1749.

Farley, C. A. 1969a. Sarcomatoid proliferative disease in a wild population of edible mussels (*Mytilus edulis*). J. Natl. Cancer Inst. 4:509–516.

Farley, C. A. 1969b. Probable neoplastic disease of the hematopoietic system in oysters, *Crassostrea virginica* and *Crassostrea gigas*. J. Natl. Cancer Inst. 31:541–555.

Farley, C. A., and A. K. Sparks. 1970. Proliferative disorders of hemocytes, endothelial cells and connective tissue cells in molluscs. Bibl. Haematol. (Basel) 36:610–617.

Frierman, E. M., and J. D. Andrews. 1976. Occurrence of hematopoietic neoplasms in Virginia oysters (*Crassostrea virginica*). J. Natl. Cancer Inst. 56:319–324.

Goner, J. J. 1971. Sea Grant Memorandum, Marine Science Center, Oregon State University.

Mix, M. 1975. Neoplastic disease of Yaquina Bay bivalve molluscs. Proc. 13th Annu. Hanford Biol. Symp. 1:369–386.

N. Y. Academy of Sciences. Conference on Aquatic Pollutants and Biological Effects, with Emphasis on Neoplasia, September 27–29, 1976, Ann. N.Y. Acad. Sci. Vol. 298. 604 pp.

QUESTIONS AND ANSWERS

M. MIX: Did you find any seasonality or any variance in the prevalence rate from one month to the next?

R. BROWN: It is difficult for me to answer that question because in most cases we examined one site only one time and therefore could only describe prevalence at the time of sampling. We tried to look at as many sites in the first year of the study as possible to give us some background data. In the second year of our study we selected areas of high and low incidence of gonadal and hematopoietic neoplasms to determine whether the disease was increasing or decreasing in the population. Upon our return to the Searsport and Quonset locations, we found the disease was still prevalent in the population and in fact had increased slightly.

D. LISK: In mammalian neoplasia there is a latency period from time of exposure to the carcinogen to manifestation of the tumor. Did you correlate any of the spills on chronic exposures with the incidence of disease?

R. BROWN: No. I think that one of the pitfalls of the ecological approach is that in many instances we really do not know the details of the ecological situation before a certain environmental insult has occurred. In a study conducted by the Environmental Protection Agency in Searsport, where the sampling began immediately, gonadal neoplasms were found about 6 to 9 months after the spill.

G. CHOULES: Have you tried injecting tumor cell extracts into normal molluscs to determine the transplantability of the tumors?

R. BROWN: I have tried transplanting whole cells from one mollusc to another, and, in one instance, one animal out of the five inoculated died a week following injection. At autopsy, we were able to identify tumor cells in sufficient numbers to indicate that these cells proliferated following transplantation. Attempts to reproduce these results have not been successful to date.

C. ANDERSON: Have you been able to determine the primary site of these neoplasms?

R. BROWN: This is a difficult question. A number of these tumors have been classified cytologically as having a hematopoietic origin, although no hematopoietic organ has yet been unequivocally identified. The gonadal neoplasms appear to arise from the germinal epithelium of the gonadal follicle.

J. WEIS: In your table you showed that *Mercenaria* have a lower incidence of tumors. There have been some reports concerning the presence of antitumor substances in *Mercenaria* (Schmeer, 1966; Schmeer *et al.*, 1966)[1]. Do you think that this particular clam may be more resistant to the development of neoplasms because it contains an antitumor substance?

R. BROWN: I don't know if the mercenene, as the antitumor substance is called, is as active against tumor cells of molluscan origin as against those of mammalian origin as Dr. Schmeer reported.

G. MIGAKI: In the gonadal neoplasms, are there any histological differences between testicular and ovarian tumors?

R. BROWN: No. The appearance is the same regardless of sex.

[1]Schmeer, M. R. 1966. Mercenene: growth inhibiting agent of *Mercenaria* extracts—further chemical and biological characterization. Ann. N.Y. Acad. Sci. 136:213–218.
Schmeer, M. R., D. Horton, and A. Tanimura. 1966. Mercenene, a tumor inhibitor from *Mercenaria:* purification and characterization studies. Life Sci. 5:1169–1178.

Benzo[a]pyrene Body Burdens and the Prevalence of Proliferative Disorders in Mussels (*Mytilus edulis*) in Oregon

MICHAEL C. MIX, STEVE R. TRENHOLM, and KEITH I. KING

Recently many investigators have advocated the use of bivalve molluscs for monitoring marine environments to detect and quantitate various pollutants, including chemical carcinogens. Indigenous populations of bivalves seem to be ideal subjects for evaluating carcinogenic loads in the marine environment. They are sedentary and, excepting the larval sojourn, spend their lives in the same location; they inhabit waters that are polluted by domestic sewage, petroleum, petroleum by-products, and industrial waste products; they tend to concentrate in their tissues the noxious substances in the environment; and they are relatively easy to locate and sample. Moreover, it is possible to obtain or locate spat (newly attached young) enabling investigators to measure the rate of uptake and incorporation of carcinogens in bivalve populations starting with young animals. Because of their ubiquitous distribution, data can be compared with those collected by other workers, and large numbers of shellfish can be analyzed and sampled (Mix *et al.*, 1977a).

In mussels (*Mytilus edulis*) the residence time of certain carcinogens is quite long. That makes them particularly attractive for use as bioaccumulators in monitoring programs for benzo[a]pyrene (BP) and other carcinogens (Dunn and Stich, 1976).

During the past decade, there has been a plethora of reports describing presumptive neoplastic cellular proliferative disorders in marine bivalve molluscs from several bays in the United States. The most common type of reported disorder is thought to be associated with abnormal leukocyte renewal (Mix *et al.*, 1977b), and many pathologists consider them to be similar to sarcomas of higher animals (Rosenfield, 1976). Farley (1969) first described such a disorder in mussels from Yaquina Bay, Oreg. Since that time, Mix and his colleagues have conducted research on a similar, perhaps identical, condition in native oysters (*Ostrea lurida*) and have reviewed earlier studies on mussels, oysters, and clams from Yaquina Bay (Mix, 1975a,b; Mix, 1976a; Mix *et al.*, 1977b).

To date, studies to determine etiologic agents (chemical pollutants, viruses, or genetic factors) of the proliferative disorders in shellfish have yielded equivocal data. Kraybill (1976) believed that, to assess the carcinogenic risk of chemicals as inducers of neoplasms in aquatic organisms, a systematic survey must be made to identify and quantitate carcinogens in waterways and to determine the body burdens or concentrations in aquatic animals and their relevance to tumor incidence in these species.

During the past year, we have been engaged in a research project designed to determine the concentrations of BP in populations of indigenous bivalves from Oregon estuaries and to establish the prevalence of cellular proliferative disorders in these populations.

This paper describes the body burdens of BP and the prevalence of proliferative disorders in mussels from four sites in Yaquina Bay, Oreg., and discusses the relevance of these findings.

METHODS AND MATERIALS

A description of our ongoing study of body burdens of BP in six species of indigenous bivalve molluscs from four Oregon bays has been published (Mix *et al.*, 1977a). Below, we describe chemical analysis for BP and histological analysis for the presence of cellular proliferative disorders in mussels from four sites in Yaquina Bay.

Site Selection

After preliminary analyses to determine BP concentrations in mussels from 14 Yaquina Bay sites, four sites were selected for further investigations (Figure 1). Mussels from two of the sites, Y1M and Y12M, initially contained very low or nondetectable levels of BP (<1 μg/kg), while those from the other two sites, Y2M and Y4M, had substantial body burdens of BP (>15 μg/kg). Mussels from all sites were removed from creosoted pilings that had undergone varying degrees of weathering. BP levels in wood from these pilings were also determined during preliminary studies. The four sites were:

Y1M. This site consists of a number of weathered, heavily creosoted pilings that formerly supported an old railroad trestle. The pilings are located slightly south of the main channel and are subjected to heavy tidal currents.

Y2M and Y4M. Mussels from these sites were removed from creosoted pilings that support two fish processing plants along the bayfront along which there are also several marinas. Tidal flows in the bayfront area are minimal.

Y12M. Mussels were removed from creosoted pilings that support a marker buoy in the main channel. This site is also subjected to heavy tidal flows.

Sampling and Histopathological Examination

Mussels were collected monthly or bimonthly from each of the four sites. Immediately after collection, they were separated according to site, placed in labeled plastic bags, put on ice in coolers, and transported to our laboratory in Corvallis. Mussels to be examined

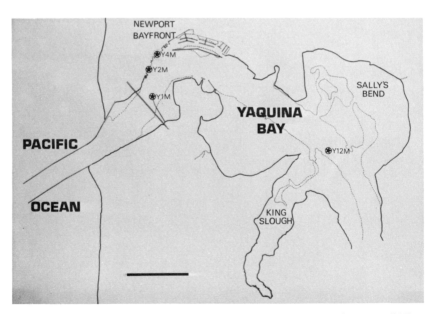

FIGURE 1 Yaquina Bay, Oreg. Only the lower bay is shown. The continuous, solid lines indicate the high tide mark. The discontinuous, lighter lines indicate low tide levels. See text for a description of the four sites (Y1M, Y2M, Y4M, and Y12M), from which mussels were sampled during this study. Bar = 1 km.

histologically were placed in Davidson's fixative (3:3:2:1:1—95% ethanol:seawater:formalin:glycerol:acetic acid added just prior to use), processed in the usual way, and sectioned at 6 μm. Tissue slides, prepared from each specimen, were examined microscopically by two or three investigators to determine if they possessed the large cells that characterize cellular proliferative disorders.

Chemical Analysis for Benzo[a]pyrene

Upon arrival at the laboratory, mussels were removed from their shells. The pooled sample from each site was then weighed and subsequently stored in a plastic bag at −20°C until BP analysis.

Aliquots of mussel samples from each site were analyzed according to the method of Dunn (Dunn, 1976; Dunn and Stich, 1976). A 30- to 40-g sample was digested by refluxing in an ethanol–potassium hydroxide solution. Following digestion, the ethanol–potassium hydroxide supernatant was extracted with 2-4 trimethylpentane (TMP) and the organic phase passed through a column of partially deactivated florisil. The polycyclic aromatic hydrocarbons (PAH's) were eluted with benzene. After removal of the benzene, the eluate was cleaned up by dimethyl sulfoxide (DMSO) extraction in TMP.

BP was isolated by preparative thin layer chromatography on 20% acetylated cellulose, made to volume in hexadecane, and its concentration was determined by spectrophotofluorimetry. Recovery of BP by the extraction procedure was determined by spiking the original digestion with an aliquot of tritiated BP and counting an aliquot of the final hexadecane solution. Analysis of wood samples from the pilings was accomplished similarly (see Mix *et al.*, 1977a).

RESULTS

Figure 2 shows the large, atypical cells that characterize the putative proliferative disorder in mussels. Recently, Michael C. Mix has begun detailed microscopic (light, phase, and transmission electron microscope [TEM]) studies in collaboration with Drs. J. Hawkes and A. Sparks of the National Marine Fisheries Service in Seattle, Wash. Examination of thin, 1-μm, plastic sections revealed two distinctly different types of large cells that are associated with the disorders (Figure 3).

One type, termed type 1 cells, possesses hyperchromatic nuclei with dense, scattered chromatin and a scant amount of basophilic cytoplasm. Nuclear diameters range from 7 to 9 μm. They are

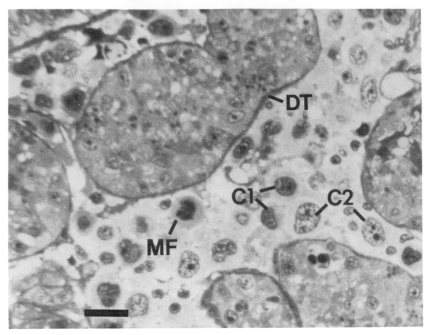

FIGURE 2 A thin, 1 μm plastic section (Richardson's stain) showing vesicular connective tissue with the atypical large cells that characterize the presumptive cellular proliferative disorders in mussels. DT = digestive tubules; Cl = cell type 1; C2 = cell type 2; MF = mitotic figure. See text for a description of the two cell types. Bar = 20 μm.

typically round or oval. Less frequently, they are bilobed, polymorphonuclear, or crenulated. Nucleoli, if present, are limited to one or, rarely, two. The second, or type 2, cells possess large (7 to 14 μm in diameter), pale-staining pleomorphic nuclei with sparse, scattered, or clumped chromatin. They contain 0 to 3 nucleoli. Their cytoplasm is very pale-staining and somewhat more abundant than in type 1 cells. The problem of whether or not these cells are indicative of a neoplastic disorder has not yet been resolved (Mix *et al.*, 1977b).

Table 1 summarizes the data on BP body burdens and the prevalence of cellular proliferative disorders in the four mussel populations.

Cellular proliferative disorders were common in mussels from the Y2M and Y4M sites, but none were found in the Y1M or Y12M groups. Mussels from the two former sites were consistently contaminated with substantial levels of BP while those from the latter two sites, with one exception (Y1M on December 16, 1976), contained low or very low (background) levels of BP.

Although our data on the prevalence, incidence, and seasonal aspects are incomplete, they seem to be at variance with reports by Farley (1969) and Farley and Sparks (1970), who reported that peak prevalences (12%) of the disorders in mussels and oysters (*O. lurida*) occurred in December and "early cases" were observed in the mussels in September. Thus far, there is no conclusive evidence that the cellular disorders are or are not seasonal with respect to the incidence or prevalence in the mussel populations. In view of the fact that the Oregon winter of 1976–1977 was the driest on record and no significant rainfall occurred until mid-February 1977, it is possible that the absence of fresh water may have modified the seasonal patterns normally associated with the disorders.

DISCUSSION

Although it is not yet possible to formulate any conclusions, it seems significant that no mussels with low body burdens of BP from the Y1M

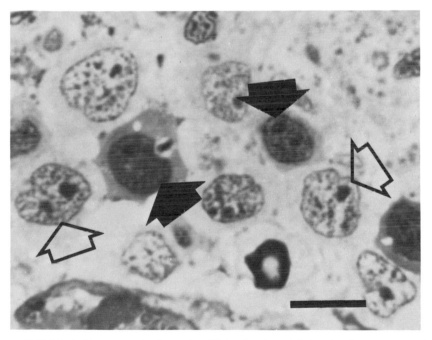

FIGURE 3 A thin, 1 μm plastic section (Richardson's stain) showing the two types of atypical large cells in mussels. Closed, black arrows = cell type 1; open arrows = cell type 2. See text for a description of the two cell types. Bar = 10 μm.

TABLE 1 Body Burdens of Benzo[a]pyrene (BP) and the Prevalence of Proliferative Disorders in Mussels (Mytilus edulis) from Yaquina Bay, Oregon

Mean Concentration of Benzo[a]pyrene (BP) in Mussels, Number of Mussels with Proliferative Disorders/Total Number Examined, and Percent of Mussels with Proliferative Disorders for Each Date Sampled

Date Sampled	Site Y1M			Site Y2M			Site Y4M			Site Y12M		
	BP, μg/kg Wet Weight	No.	%	BP, μg/kg Wet Weight	No.	%	BP, μg/kg Wet Weight	No.	%	BP, μg/kg Wet Weight	No.	%
6/15/76	0.12	0/10	0.0	30.10	0/10	0.0	15.00	1/10	10.0	0.44	0/10	0.0
7/22/76	4.72	0/10	0.0	67.88	0/10	0.0	6.73	0/10	0.0	0.70	0/10	0.0
8/22/76	2.73	0/45	0.0	14.22	0/42	0.0	—[a]	6/161	3.7	—	—	—
9/24/76	0.68	—	—	33.84	—	—	6.88	—	—	0.82	—	—
10/21/76	0.60	0/194	0.0	—	—	—	—	16/199	8.0	—	—	—
11/16/76	0.62	—	—	40.20	12/47	25.5	8.95	13/90	14.4	0.45	0/100	0.0
12/16/76	8.41	0/101	0.0	12.23	2/46	4.3	7.47	13/51	25.5	—	0/97	0.0
2/13/77	3.77	0/100	0.0	32.97	11/98	11.2	170.13	7/50	14.0	0.46	0/95	0.0
4/8/77	1.72	0/43	0.0	21.89	—	—	12.67	—	—	0.50	—	—
TOTAL		0/503	0.0		25/253	9.9		56/571	9.8		0/312	0.0

[a] Not measured.

and Y12M sites have contained the large cells, whereas substantial numbers have been encountered in mussels collected from the Y2M and Y4M sites with high body burdens of BP. These analyses must be continued indefinitely to determine the precise relationship, if any, between BP levels and the cellular disorders.

One of the major concerns connected with industrial and urban contamination of productive coastal waters is the occurrence, in many petroleum products, of substantial quantities of potent carcinogens such as BP and benzanthracene (Dunn and Young, 1976). The sources of BP in mussels from the four Yaquina Bay sites have not yet been established. A recent paper (Mix *et al.*, 1977a) reported the following levels of BP in creosoted wood from three of the four Yaquina Bay sites: Y1M = 265,514 μg/kg, Y2M = 137,204 μg/kg, and Y4M = 107,097 μg/kg. Mussels from the Y1M site contained low concentrations of BP, although they had been removed from creosoted pilings with the highest BP concentrations. However, both Y1M and Y12M are situated in areas with strong tidal currents. Therefore, BP leached from these pilings into the surrounding water may be removed before it can be incorporated into the tissue. Y2M and Y4M are protected somewhat by a breakwater and are not subjected to strong tidal fluxes. A second significant difference between the low and high BP sites is that Y2M and Y4M are near a number of large marinas that may contribute petroleum products to the surrounding environment. A third potential source of BP is the waste products that are routinely discharged into the bayfront area from fish-processing plants. There may also be additional, as yet unidentified, sources of BP.

A salient question that remains to be answered is whether BP can cause neoplasia or the malignant equivalent in bivalve molluscs. Many chemical carcinogens must be metabolized to reactive substances prior to eliciting their carcinogenic effects (Miller and Miller, 1976). Polycyclic aromatic hydrocarbons (PAH's) are metabolically activated by the complicated microsomal enzyme systems variously known as mixed-function oxidases (MFO), nonspecific drug metabolizing enzymes, cytochrome P-450 enzymes, benzo[*a*]pyrene hydroxylase, and aryl hydrocarbon hydroxylase (AHH) (Heidelberger, 1976). Such compounds must also be metabolically altered before they can be excreted effectively. Both activation and excretion pathways are initiated by the microsomal MFO system (Philpot *et al.*, 1976).

The evidence accumulated to data indicates that few, if any, bivalve molluscs possess such enzyme systems (e.g., Neff *et al*, 1976; Dunn and Stich, 1976). Thus, the absence of MFO activity may lead to the retention of unaltered BP in mussels. Studies (e.g., Di Salvo *et al.*,

1975; Neff *et al.*, 1976; Dunn and Stich, 1976) to determine the rates of BP uptake and elimination in mussels and clams indicate that BP is stored in two compartments. Most BP is probably contained within a fluid compartment and, after removal from a contaminated system, the loss from this compartment would be quite rapid (hours to days). Loss from a tissue compartment would be considerably slower, possibly requiring weeks or months for complete depuration. If metabolic conversion does not occur, it is difficult to conceive of a physiological mechanism by which BP or other carcinogenic PAH's could cause the proliferative cell disorders in these organisms.

However, BP may be associated with the cellular disorders in a different way. It became evident during our analyses that BP is only one of many PAH's in the mussels analyzed during this study. We consistently found 6 to 10 fluorescent bands on the thin-layer chromatography (TLC) plates that may represent individual PAH or groups of PAH's. There is also a direct correlation between the concentration of BP and the number and size of the fluorescent bands. Therefore, BP body burdens may represent only a small fraction of the total chemical load associated with PAH and possibly other anthropogenic environmental insults. The cellular responses of mussels to chronic chemical contamination are not known. Mix *et al.* (1977b) have speculated that the large, apparently abnormal, cells in various shellfish species may be the result of physiological hyperplasia (initiated by unknown factors) of the leukocyte proliferative compartments coupled with the failure of resultant cells to differentiate normally. Perhaps, then, the large cells represent an abnormal response to continuous chemical insult. Unfortunately, so little is known about bivalve molluscan leukocyte renewal (Mix, 1976b) that it is not yet possible to test this hypothesis.

It is also possible that BP and PAH may somehow react with other agents (chemical, physical, and/or viral) to elicit a proliferative response in mussels. Most experimental data on carcinogenic agents have revealed multiple agents, either in sequence or in combination (Bingham *et al.*, 1976). Finally, of course, BP may not be associated in any significant way with the large atypical cells that characterize these conditions.

We are convinced that bivalve molluscs are valuable organisms for monitoring the marine environment for chemical pollutants. Histopathological disorders, particularly proliferative disorders, in bivalve populations may indicate environmental stress. Our results suggest that there may be a relationship between the body burdens of BP and the presence of abnormal, possibly neoplastic, cells in mussel populations. As more data become available, answers to the many difficult questions will be forthcoming.

SUMMARY

Levels of benzo[*a*]pyrene (BP) in economically important shellfish from Oregon estuaries were studied. Shellfish from the more polluted areas of Yaquina Bay, Oreg. (sites Y2M, Y4M, Figure 1), contained significant levels of BP (>15 μg/kg), while those from cleaner areas (sites Y1M, Y12M, Figure 1) contained very low or undetectable levels of BP. Histological studies have shown that nearly 10% of the mussels with significant BP body burdens have apparent proliferative disorders compared to none for those sampled from clean sites.

ACKNOWLEDGMENTS

We would like to express our gratitude to Dr. John Couch for his encouragement and support, and to Ron Riley, Randy Schaffer, Debbie Parker, Diane Bunting, and Tim Corkill, who provided valuable technical assistance. Epon sections used in Figures 2 and 3 were prepared by Dr. Joyce Hawkes.

This research was supported by grant # R804427010 from the Environmental Protection Agency, Gulf Breeze, Fla.

REFERENCES

Bingham, E., R. W. Niemeier, and J. B. Reid. 1976. Multiple factors in carcinogenesis. Ann. N.Y. Acad. Sci. 271:14–21.

Di Salvo, L. H., H. E. Guard, and L. Hunter. 1975. Tissue hydrocarbon burden of mussels as potential monitor of environmental hydrocarbon insult. Environ. Sci. Technol. 9:247–251.

Dunn, B. P. 1976. Techniques for determination of benzo(α)pyrene in marine organisms and sediments. Environ. Sci. Technol. 10:1018–1021.

Dunn, B. P., and H. F. Stich. 1976. Release of the carcinogen benzo(α)pyrene from environmentally contaminated mussels. Bull. Environ. Contam. Toxicol. 15:398–401.

Dunn, B. P., and D. R. Young. 1976. Baseline levels of benzo(α)pyrene in Southern California mussels. Mar. Pollut. Bull. 7:231–233.

Farley, C. A. 1969. Sarcomatoid proliferative disease in a wild population of blue mussels (*Mytilus edulis*). J. Natl. Cancer Inst. 43:509–516.

Farley, C. A., and A. K. Sparks. 1970. Proliferative diseases of hemocytes, endothelial cells and connective tissue cells in mollusks. Bibl. Haematol. 36:610–617.

Heidelberger, C. 1976. Studies on the mechanisms of carcinogenesis by polycyclic aromatic hydrocarbons and their derivatives. Pp. 1–8 in R. I. Freudenthal and P. W. Jones, eds. Carcinogenesis, Vol. 1. Polynuclear Aromatic Hydrocarbons: Chemistry, Metabolism and Carcinogenesis. Raven Press, New York.

Kraybill, H. F. 1976. Distribution of chemical carcinogens in aquatic environments. Prog. Exp. Tumor Res. 20:3–34.

Miller, E. C., and J. A. Miller. 1976. The metabolism of chemical carcinogens to reactive electrophiles and their possible mechanisms of action in carcinogenesis. Pp. 737–762 in C. E. Searle, ed. Chemical Carcinogens. ACS Monograph 173. American Chemical Society, Washington, D.C.

Mix, M. C. 1975a. The neoplastic disease of bivalve molluscs. Pp. 369–386 in J. C. Hampton, ed. The Cell Cycle in Malignancy and Immunity. NTIS No. CONF-731005. National Technical Information Service, Springfield, Va.

Mix, M. C. 1975b. Proliferative characteristics of atypical cells in native oysters (*Ostrea lurida*) from Yaquina Bay, Oregon. J. Invert. Pathol. 26:289–298.

Mix, M. C. 1976a. A review of the cellular proliferative disorders of oysters (*Ostrea lurida*) from Yaquina Bay, Oregon. Prog. Exp. Tumor Res. 20:275–282.

Mix, M. C. 1976b. A general model for leucocyte cell renewal in bivalve molluscs. Mar. Fish. Rev. 38:37–41.

Mix, M. C., R. T. Riley, K. I. King, S. Trenholm, and R. Schaffer. 1977a. Chemical carcinogens in the marine environment. Benzo(*α*)pyrene in economically important bivalve mollusks from Oregon estuaries. Pp. 421–431 in D. A. Wolfe, ed. Fate and Effects of Petroleum Hydrocarbons in Marine Ecosystems and Organisms. Pergamon Press, New York.

Mix, M. C., H. J. Pribble, R. T. Riley, and S. P. Tomasovic. 1977b. Neoplastic disease in bivalve mollusks from Oregon estuaries with emphasis on research on proliferative disorders in Yaquina Bay oysters. Ann. N.Y. Acad. Sci. 298:356–373.

Neff, J. M., B. A. Cox, D. Dixit, and J. W. Anderson. 1976. Accumulation and release of petroleum-derived aromatic hydrocarbons by four species of marine animals. Mar. Biol. 38:279–289.

Philpot, R. M., M. O. James, and J. R. Bend. 1976. Metabolism of benzo(*α*)pyrene and other xenobiotics by microsomal mixed-function oxidases in marine species. Pp. 184–199 in AIBS Symposium Volume: Sources, Effects and Sinks of Hydrocarbons in the Aquatic Environment. American Institute of Biological Sciences, Rosslyn, Va.

Rosenfield, A. 1976. Recent environmental studies of neoplasms in marine shellfish. Prog. Exp. Tumor Res. 20:263–274.

QUESTIONS AND ANSWERS

D. SCARPELLI: I am pleased, Dr. Mix, that you broached the question of the significance of 3,4-benzo[*a*]pyrene. This suggests that there may be a relationship between the presence of benzo[*a*]pyrene in the environment and proliferative lesions in molluscs. It is well established that most carcinogens are activated by mixed oxygenase enzyme systems, and that these enzymes in turn alkylate and damage DNA. This causes a genetic error that is recognized by a loss of growth control that leads to the development of neoplasms. What is disturbing is that there is no evidence that molluscs possess enzymes capable of activating carcinogens. If the enzymes are not there and if there is a causal relationship between the presence of carcinogens and the proliferative lesions, then we have to identify other mechanisms giving rise to lesions. This is a very interesting challenge.

J. PAYNE: In our 4 years of studying aryl hydroxylase activity, we have not been able to demonstrate metabolism of these compounds in any of the four species of bivalves that we studied.

R. ANDERSON: Have you examined other taxa?

J. PAYNE: Yes, besides teleosts we have found activity in crustaceans, several species of gastropods, pelecypods, echinoderms, and annelids from the North Atlantic.

R. ANDERSON: We have been addressing the problem of aryl hydrocarbon hydroxylase activity in marine invertebrates, and have also had great difficulty in demonstrating benzo[*a*]pyrene hydroxylase in bivalves. However, we have recently developed an assay that demonstrates activity of this enzyme in insects (Anderson, 1978a),[1] which we have adapted for the study of oyster benzpyrene hydroxylase (Anderson, 1978b).[2] When studying invertebrates, the methods developed for vertebrates should not be followed too closely. For example, the cofactor reduced nicotinamide adenine dinucleotide phosphate (NADPH), which is required for all mixed function oxidases, inhibits the enzyme in the American oyster (*Crassostrea virginica*) when present at the levels used in assays for vertebrate enzymes. The fluorescent method is most commonly used for detecting benzo[*a*]pyrene metabolites. We have been able to increase the sensitivity of enzyme detection by using ^{14}C-labeled or tritium-labled benzo[*a*]pyrene as substrate. We should learn more about the generation of 9-10 epoxide of the 7-8 dihydrodiol of benzo[*a*]pyrene by *Crassostrea* microsomes, since it has been suggested that this epoxide may be carcinogenic in vertebrates (Sims *et al.*, 1974).[3]

D. SCARPELLI: How would you compare the activity of benzo[*a*]pyrene-hydroxylase in higher vertebrates with the enzyme system in molluscs and other marine invertebrates?

R. ANDERSON: The specific activity of the enzyme in vertebrates is higher than in invertebrates; in certain strains of laboratory rodents it is perhaps six to ten times more active.

D. SCARPELLI: How does the activity compare with the enzyme present in fish liver?

R. ANDERSON: It is less.

J. PAYNE: Organisms such as gastropods can have one-thousandth of the activity of rat livers and can still be detected easily by fluorometry. I am wondering how much of the 3-hydroxy derivative of benzo[*a*]pyrene you are detecting.

R. ANDERSON: After 60 min incubation, about 0.2% to 0.3% of the benzo[*a*]pyrene present in the medium has been converted to phenolic derivatives. It appears that quinones are the major metabolites produced by oyster aryl hydrocarbon hydroxylase *in vitro*.

S. NIELSEN: I don't understand why Drs. Mix, Brown, and Scarpelli seem to be apologetic about using the terms myeloproliferative or lymphoproliferative disorders. The more species in which we study neoplastic diseases the

[1] Anderson, R. S. 1978a. Aryl hydrocarbon hydroxylase in an insect, *Spodoptera eridanea* Cramer. Comp. Biochem. Phys. 59C:87–93.

[2] Anderson, R. S. 1978b. Benzo[*a*]pyrene metabolism in the American oyster, *Crassostrea virginica*. Ecological Research Series, U.S. Environmental Protection Agency, EPA 600/3-78-009.

[3] Sims, P., P. L. Grover, A. Swaisland, K. Pal, and A. Hewer. 1974. Metabolic activation of benzo(*a*)pyrene proceeds by a diol-epoxide. Nature 252:326–328.

clearer it becomes that the distinction between cancerous and normal tissue is not clear-cut. There is a big gray zone between the two. I do not think there should be any reluctance to accept what we see in clams, mussels, and certain mammalian species as being on the borderline between normal and neoplastic tissue. We have myeloproliferative, immunoproliferative, and lymphoproliferative disorders that are neither frank cancers nor inflammatory lesions.

M. MIX: I certainly didn't mean to sound apologetic. I am simply being cautious since we have often been abused by others less knowledgeable about these conditions in shellfish.

S. NIELSEN: What is the alternative to the creosote treatment as a wood preservative for pilings?

M. MIX: I do not know. We do not have conclusive evidence that creosote is the sole source of benzo[*a*]pyrene in mussels.

D. SCARPELLI: We should be cautious until we know more about the true nature of these lesions; they are certainly proliferative but they have not been shown unequivocally to be neoplasms, much less malignant ones.

J. COUCH: An oft-mentioned possibility is that we have some undescribed rare protistan infection that produces a cell that could be confused with a neoplastic cell of the host. Those of us who have worked extensively with the proliferative lesions and pathogens of molluscs feel that this is not the case. However, the question still persists and will not be answered definitively until someone produces proliferative disorders in molluscs under experimental conditons.

The American Oyster
(*Crassostrea virginica*) as an
Indicator of Carcinogens in the
Aquatic Environment

JOHN A. COUCH, LEE A. COURTNEY,
JAMES T. WINSTEAD, and STEVEN S. FOSS

Synthetic organic chemicals, agricultural chemicals, heavy metals, and petroleum enter the environment as pollutants generated by human activity. Some oncologists now believe that most human neoplasms are related to the presence of carcinogens in the environment (Cairns, 1975). Certain pollutants are actual or potential carcinogens, which may contact aquatic organisms that will accumulate, metabolize and translocate the pollutant or its metabolites (Kraybill, 1975). Results of several studies on the effects, bioaccumulation, and fate of pollutant chemicals in aquatic species have been published (Pimentel, 1971). One potentially serious, but neglected, aspect is the long-term tissue and cellular response of aquatic species that may reflect or indicate the presence of pollutants that are frank or suspect carcinogens.

Evidence that aquatic animal species are susceptible to neoplasia is abundant. However, the etiologies of spontaneous neoplasms in feral aquatic animals are unknown, and the significance of the neoplasms to ecosystems and human health is uncertain (Dawe, 1969). Because some proliferative cellular diseases in fishes and molluscs have been epizootic (Scarpelli and Rosenfield, 1976), the experimental determinations of their etiological agents are necessary. Possible candidate etiologies are chemical pollutants, viruses, and genetic factors.

The selection of aquatic species for use as laboratory carcinogen test indicators is difficult. Many aquatic species are not adaptable or acclimatable to long-term laboratory exposures and manipulations. Aspects of nutritional requirements, behavior, natural disease, life cycle, and hydrographics limit the number of species amenable even to flowing-water aquarium systems, much less to the static aquarium conditions that are necessary in many inland or metropolitan research centers.

Table 1 lists some aquatic species that are considered suitable for laboratory experimentation. Each of these species has one or more or all of the following qualities that make it a probable candidate for carcinogenesis studies: It can be maintained for relatively long periods under laboratory conditions; it can be manipulated experimentally; it can be reared from egg to adult in the laboratory; it has a history of spontaneous or experimentally induced neoplasia; it is representative of a broad distribution; and it has shown metaplastic, anaplastic, or hyperplastic tissue responses to chemical exposures.

Methods for studying effects of carcinogens in aquatic species must be carefully planned to derive maximum information. Short- or long-term exposures of aquatic organisms to carcinogens require safe and adequate holding facilities (properly isolated and ventilated raceways, aquaria, ponds, or glassware) that insure a stable, healthy background for select test species. There are methods of applying chemicals to exposure water that ensure continual direct or intermittent contact of test chemical with test animals.

In static systems, the chemical is dissolved in solvent or directly in water to attain desired concentration. In flow-through systems, concentrations of stock solution are introduced by controlled injection[1] into systems with known water flow rate to maintain desired ambient concentrations of toxicants. Food, somatic injection, or implant can also be used as vehicles for desired doses of chemical.

The experimental exposure regimens for aquatic species given in Figure 1 (A–D) cover several of the contemporary hypotheses for chemical carcinogenesis. Figure 1A represents long-term continuous exposure of test species to sublethal concentrations of chemical. This method accounts for situations in which subjects are under chronic chemical insult, much as would be the case in low-level, chronically polluted rivers or estuaries. The regimen in Figure 1B could be used to simulate a situation in which an aquatic community receives a single dose of chemical, as in accidental spills of oil or industrial wastes. In Figure 1C aquatic species might receive repeated but discrete exposures to a chemical in order to reproduce regular periodic ocean dumping of municipal or industrial wastes or seasonal pesticide spraying. Figure 1D accounts for a theory of chemical carcinogenesis (Berenblum, 1970)

---

[1] Several injection-devices are commercially available. These devices use gear-driven plungers in injection syringes. Delivery diluters can be built.

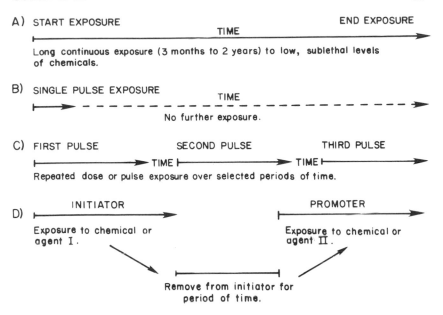

A) START EXPOSURE   END EXPOSURE

TIME

Long continuous exposure (3 months to 2 years) to low, sublethal levels of chemicals.

B) SINGLE PULSE EXPOSURE

TIME

No further exposure.

C) FIRST PULSE   SECOND PULSE   THIRD PULSE

TIME   TIME

Repeated dose or pulse exposure over selected periods of time.

D)   INITIATOR   PROMOTER

Exposure to chemical or   Exposure to chemical or
agent I.   agent II.

Remove from initiator for
period of time.

FIGURE 1 Some possible experimental exposure regimens of aquatic species to test chemicals for carcinogenic activity.

that involves an initiating agent followed by a promoting agent. Under aquatic conditions, this sequence could occur if exposure to pollutant with initiating carcinogenic qualities (e.g., a mutagenic substance) was followed by exposure to a second pollutant with tumor-promoting qualities (see Prehn, 1971).

This paper describes the basis for, and the potential laboratory use of, the American oyster (*Crassostrea virginica*) as an indicator of carcinogens in the aquatic environment. The oyster was selected because it fulfills most of the above-listed requirements for an aquatic species to be used in carcinogen studies and because we are familiar with pathology of oysters.

MATERIALS AND METHODS

To date, we have designed and implemented an experimental exposure system for determining the rates of uptake, tissue accumulation, metabolism, and effects of two known mammalian carcinogens in relatively large numbers of oysters. The system consists of four tables that receive flowing estuarine water (10‰ to 25‰ salinity). The

TABLE 1   Aquatic Species Considered Suitable for Carcinogenesis Experiments. Many of These Species Are North American Forms; However, Comparable Species Are Available in Europe, the Soviet Union, and the Orient

Species	Container[a]	Sample Reference on History of Tumor Studies in Species
Freshwater		
Guppy (*Lebistes reticulatus*)	TS	Pliss and Khudoley, 1972
Zebra danio (*Brachydanio rerio*)	TS	Stanton, 1969
Platyfish (*Xiphophorus maculatus*)	TS	Anders, 1967
Swordtail hybrids (*X. helleri*)	TS	Anders, 1967
Rainbow trout (*Salmo gairdner*)	RW, TFT	Halver and Mitchell, 1967
Brook trout (*Salvelinus fontinalis*)	RW, TFT	Wellings, 1969
Yellow bullhead (*Ictalurus natalis*)	TFT, TS	Harshbarger and Couch, in preparation
Channel catfish (*I. punctatus*)	TFT, TS	None
Blue gill (*Lepomis macrochirus*)	TFT, TS	None
Mussel (*Anodonta* sp.)	TFT	Pauley, 1969
Snail (*Biomphalaria glabrata*)	TS	Richards, 1973
Crayfish (*Procambarus clarkii*)	TS, TFT	Harshbarger et al., 1971
Flatworms (*Dugesisa* sp.)	TS, FBS	Foster, 1969
Hydra (*Hydra littoralis*)	TS, FBS	Lenhoff et al., 1969
Marine		
Sheepshead minnow (*Cyprinodon variegatus*)	TFT	None
Croakers, several species	TFT	Harshbarger, 1972, 1974
Dover sole (*Microstomus pacificus*)	TFT	Mearns and Sherwood, 1974
Mosquito fish (*Gambusia affinis*)	TFT, TS	None
Oysters (*Crassostrea* sp.; *Ostrea* sp.)	TFT	Couch, 1969; Farley, 1969
Soft clam (*Mya arenaria*)	TFT	Barry and Yevich, 1972
Octopus (*Octopus vulgaris*)	TFT	Rungger et al., 1971
Cuttlefish (*Sepia* sp.)	TFT	Jacquemain et al., 1947
Shrimp (*Palaemonetes* sp.)	TFT, TS	None

[a] TS = Tank or aquaria, static water, but oxygenated or aerated.
RW = Raceways.
TFT = Tank or aquaria with flow-through water supply.
FBS = Finger bowl, static water, possibly oxygenated or aerated.

water is pumped from Santa Rosa Sound near Pensacola, Fla., travels through an overhead delivery trough, and is siphoned to each water table. Carcinogen-contaminated water was delivered via overflow stand pipes (from carcinogen-receiving water tables) to an effluent-holding pond (Figure 2) and water from the control table was returned to Santa Rosa Sound.

Approximately 1,500 oysters were taken from Apalachicola Bay and acclimated to the holding tanks. A base sample of 150 was examined

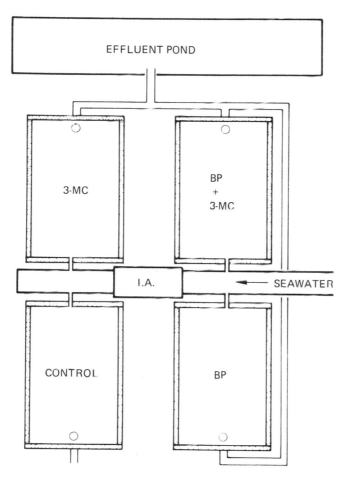

FIGURE 2 Scheme of control and exposure water tables for oysters. BP = benzo[*a*]pyrene exposure; 3-MC = 3-methylcholanthrene exposure; BP + 3-MC = combination exposure; I.A. = injection apparatus.

histologically. Benzo[*a*]pyrene (BP) and 3-methylcholanthrene (3-MC) dissolved in triethylene glycol were each injected into the delivery water of separate tables each containing 300 oysters (Figure 3). In a combination exposure, BP and 3-MC were both injected into the water entering a third table, which also contained 300 oysters. The fourth table, the control table, also containing 300 oysters, received only triethylene glycol in estuarine water. Exposures were made according to a pulse or intermittent schedule to simulate regimen C in Figure 1. The test carcinogens were injected into the system 5 days a week, 8 h a day. The nominal exposure concentration was set at 1 μg of carcinogen per liter of water (1.0 μg/l) from July 14, 1976, to March 22, 1977. At this time the nominal concentration was increased to 5.0 μg/l and exposures continued at this concentration until September 8, 1977, at which time exposure was stopped. Surviving oysters were maintained and depurated in flowing seawater and observed for cellular changes. During the

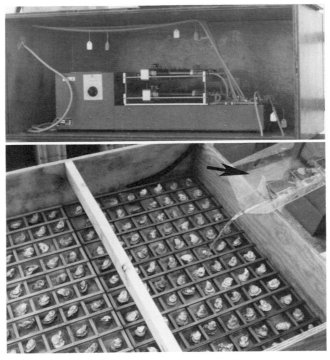

FIGURE 3 Injection apparatus and water table with oysters; carcinogen is injected directly into influent estuarine water which flows into mixing baffle (arrow).

exposure, oysters were able to obtain food from the natural estuarine water delivered to their holding tables. Flow rate of water in each table (700 l/h) was sufficient to support new shell growth in the oysters for the entire test period.

Exposed and control oysters were examined weekly for mortality, growth, and fouling. During the course of the test, a total of seven samples were taken for chemical and histopathological analysis, including a base sample prior to exposure and a depuration sample 3 months after exposure had been terminated. Twelve oysters were taken from each table, for a total of 48 oysters per sample period.

Oyster tissues (1-cm cross section through visceral mass) were fixed in Davidson's fixative, paraffin processed, sectioned at 7 μm, and stained with Harris hematoxylin and eosin. These oyster sections were studied with light microscopy. The remaining tissues from each sample were extracted and analyzed with high-pressure liquid chromatography for the parent compound of polycyclic aromatic hydrocarbons used. Oysters naturally growing in the effluent stream in the receiving pond were analyzed for BP and 3-MC concentrations after 8 months of exposure. Individual oysters after 5.5 months of exposure to BP and 3-MC and controls were analyzed for BP hydroxylase activity by Anderson (1978).

Another experiment to determine rates of uptake and distribution of BP and 3-MC in different tissues of oysters was completed using radiolabeled BP and 3-MC. Two cleaned oysters were placed in each of three 32-l glass aquaria. To one aquarium was added 3 μg/l of ^{14}C-BP (SA 51.0 μCi/μmol) dissolved in a benzene–ethyl alcohol solution (1:50); to a second was added 3.0 μg/l ^{14}C-3-MC (SA 60.2 μCi/μmol) dissolved in benzene–ethyl alcohol solution; to a third was added only the benzene–ethyl alcohol solution in volume equal to that used in the first and second aquaria. To a fourth aquarium, without oysters, 3.0 μg/l ^{14}C-BP in benzene–ethyl alcohol was added. One oyster was taken from each aquarium after 7 days of exposure. Scintillation counts were made on oyster visceral mass samples to determine degree of radioactivity and, hence, degree of accumulation of radiolabeled carcinogen. The remaining oysters from each aquarium were taken after 14 days of exposure and radioactivity was measured. One oyster exposed to ^{14}C-BP for 7 days was allowed to depurate in clean estuarine water for 7 days. Then radioactivity was determined in that oyster. Water samples were taken periodically from the exposure aquaria and the aquarium without oysters to determine rates of loss of labeled carcinogen from water with and without oysters. Histological sections (7 μm) were taken from the visceral masses of each oyster exposed to ^{14}C-BP prior to scintillation counting and prepared for autoradiography according to the liquid emulsion coating technique of Lillie (1965). Emulsions were exposed for 3, 5, and 7 weeks, developed, and examined for comparative distribution of labeled BP in tissues.

RESULTS

Concentration of Carcinogens in Oysters Exposed Intermittently for 14 Months

Chemical analyses data on uptake and accumulation of BP and 3-MC in oysters are presented in Figure 4. The relatively heavy concentrations of both compounds found in the oysters from the September 1976 sample (193-h exposure) can be explained by the occurrence of an accidental injection at the beginning of the test (July 14, 1976) of relatively high concentrations of carcinogen stock solutions. This occurred when the injection apparatus malfunctioned, resulting in a much faster rate of injection than planned. Subsequent to correcting this malfunction, the desired nominal rate of injection (1 μg/l) was maintained until March 22, 1977 (Figure 4).

In the December 1976 sample (640-h exposure), there were no carcinogens in the oysters analyzed. This complete loss in the second sample of carcinogen from tissues is difficult to explain. There are three probable explanations:

1. Heavy growth of winter algae in exposure tables may have resulted in rapid adsorption and perhaps metabolism of carcinogen by algae, resulting in a lowering of levels of carcinogen available to the oysters.

2. Oyster microsomal enzymes may have been induced and the carcinogens metabolically eliminated (we analyzed only for parent compound, not metabolites); the high levels detected subsequently may be due to the higher levels of carcinogen administered at that time.

3. Errors were made in the chemical analyses of this sample (replicate samples gave the same results).

The next sample, which was taken in March 1977 (1,229-h exposure), revealed carcinogens in all exposed oysters (Figure 4). However, the concentrations did not show a trend of rapid increase. There was only a relatively low increase in the BP-exposed oysters. These oysters had accumulated 209 times the ambient exposure concentration after the December 1976 sampling. Oysters exposed to BP and 3-MC combined showed a tremendous loss of both carcinogens.

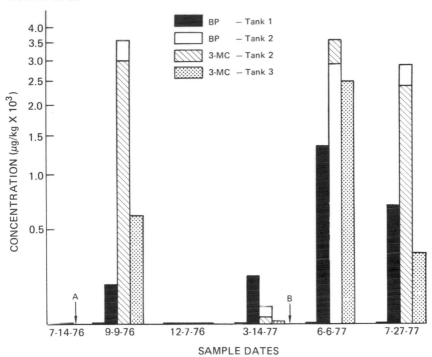

FIGURE 4 Uptake of BP and 3-MC by oysters over 12 months of exposure. Combined open and lined histograms represent results from combination-exposure water table. Point A represents time of accidental "spill" into water tables at beginning of tests. Point B represents time at which concentrations of PAH's were increased from 1 μg/l to 5 μg/l.

Following the March sample, ambient exposure concentration was increased to 5 μg/l (on March 22, 1977). Subsequent samples (June and July 1977) showed corresponding increase in tissue concentrations (Figure 4). Exposure was terminated on September 8, 1977, and a final sample taken in December 1977, after 3 months of depuration, revealed no trace of the parent compounds in any of the test organisms.

Oysters taken from the effluent pond after 6 months of exposure to the mixed effluent had from 13.1 μg/kg to 84.4 μg/kg 3-MC and from a trace to 36.4 μg/kg BP. Those oysters taken nearest the effluent stream had the highest concentrations.

There was aryl hydrocarbon hydroxylase (AHH) activity in hepatopancreas homogenates of oysters exposed 5.5 months to BP or 3-MC. Enzymatic activity was not detected in nonexposed animals (Anderson, 1978). Activity in homogenates from BP-exposed animals was 0.18 nmol BP metabolites per milligram of protein per 10 min.

Similar activity in 3-MC-exposed oysters was 0.21. These data suggest that BP hydroxylase activity may be induced by polycyclic aromatic hydrogen compounds in oysters, as it is in other higher animals. AHH activity is low in oysters compared to that of mammals, but can be consistently shown *in vitro* by the generation of radiolabeled metabolites derived from ^{14}C-BP (Anderson, 1978).

HISTOPATHOLOGICAL EFFECTS OF CARCINOGENS

Oysters in all four water tables had significant mortality in August, September, and October 1976 due to relatively heavy infections by *Dermocystidium marinum,* a common, natural protistan pathogen of oysters of the Gulf of Mexico and Atlantic coast of the United States. Approximately 30% of all oysters in the test was lost during this period. Histological examination determined the extent and nature of the protistan lesions in individual moribund oysters. Because of our extensive experience with natural oyster diseases, we can discriminate

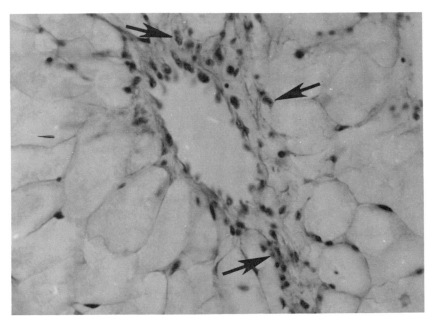

FIGURE 5 Normal blood vessel in mantle of control oyster; note relatively few leukocytes associated with endothelium and in blood sinuses between vesicular connective tissue cells (arrows). (×430.)

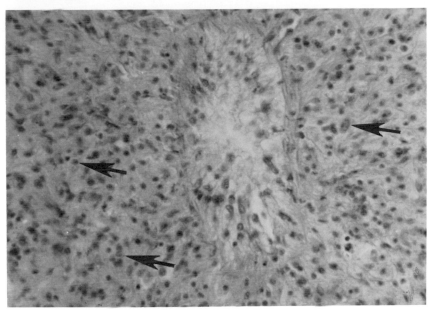

FIGURE 6 Perivascular lesion that resembles inflammatory focus from oyster exposed to 1 μg/l of 3-methylcholanthrene for 1,229 h; note pleomorphic nuclei of cells surrounding blood vessel (arrows). (×430.)

between histological effects of the carcinogens and those of most natural agents on oysters in our region.

Samples of the September 1976 (193-h exposure) and the December 1976 (640-h exposure) oysters exhibited no histopathological changes that could be attributed to the experimental exposures. In these and control animals the tissues were normal (Figure 5). However, in eight of 12 oysters from the 3-MC-exposure (1,229 h of intermittent exposure to 1.0 μg/l 3-MC), there was found a lesion suggesting possible chemical effects. The lesion consisted of incipient perivascular inflammatory foci in the mantle (Figures 6 and 7). Large numbers of at least two different blood cell types made up the lesion surrounding and radiating from the larger blood vessels of the respiratory and somatic mantle. Occasional mitotic figures were found in relatively large cells (nuclear diameter = 3.5 to 6 μm) that predominated in the lesions (Figure 7). This cell type closely resembled those cells constituting the pathologic cell type in sarcomoid neoplasms in feral oysters and clams described by Couch (1969), Farley (1969), and Barry and Yevich (1972) (compare Figures 6, 7 and 8). Both cell types

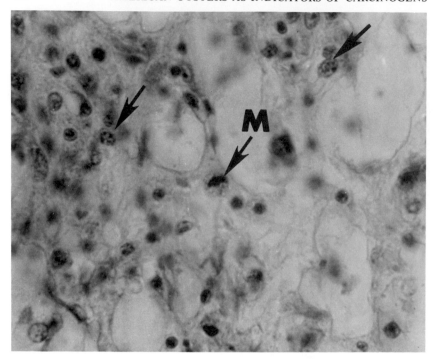

FIGURE 7 Higher magnification of periphery of lesion similar to that illustrated in Figure 6; note mitotic figure (metaphase arrow-M) and other similar nondividing cells (arrows). (×1,000.)

possess pleomorphic nuclei, mitotic activity, and large size relative to normal oyster leukocytes. The extent of lesions in the 3-MC-exposed oysters was not as great as those in the feral molluscan neoplasms. It would be premature, therefore, to consider these lesions in the 3-MC-exposed oysters as unequivocal neoplasms. Further study of other surviving oysters that were exposed to 3-MC will be necessary before the real nature of the perivascular lesions is known.

Accumulation and Tissue Distribution of ^{14}C Benzo[a]Pyrene and ^{14}C 3-Methylcholanthrene

Scintillation count analyses of tissues from oysters exposed for 7 or 14 days to labeled carcinogens in aquaria are given in Table 2. Oysters exposed for 336 h (14 days) to 3.0 μg/l of ^{14}C-labeled compounds accumulated less or only slightly more carcinogen than oysters exposed for only 168 h (7 days). Concentration factors (concentration in tissues/initial concentration in aquarium water) for different oysters ranged from 204× to 264× in 7- to 14-day exposures to 3.0 μg/l of each carcinogen (Table 2).

Comparison of loss of radioactive label in water samples from aquaria with oysters and without oysters showed a more rapid loss in the aquarium with oysters (lower line in Figure 9) than in the aquarium without oysters (upper line in Figure 9). However, the rate of loss leveled off in both types of aquaria approximately 100 hours after introduction of the compound into the system.

Data on depuration of oysters exposed 7 days then depurated in clean water for 7 days show a 66.8% loss of activity in tissue samples (Table 2).

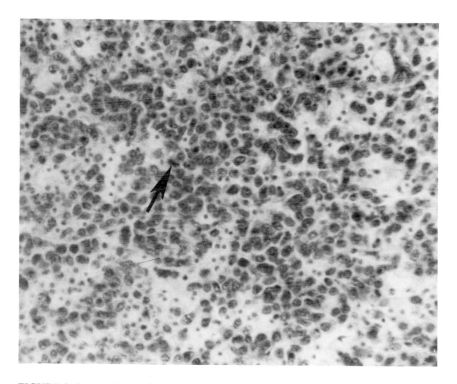

FIGURE 8 Center of neoplastic lesion from a feral oyster suffering from sarcoma-like disease; note cell type morphology and mitotic activity (arrow). (×650.)

TABLE 2   Concentration of ¹⁴C Carcinogens in Oysters

¹⁴C-Labeled Compound	Exposure at 3 μg/l, h	Total PAH[a] in Oyster, μg	Weight of Oyster, μg	Concentration in Oyster, μg/g	Concentration Factor[c] ($\times$ = times)
Benzo[a]pyrene	168	7.169	9.80	0.732	244$\times$
Benzo[a]pyrene	336	4.750	6.70	0.709	236$\times$
Benzo[a]pyrene	336	1.746	2.40	0.727	242$\times$
3-Methyl-cholanthrene	168	3.711	6.05	0.613	204$\times$
3-Methyl-cholanthrene	336	4.046	5.10	0.793	264$\times$
Benzo[a]pyrene	168[b]	7.10	1.727	0.243	66.8%[d]

[a] PAH = polycyclic aromatic hydrocarbons.
[b] 168-h exposure followed by 168-h depuration.
[c] Concentration factors were calculated by dividing final concentration of label in oyster tissue by concentration originally put in aquarium (conversion of specific activity of labeled carcinogen to molar quantities was necessary).
[d] % radioactivity lost during depuration.

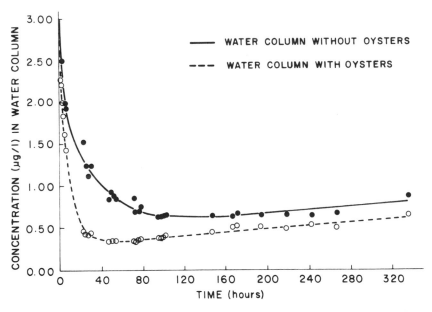

FIGURE 9 Graph of loss of ¹⁴C-labeled benzo[a]pyrene from water in 32-l aquarium with three oysters (lower curve) and from water in similar aquarium without oysters (upper curve).

These results suggest that oysters take up and accumulate relatively small amounts of BP and 3-MC in contrast to relatively greater uptake of many pesticides (e.g., Parrish *et al.*, 1976; Schimmel *et al.*, 1976). They then quickly either excrete or metabolize (or both) the compounds when returned to clean water.

Radioautographic analyses of tissue sections from the oysters exposed for 7 and 14 days to labeled carcinogens indicate that the distal portions of the digestive tubules in the digestive gland are the chief sites of uptake (Figure 10). Compare this with a histological section from a control oyster shown in Figure 11. Some label occurs in the vesicular connective tissues and gonads surrounding the digestive tubules after 7 to 14 days, but transport to the connective tissue and thence to gonadal tissues probably follows initial absorption by the digestive epithelial cells. Some oyster ova became labeled after 7 to 14 days exposure. Whether or not uptake occurs directly from fluid of the gonoducts or indirectly via the vesicular connective tissue blood sinuses is unknown.

DISCUSSION

The data presented in this report indicate that oysters are capable of concentrating known mammalian carcinogens under long- and short-term experimental exposure regimens. There is some indirect evidence

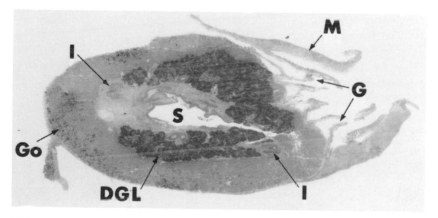

FIGURE 10 Radioautograph of section of oyster exposed for 7 days to ^{14}C-BP in aquarium; emulsion was exposed for 3 weeks prior to development; note radioactive exposed emulsion and distribution of ^{14}C label as indicated by beta tracks in emulsion (arrows); heavily labeled (DGL) circular areas are epithelial layers of digestive gland tubules. (×4.8.) Go, gonads; I, intestine; G, gills; DGL, digestive gland; S, stomach; M, mantle.

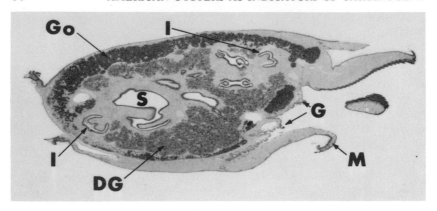

FIGURE 11 Histological section from control oyster for orientation of reader; vesicular connective tissue is light area among circular epithelial layers (arrows). Sections were stained with hematoxylin and eosin. (×4.8.) Go, gonads; I, intestine; G, gills; DG, digestive gland; S, stomach; M, mantle.

from these data and direct evidence from the work of Anderson (1978) that oysters have at least a minimal capacity to metabolize BP. We have not studied the possible role that polycyclic aromatic metabolites may play in oysters. Accumulation of BP and 3-MC by oysters in our studies appeared to be dependent upon the amount of compound available to the oysters, but oysters did not accumulate polycyclic aromatics as rapidly as they have organochlorine pesticides.

Incipient histopathological changes in oysters exposed intermittently to 3-MC for 1,229 actual hours of exposure are suggestive of possible pre-neoplastic or at least inflammatory responses. Survivors of these and longer exposures (14 months) will be examined to determine, if possible, the fate of the 3-MC-related lesions.

Radioautographic studies demonstrated the uptake of both BP and 3-MC by the digestive glands of oysters. No significant uptake was noted for external gill or mantle epithelium. Some ova were heavily labeled with ^{14}C-BP, suggesting that, in the early stages of their life cycles, oysters could be exposed to mutagenic and carcinogenic compounds.

The American oyster (C. virginica) appears to be suitable for studies of the fate and possibly the effects of some carcinogens. The fact that oysters can accumulate two known mammalian carcinogens from estuarine waters suggests a need for careful monitoring of these organisms to safeguard human and estuarine species health.

SUMMARY

The American oyster (*C. virginica*) was used as the experimental animal for chronic exposure to 3-methylcholanthrene (3-MC) and benzo[*a*]pyrene (BP) in an exposure system in which the carcinogens can be continuously injected into free flowing water at fixed rates ranging from 1 to 5 μg/l. Experiments designed to determine uptake and distribution of H³MC and H³BP showed that these are concentrated in oyster tissues in direct proportion to the dosage of carcinogen injected into the system. Residual concentrations as high as 84.4 μg/kg of MC and 36.4 μg/kg of BP were present in oysters as long as 6 months following exposure. Autoradiography showed intense localization of H³BP in distal portions of the tubules of the digestive gland and to a lesser extent in the gonadal tissues. Aryl hydrocarbon hydroxylase (AHH) activity was present in homogenates of hepatopancreas after 5.5 months of exposure to the carcinogens, in contrast to control animals in which AHH activity was quite low. In eight oysters exposed to MC, an infiltration of cells believed to be of hematopoietic origin was encountered in the mantle. Some appear to be identical in type to those which constitute sarcoma-like lesions encountered in feral oyster populations. However, it would be premature at this stage to assign any etiological significance to the experimental findings.

ACKNOWLEDGMENT

We thank Dr. Peter Schoor for his help on the chemical analysis performed in his laboratory.

REFERENCES

Anders, F. 1967. Tumor formation in platyfish–swordfish hybrids as a problem of gene regulation. Experientia 23:1–10.

Anderson, R. S. 1978. Benzo[*a*]pyrene metabolism in the American oyster *Crassostrea virginica*. Ecol. Res. Ser. U.S. EPA 600/3-78-004. Washington, D.C.

Barry, M., and P. Yevich. 1972. Incidence of gonadal cancer in the quahog, *Mercenaria mercenaria*. Oncology 26:87–96.

Berenblum, I. 1970. The study of tumors in animals. In H. W. Florey, ed. General Pathology. W. B. Saunders Co., Philadelphia.

Cairns, J. 1975. The cancer problems. Sci. Am. 233:64–80.

Couch, J. 1969. An unusual lesion in the mantle of the American oyster, *Crassostrea virginica*. Natl. Cancer Inst. Monogr. 31:557–562.

Dawe, C. J. 1969. Phylogeny and oncogeny. Neoplasms and related disorders of invertebrate and lower vertebrate animals. Natl. Cancer Inst. Monogr. 31:1–40.

Farley, A. 1969. Probable neoplastic disease of the hematopoietic system in oysters, *Crassostrea virginica* and *Crassostrea gigas*. Natl. Cancer Inst. Monogr. 31:541–555.

Foster, J. 1969. Malformations and lethal growths in planaria treated with carcinogens. Natl. Cancer Inst. Monogr. 31:683–691.

Halver, J., and I. Mitchell. 1967. Trout Hepatoma Research Conference Papers. U.S. Bureau of Sport Fisheries and Wildlife Research Report (U.S.) Res. Rep. 70. 199 pp.

Harshbarger, J. 1972. Work of the registry of tumors in lower animals with emphasis on fish neoplasms. Pp. 285–303 in Symp. Zool. Soc. Lond., No. 30. Academic Press, New York.

Harshbarger, J. 1974. Activities report of the Registry of Tumors in Lower Animals 1965–1973. Smithsonian Institution, Washington, D.C. 141 pp.

Harshbarger, J., and J. Couch. In preparation.

Harshbarger, J., G. Cantwell, and M. Stanton. 1971. Effects of n-nitroso-dimethylamine on the crayfish, *Procambarus clarkii*. Pp. 425–430 in Proc. IV Int. Colloq. Insect Pathol. College Park, Md.

Jacquemain, R., A. Jullien, and R. Noel. 1947. Sur l'action de certains corps cancerigenes chez les Cephalopodes. C. R. Acad. Sci. Paris 225:441–443.

Kraybill, H. 1975. The distribution of chemical carcinogens in the aquatic environment. Pp. 3–34 in F. Homburger, ed. Neoplasms in Aquatic Animals. Progress in Experimental Tumor Research, Vol. 20. S. Karger, New York.

Lenhoff, H., C. Rutherford, and D. Heath. 1969. Anomalies of growth and form in *Hydra*: polarity, gradients, and a neoplasia analog. Natl. Cancer Inst. Monogr. 31:709–737.

Lillie, R. D. 1965. Histopathologic Technic and Practical Histochemistry. McGraw-Hill Book Co., New York. 648 pp.

Mearns, A., and M. Sherwood. 1974. Environmental aspects of fin erosion and tumors in Southern California Dover sole. Trans. Am. Fish. Soc. 103:799–810.

Parrish, P. R., S. C. Schimmel, D. J. Hansen, J. M. Patrick, and J. Forester. 1976. Chlordane: Effects on several estuarine organisms. J. Toxicol. Environ. Health 1:485–494.

Pauley, G. 1969. Neoplasia in mollusks. Natl. Cancer Inst. Monogr. 31:509–529.

Pimentel, D. 1971. Ecological Effects of Pesticides on Non-target Species. Stock No. 4106-0029. U.S. Government Printing Office, Washington, D.C. 220 pp.

Pliss, G., and V. Khudoley. 1972. The use of aquarium fish for experimental cancer studies. In Problems of Prophylaxis of Environmental Pollution by Carcinogenic Agents. Tallin. 30–35. (Russian)

Prehn, R. 1971. Neoplasia. Pp. 191–241 in LaVia and Hill, eds. Principles of Pathobiology. Oxford University Press, New York.

Richards, C. 1973. Tumors in the pulmonary cavity of *Biomphalaria glabrata*: genetic studies. J. Invert. Pathol. 22:283–289.

Rungger, D., M. Rastelli, E. Braendle, and R. G. Malsberger. 1971. A viruslike particle associated with lesions in the muscles of *Octopus vulgaris*. J. Invert. Pathol. 17:72–80.

Scarpelli, D., and A. Rosenfield, eds. 1976. Molluscan pathology. Mar. Fish. Rev. 38:1–50.

Schimmel, S. C., J. M. Patrick, and J. Forester. 1976. Heptachlor: Toxicity to and uptake by several estuarine organisms. J. Toxicol. Environ. Health 1:955–965.

Stanton, M. 1969. Diethylnitrosamine-induced hepatic degeneration and neoplasia in the aquarium fish *Brachvdanio rerio*. J. Natl. Cancer Inst. 1:117–130.

Wellings, S. R. 1969. Neoplasia and primitive vertebrate phylogeny: echinoderms, pre-vertebrates, and fishes. Natl. Cancer Inst. Monogr. 31:59–128.

QUESTIONS AND ANSWERS

M. MIX: I am interested in the curve showing benzo[*a*]pyrene in water alone and in water with oysters. Have you ever conducted studies where you simply put oysters in water and then removed them? We have done this with naphthalene and other compounds, and the curves are remarkably similar for water containing oysters and for water in which oysters were placed briefly and then removed. In other words, we think that there is possibly a substantial amount of bacterial action and not just adsorption to glass and particulate materials.

J. COUCH: I am sure there is even though we clean the oyster shells before placing them in a test situation. Considerable flora and fauna live in the gills, mantle cavity, and in the digestive tract of these animals. These organisms could affect the uptake and metabolism of carcinogens. We have not done what you mentioned, but you should take note that the curve leveled off both with and without oysters in the tanks. This indicates to us that oysters can take the compound out of the water at a certain rate. The oyster is able to control either its accumulation or its metabolism of the compound. At this point, we do not know which. We have repeated this experiment three times and have obtained identical curves. When the curves level off the concentration approaches that in a tank without oysters. Therefore, in a glass-lined aquarium or tank, you are probably seeing the results of an equilibrium between adsorption and desorption on the walls. This is the first study in which I have seen perivascular cellular proliferation of this nature in response to any experimental stimulus (i.e., methylcholanthrene) among the many oysters I have examined in histologic section.

G. MIGAKI: Dr. Anderson, would you care to comment on the data on benzo[*a*]pyrene hydroxylase activity?

R. ANDERSON: I would like to emphasize the preliminary nature of these data, and also to indicate that one should not assume that the control value is actually zero. We have found subsequently that the method we used to preserve and ship tissues for analysis was not optimal. Apparently the enzyme systems are labile, and a certain amount of activity is lost in rapid freezing. Using the best available method, the activity for aryl hydrocarbon hydroxylase in a normal oyster would be about 0.5 nmol/mg of protein/10 min. It is well established that these enzyme systems are easily induced by polycyclic hydrocarbons (PCB's) and other insecticides in the aquatic environment. Neither the genetic background of these oysters nor their history of exposure to environmental pollutants is known. Consequently, one should expect considerable enzymatic variation among animals due to these variables.

M. LIPSKY: What activation temperature was used in the assays and was there an increase in protein? Also, precisely what fraction was used for the assay, the postmitochondrial supernate, or the pure microsomal fraction?

R. ANDERSON: The optimal *in vitro* temperature is about 37°C, which is considerably higher than that encountered *in vivo*. Apparently the enzyme is thermostable. By raising the temperature of the incubation mixture one can increase the activity to a point above which it is inactivated. This occurs at 45°C. Enzyme activity can be demonstrated both with whole homogenate and with the microsomal fraction. We did not detect any increase in protein concentration in hepatopancreas preparations of oysters induced by previous exposure to polycyclic aromatic hydrocarbons.

J. WEIS: A bivalve that is subject to heavy pollution is *Modiolus,* the large mussel. Have any neoplasms been described in this species?

J. COUCH: I know of no work on *Modiolus.*

R. BROWN: I have collected and examined a few of the mussels when they were found near *Mya.* Although the numbers are too small to be significant, I did not find any lesions in the animals examined.

Studies on Aryl Hydrocarbon Hydroxylase, Polycyclic Hydrocarbon Content, and Epidermal Tumors of Flatfish

W. T. IWAOKA, M. L. LANDOLT, K. B. PIERSON,
S. P. FELTON, and A. ABOLINS

Three generalizations have been made about cutaneous papillomas in fish: they most frequently occur in bottom feeding species; they have focal geographic distribution; and their distribution is usually associated with urbanization and industrialization (Dawe and Harshbarger, 1975). All of these factors suggest the presence of one or more environmental oncogens.

The occurrence of superficial tumors in flatfish along the Pacific Coast of North America was noted as early as the late 1800's by fishermen in the Gulf of Alaska. Subsequent descriptions of the tumors were published by Schlumberger and Lucké (1948) and by Nigrelli *et al.* (1965); however, it was not until the 1960's that extensive research into the characterization of the tumors was undertaken, principally by Wellings and Chuinard (1964). The three distinct morphological types of epidermal tumors in flatfish were described by Wellings *et al.* (1964) as an angioepithelial nodule, an angioepithelial polyp, and an epidermal papilloma (Figure 1).

The distribution of flatfish with tumors is geographically focal. The highest incidences occur near heavily industrialized areas (Cooper and Keller, 1969; Miller and Wellings, 1971), such as Puget Sound, Wash., which contain a variety of organic pollutants in varying concentrations (Dexter and Pavlou, 1976). Many of these compounds or their metabolites are carcinogenic. Some of them seem to occur naturally in the environment, while others have been introduced by humans (Dexter and Pavlou, 1976).

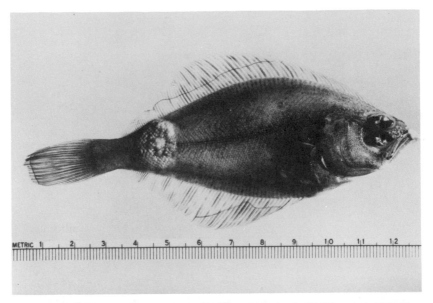

FIGURE 1 Representative examples of epidermal tumors found on flatfish. Above: flathead sole (*Hippoglossoides elassodon*). Below: English sole (*Parophrys vetulus*).

MATERIALS AND METHODS

One hundred adult flatfish were collected during May through July 1976, from one location on the west coast of Vancouver Island, British Columbia, and from two locations in Puget Sound. The Canadian site was the Bamfield Marine Station on Barkley Sound, a relatively pristine area far removed from industrial or municipal effluent (Dunn and Stich, 1975). The Puget Sound locations included Alki Point, which is the site of the City of Seattle sewage outfall, and Golden Gardens Park, which is adjacent to a large marina and which receives significant amounts of petroleum hydrocarbons. The five species of fish collected were rock sole *(Lepidopsetta bilineata)*, C-O sole *(Pleuronichthys coenosus)*, starry flounder *(Platichthys stellatus)*, English sole *(Parophrys vetulus)*, and flathead sole *(Hippoglossoides elassodon)*.

Each fish was assigned an accession number, was measured, and was given a complete postmortem examination. At necropsy, representative tissues were preserved for histopathological examination. The livers were immediately processed for aryl hydrocarbon hydroxylase (AHH) assay, and the remaining tissues were frozen at -20°C for hydrocarbon analysis.

Tissues taken at necropsy included samples from visceral organs, integument, and from the central nervous and reticuloendothelial systems. Following routine histological procedures, tissues were processed, stained with hematoxylin and eosin, and examined histologically. AHH activity was assayed, using a modification of a method of Nebert and Gelboin (1968) and expressed as fluorescent units (FU) of alkali-extractable 8-hydroxybenzo[*a*]pyrene/min/mg of protein. Protein content was determined by the method of Lowry *et al.* (1951).

Samples of fish tissue (up to 100 g) were analyzed for polycyclic hydrocarbons essentially following the extraction procedure of Grimmer and Bohnke (1975). Gas-liquid chromatographic (GLC) analyses were performed on a 183-cm × 0.3-cm OD column packed with 10% OV101 on 80/100 mesh Chromosorb W., using temperature programming.

Following review of the enzyme data, 20 Puget Sound rock sole with high, low, and intermediate levels of AHH activity were examined microscopically and chemically. All data obtained from these fish were then combined for analysis in a computerized pattern recognition program (Kowalski, 1975).

RESULTS AND DISCUSSION

The levels of AHH activity in the Puget Sound fish of a given species and age class appeared to be quite variable, ranging from <20 to >180 FU/min/mg of protein, with a mean of 60 FU/min/mg of protein. The flathead sole from Bamfield also had some variability in enzyme activity—from 10 to 60 FU/min/mg of protein. However, a mean activity of only 20 FU/min/mg of protein made them perceptibly different from the rock sole taken near Seattle. The AHH levels in Alki Point rock sole and Golden Gardens rock sole were indistinguishable.

A progressive increase in AHH activity was observed throughout the summer sampling period. Perhaps this increasing activity reflected increasing ambient temperature. Dewaide and Henderson (1970) reported a similar rise in hepatic microsomal activity with temperature in a European freshwater fish, *Leuciscus rutilus*.

Recently a number of investigators have proposed the use of piscine AHH levels as a direct index of aquatic pollution (Payne and Penrose, 1975). The preliminary study reported in this paper has revealed, however, that a single enzyme determination is not sufficient, since flatfish AHH activity increased markedly with increasing ambient temperatures.

The two groups of Puget Sound fish could not be separated from one another histopathologically. Comprising the 50 fish collected at each of the two sites were English sole *(Parophrys vetulus)*, C-O sole *(Pleuronichthys coenosus)*, starry flounder *(Platichthys stellatus)*, and rock sole *(Lepidopsetta bilineata)*. In the Golden Gardens collection, 1 of 14 C-O sole, 1 of 10 English sole, and 1 of 3 starry flounder had proliferative lesions on the caudal peduncle, dorsal fin, and gular area, respectively. In each case, the lesion was a single hyperemic focus with a raised granular surface. Histologically, each consisted of papillomatous proliferations of epithelial cells with minimal amounts of connective tissue stroma. Characteristic "X cells" were present in the lesions. The tumors appeared to be consistent with published descriptions of flatfish epidermal papillomas (Wellings *et al.*, 1964). From the Alki Point collection, one of nine C-O sole had a single proliferative lesion in the opercular region. This lesion was histologically indistinguishable from the tumors seen in the Golden Gardens fish. Also from Alki Point, 1 of 35 rock sole had a single renal lesion which, histologically, proved to be a nephroblastoma.

To allow for comparative studies, a single species, the rock sole, was selected for biochemical, tissue residue, and histological studies. Twenty-two rock sole were collected at Golden Gardens and 35 at Alki Point. With the exception of the nephroblastoma, the fish appeared to be in general good health except for random visceral lesions and to have adequate nutritional status.

Proliferative skin disorders were noted in 4 of the 51 fish taken from Bamfield. These lesions were frequently multiple and appeared grossly as opalescent white granular lesions. Histologically, they consisted of hypertrophic dermal connective tissue cells covered by a nonhyperplastic layer of epithelial cells. Contained within the hypertrophic dermal cells were basophilic cytoplasmic inclusions. These lesions were considered to be consistent with those of the viral disease, lymphocystis. The fish from Bamfield were in general good health, but a few fish exhibited random visceral lesions similar to those in the Puget Sound specimens. No tumors were noted.

Gas chromatographic analyses of the extracts obtained from the Puget Sound flatfish showed a number of peaks, with their areas corresponding to concentrations ranging from <1 mg/kg up to 50 mg/kg in the fish. Phenanthrene was used as an external standard and for quantitating the peaks in the chromatogram. Retention times and peak areas were calculated for each peak in the chromatogram, but no attempt was made to identify any of the peaks in the chromatogram.

The retention time of each GLC peak, each peak area, the enzyme activity, and the histopathological data (a total of 102 data points or variables) for each fish analyzed were then entered into the pattern recognition program. One goal was to determine if any of the variables could be correlated with each other or if similar subgroupings of data points (e.g., all the enzyme activity data) could be correlated to other similar subgroupings of data (e.g., GLC retention times or histopathological data). The second goal was to learn if any of the variables used in the program would allow the fish to be separated into two groups on the basis of their collection sites.

There were a number of high correlations from the data; most of them were GLC peak retention times correlating to other retention times or with certain GLC areas. No correlation was observed when histopathological data were compared to enzyme activity levels or to the GLC data.

A number of data points could be used to separate the fish on the basis of their collection sites. From two data points (a GLC retention time and a GLC peak area) 80% of the flatfish could be correctly classified by collection site. If three data points were used (two GLC retention times and one peak area), the flatfish could be classified with 100% accuracy (Figure 2).

The flatfish were separated into two groups: one group had two more GLC peaks per fish than the other group. The fish with the "extra" peaks were those collected next to the large marina. The compounds represented by these two peaks are still unknown. They may be innocuous compounds stored by the organisms. Results suggest that these compounds did not seem to increase hepatic enzyme activity or induce pathological changes.

With this program, the capture sites of other flatfish taken from

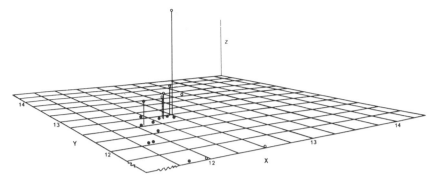

FIGURE 2 Three dimensional plot showing how the pattern recognition program separated the two groups of flatfish according to their collection site. X axis = GLC retention time$_1$ (cm). Y axis = GLC retention time$_2$ (cm). Z axis = peak area (cm^2). ● = Alki Point flatfish. ○ = Golden Gardens flatfish.

either of the two sites during the original collections could be determined by analysis of GLC data.

No correlations were found between enzyme activity levels or histopathological data and the site of collection of these flatfish. Since the fish from both Puget Sound locations appeared fairly healthy, no meaningful difference was expected. Although the anticipated correlation between hydrocarbon occurrence and enzyme activity was not found, it may exist. Genetic diversity or other factors may be of sufficient magnitude to mask it.

The importance of pattern recognition is to provide a preliminary sorting of all the data collected so that certain exhaustive and expensive work, such as that of compound identification, can be minimized. Thus, only the data points that showed high correlation need to be investigated further. As mentioned in the text, pattern recognition may provide unexpected and meaningful insights into the data that otherwise might have remained unnoticed because of the researcher's particular working hypothesis or bias. These advantages are of particular importance to analytical chemists who are providing data for biological monitoring programs.

SUMMARY

A preliminary survey on histopathology, liver aromatic hydrocarbon content, and aryl hydrocarbon hydroxylase levels of flatfish from several selected sites in Puget Sound and Canadian waters was made to determine how well the flatfish qualifies as a wildlife monitor of pollution. No significant correlations were found, although certain GLC and other data indicate that further study is needed.

ACKNOWLEDGMENT

This research was supported in part by the institutional cancer grant IN-26 from the American Cancer Society.

REFERENCES

Cooper, R. C., and C. A. Keller. 1969. Epizootiology of papillomas in English sole, *Parophrys vetulus*. Natl. Cancer Inst. Monogr. 31:173–185.

Dawe, C. J., and J. Harshbarger. 1975. Neoplasms in feral fishes: Their significance to cancer research. Pp. 871–894 in W. Ribelin and G. Migaki, eds. Pathology of Fishes. University of Wisconsin Press, Madison.

Dewaide, J. H., and P. T. Henderson. 1970. Seasonal variation of hepatic drug metabolism in the roach, *Leuciscus rutilus* L. Comp. Biochem. Physiol. 32:489–497.

Dexter, R. N., and S. P. Pavlou. 1976. Partitioning characteristics of trace organic compounds in Puget Sound. Presented at the Fifth Technical Conference on Estuaries of the Pacific Northwest, Oregon State University, Corvallis, Oregon, April 1–2, 1976.

Dunn, B. P., and H. F. Stich. 1975. The use of mussels in estimating benzo(a)pyrene contamination in the marine environment. Proc. Soc. Exp. Biol. Med. 150:49–51.

Grimmer, G., and H. Bohnke. 1975. Polycyclic aromatic hydrocarbon profile analysis of high protein foods, oils and fats by gas chromatography. J. Assoc. Off. Anal. Chem. 58:725–733.

Kowalski, B. 1975. Measurement analysis by pattern recognition. Anal. Chem. 47:1152A–1162A.

Lowry, O. M., N. J. Rosenbrough, A. L. Farr, and R. J. Randall. 1951. Protein measurement with the Folin phenol reagent. J. Biol. Chem. 193:265–275.

Miller, B. S., and S. R. Wellings. 1971. Epizootiology of tumors on flatfish sole *(Hippoglossoides elassodon)* in East Sound, Orcas Island, Washington. Trans. Am. Fish. Soc. 100:247–266.

Nebert, D. W., and M. V. Gelboin. 1968. Substrate-inducible microsomal aryl hydroxylase in mammalian cell culture. J. Biol. Chem. 243:6242–6249.

Nigrelli, R. F., K. S. Ketchen, and G. D. Rugguri. 1965. Studies on virus diseases in fishes, epizootiology of epithelial tumors in the skin of flatfishes of the Pacific Coast, with special reference to the sand sole *(Psettichthys melanostictus)* from Northern Hecate Strait, British Columbia, Canada. Zoologica (N.Y.) 50:115–122.

Payne, J. F., and W. R. Penrose. 1975. Induction of aryl hydrocarbon (benzo[a]pyrene) hydroxylase in fish by petroleum. Bull. Environ. Contam. Toxicol. 14:112–115.

Schlumberger, M. G., and B. Lucké. 1948. Tumors of fishes, amphibians and reptiles. Cancer Res. 8:657–712.

Wellings, S. R., and R. G. Chuinard. 1964. Epidermal papillomas with virus-like particles in flathead sole, *Hippoglossoides elassodon*. Science 146:932–934.

Wellings, S. R., R. G. Chuinard, R. T. Gourley, and R. A. Cooper. 1964. Epidermal papillomas in the flathead sole, *Hippoglossoides elassodon*, with notes on the occurrence of similar neoplasms in other pleuronectids. J. Natl. Cancer Inst. 33:991–1004.

Wellings, S. R., M. E. Bens, and R. G. Chuinard. 1965. A comparative of skin neoplasms in four species of pleuronectid fishes. N.Y. Acad. Sci. 126:479–501.

QUESTIONS AND ANSWERS

R. ANDERSON: Is it not correct that the incidence of tumors in flatfish on the East Coast is very low?

W. IWAOKA: Yes, you are correct.

R. ANDERSON: Wouldn't this imply that there is something more than environmental pollution involved in the genesis of these lesions, since levels of pollution on the East Coast are quite high? Possibly there is an interaction with some other causative factor, such as a virus.

W. IWAOKA: Yes. Viruses, bacteria, and parasites have been mentioned in this context. Electron microscopic examination of the tumor tissue has not, however, revealed the presence of recognizable parasites or bacteria, although there is a question as to whether the "X cells" represent host or foreign tissue.[1] Virus-like particles have been described, but transmission of the tumor has so far been unsuccessful.[2,3] On the other hand, many workers cite the occurrence of natural oil seepages in the Pacific as being the unique factor associated with these tumors.[4] Natural oil seepages are more or less absent along the East Coast. The oil seepage theory seems to be supported by the pattern of worldwide tumor distribution.[4,5]

R. ANDERSON: We sometimes lose sight of the fact that the purpose of aryl hydrocarbon hydroxylase is to detoxify naturally occurring xenobiotics and other chemicals. Therefore, a high level of aryl hydrocarbon hydroxylase does not necessarily imply increased carcinogen activation. For example, if these enzymes are induced in certain strains of mice prior to the application of a carcinogen, they will be protected since the rate of detoxification of carcinogens is increased. The expression of chemical carcinogenesis is probably a result of the balance between detoxification mechanisms on the one hand and activation mechanisms on the other.

[1]Brooks, R. E., G. E. McArn, and S. R. Wellings. 1969. Ultrastructural observations on an unidentified cell type found in epidermal tumors of flounders. J. Natl. Cancer Inst. 43:97–109.

[2]Chuinard, R. G. 1966. The natural history and pathology of epidermal papillomas of flathead sole, *Hippoglossoides elassodon,* from East Sound of Orcas Island, Washington. M.S. Thesis, University of Oregon.

[3]Wellings, S. R., and R. G. Chuinard. 1964. Epidermal papillomas with virus-like particles in flathead sole, *Hippoglossoides elassodon.* Science 14:229–242.

[4]Koons, C. B., and P. H. Monaghan. 1976. Input of hydrocarbons from seeps and recent biogenic sources. In Sources, Effects and Sinks of Hydrocarbon in the Aquatic Environment. American Institute of Biological Sciences, Rosslyn, Va.

[5]Dawe, C. J., and J. C. Harshbarger. 1975. Neoplasms in feral fishes: Their significance to cancer research. Pp. 871–894 in W. Ribelin and G. Migaki, eds. Pathology of Fishes. University of Wisconsin Press, Madison.

J. WEIS: Winter flounder *(Pseudopleuronectes americanus)* in the New York area have a very high incidence of fin rot that has not been associated with any particular causative organism, but has been associated with pollution. Do West Coast flatfish have fin rot?

W. IWAOKA: Fin rot is characteristic of many flatfish populations indigenous to the West Coast. The starry flounder *(Platichthys stellatus)* in the Duwamish River near Seattle and off the Southern California coast frequently display this condition. The cause of this condition seems to be nonspecific and does not appear to be related to tumor incidence.

M. MIX: In Yaquina Bay we have flatfish populations that migrate out to sea. They come in to spawn, remain a year, then return to the ocean. What sort of problems does this present to your study? How much do you know about the distances involved in migration, the frequency of migration, and the comingling among various populations?

W. IWAOKA: Flatfish spawn in deep water. After the larvae metamorphose, they move into shallow water near beaches and remain there for several months. After about a year, they migrate to deeper water and their whereabouts during this period are generally unknown. However, most biologists seem to agree that flatfish movement is generally confined to Puget Sound.

D. HINTON: Hasn't it been shown that the so-called "X cells" or "Z cells" of epidermal papillomas in flatfish are associated with parasites?

W. IWAOKA: "X cells" have been described as unidentified, tumor-specific cells in flatfish epidermal lesions. The "X cells" may be associated with an exogenous parasite, but there is still little known about whether these are fish cells or cells of foreign origin.

Congenital Abnormalities in Estuarine Fishes Produced by Environmental Contaminants

PEDDRICK WEIS and JUDITH S. WEIS

Estuaries, although among the most biologically productive areas, receive the brunt of civilization's destructive wastes. Heavy metal compounds in industrial effluents and insecticides in agricultural runoff are among the environmental insults to which adult estuarine organisms and their embryos are exposed. We have studied the effects of selected insecticides and heavy metals on the embryonic development of the more common estuarine fishes of the West Atlantic temperate coastal region. In so doing, we have demonstrated that pollutant levels found in estuarine environments can in some cases have a deleterious effect on developing fish.

However, substances that are embryotoxic or teratogenic for one species of fish may be tolerated by another species at much higher concentrations. Predictions of biological effects on a given species cannot, apparently, be made on the basis of studies in another species. Nor can predictions of effects on embryos be extrapolated from effects on adults of the same species nor from effects on the same species by related compounds.

Our early work involved insecticides. We tested dichlorodiphenyltrichloroethane (DDT) as a representative organochlorine insecticide; the commonly used organophosphate insecticides malathion and parathion; and carbaryl, a carbamate, on three species of fish. While the organophosphate and carbamate insecticides have environmental half-lives measured in days rather than in years, as with the organochlorine insecticides, they could still be embryotoxic or teratogenic to fish if present during sensitive developmental stages.

Of the three fish species studied, the killifish *(Fundulus heteroclitus)* was found to be a suitable test organism with which to study the effect of heavy metals, because it is adaptable to tank life, its eggs are easy to obtain and to handle, and it is a hardy species. Our experiments included inorganic salts of lead, cadmium, mercury, and methyl mercury.

MATERIALS AND METHODS

All the fish used in these experiments were collected during early summer months in the vicinity of Montauk, N.Y., at the eastern end of Long Island. The three species collected were the killifish *(F. heteroclitus)*, the Atlantic silverside *(Menidia menidia)*, and the sheepshead minnow *(Cyprinodon variegatus)*. Eggs and sperm were stripped into fingerbowls of filtered seawater (salinity — 30‰), which was maintained at 20°C. Eggs were checked for successful fertilization and initiation of cleavage. They were then divided into groups that were placed in separate fingerbowls of filtered seawater of the same salinity and temperature. Abnormal eggs were discarded. Exposure of eggs to the insecticides and heavy metals was initiated anywhere from the 2- to 4-cell stage to late gastrula. Observations were compared to descriptions of abnormal and normal development by Costello *et al.* (1957) for the Atlantic silverside, and by Armstrong and Child (1965) for killifish. Developmental stages of killifish were used to determine stages of sheepshead minnow development.

Doses varied from 0.001 mg/l (1 ppb) to 10 mg/l (10 ppm). In the insecticide experiments, control batches of embryos received 10 μl of acetone as a solvent control.

In some cases, the experimental agent was left in the rearing solution until time of hatching, but not renewed; in others, the embryos were removed after various intervals, washed in seawater, and placed in clean water for the remainder of their development.

During development, embryos were examined under a dissecting microscope and scored for axis formation, initiation of heartbeat, pigment formation, gross malformations, and general retardation. Hatching rates and swimming behavior were also recorded.

Representative embryos were fixed and embedded for microscopic examination.

RESULTS

Insecticide Experiments

Embryos of Atlantic silverside, which were exposed to DDT, malathion, and carbaryl, showed retardation of axis formation and heartbeat initiation, and optic abnormalities. In one of the five experiments, cardiovascular retardation was caused by doses as low as 0.01 mg/l of each insecticide (Weis and Weis, 1976a). The various effects were not generally dose dependent.

The optic abnormalities consisted of unilateral and bilateral microphthalmia and anophthalmia, with occasional cyclopia. Unaffected embryos hatched normally in all treatment groups, but occasional examples of scoliosis and lordosis in treated fry were observed at concentrations as low as 0.01 mg/l of each insecticide. In four experiments, the embryos were exposed at the late blastula stage of development. In one experiment, treatment began at the 2- to 4-cell stage. There were no obvious differences in effects from all five exposures when compared to experiments in which exposure was initiated at a later stage.

When killifish embryos were exposed to DDT, malathion, and carbaryl at either the 8- to 16-cell or blastula stages, there was no significant reduction in axis formation at the concentrations that affected the Atlantic silverside. However, when the experiments were repeated at concentrations of 1 and 10 mg/l, the embryos in 10 mg/l carbaryl group eventually showed complete developmental arrest at stage 22 (Weis and Weis, 1974).

To test the reversibility of the carbaryl, exposed 4.5-day-old embryos were washed and placed in clean seawater. After 2 days, 29% resumed normal development, while the majority developed cardiovascular abnormalities. The heart failed to develop normally and remained a feebly beating thin tube with rudimentary chambers incapable of circulating blood (Figure 1). Static pools of blood cells were observed around the yolk. Embryos with "tube-heart" syndrome remained smaller than controls, probably due to the lack of blood circulation.

This experiment was repeated by placing embryos in 10 mg/l carbaryl for 1-, 2-, and 3-day periods starting on day 1, 2, 3, or 4 of development. The tube-heart syndrome appeared primarily in the embryos exposed for 3 days during the third to fifth day of development. The majority of embryos were able to resume normal development. Parathion, which was introduced to the embryos at the 8- to 16-cell stage, caused

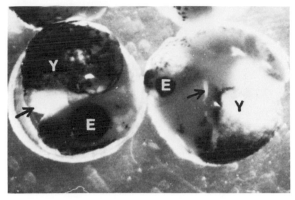

FIGURE 1 Macrophotographs of fixed killifish embryos,
indicating the normal heart (in the upside-down embryo) and
a "tube heart" (on the right), which resulted from exposures
to carbaryl at 10 mg/l. The hearts are the light structures
indicated by arrows. E = eye; Y = yolk. (×20.)

effects similar to those produced by carbaryl, i.e., developmental arrest
followed by the tube-heart syndrome after return to clean water.
Parathion was a more effective teratogen than carbaryl since retardation developed earlier than after carbaryl exposure and at a lower dose.

Cardiovascular anomalies caused by carbaryl and parathion in killifish and Atlantic silverside were similar to those observed in embryos
of the medaka *(Oryzias latipes)* (Schreiweis and Murray, 1976; Solomon, 1977) following exposure to these substances or the herbicide
2,4,5-T.

When embryos of sheepshead minnows were exposed to DDT,
malathion, or carbaryl at concentrations up to 10 mg/l, there were no
effects observed except at the 3 and 10 mg/l doses of malathion (Weis
and Weis, 1976b). These embryos were slightly retarded and started
hatching on the fourteenth rather than the thirteenth day. Their posthatching appearance and behavior were different from the controls and
the other exposed embryos. Their movements were convulsive and
uncoordinated, making swimming ineffective. Some animals were unable to maintain their postural equilibrium. Their physical appearance
(shown in Figure 2) indicated various degrees of spinal deformity in the
lateral plane. When quantitated on an arbitrary scale of 0 to 3+, the
effects indicated a dose-related response (Table 1). The physiological
disability seemed to be the basis of the problem since some fry with a
normal appearance could not swim normally.

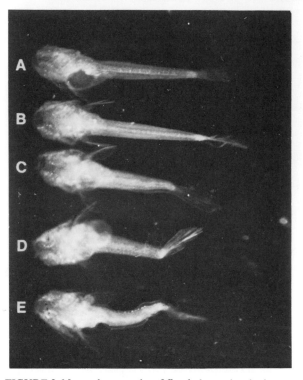

FIGURE 2 Macrophotographs of fixed sheeps-head minnow fry showing a control specimen (A), a specimen exposed to 1 mg/l malathion (B), and specimens with 1+, 2+, and 3+ responses to 3 and 10 mg/l malathion (C, D, and E, respectively). (Approximately X11.) From Weis and Weis, 1976b. (Reprinted with permission of Academic Press.)

The convulsive, uncoordinated movements of newly hatched malathion-exposed sheepshead minnows may have been due to inhibition of acetylcholinesterase and the abnormal spastic movements may have been, in turn, responsible for the skeletal malformations that many hatchlings developed. McCann and Jasper (1972) found that insecticide-treated bluegills *(Lepomis macrochirus)* suffered vertebral dislocations and fractures that were interpreted to be a result of muscle spasms. The concept that this neuromuscular problem is probably physiological rather than developmental is reinforced by Solomon (1977), who has shown that newly hatched medaka exposed to malathion will quickly develop vertebral dislocation that can be corrected by

TABLE 1   Spinal Deformity in Sheepshead Fry After Exposure to Different Concentrations of Malathion[a]

Exposure Group	Normal Fry, %	Fry with Spinal Abnormalities, %		
		1+[b]	2+[b]	3+[b]
Control	100	0	0	0
1 mg/l	100	0	0	0
3 mg/l	41	35	19	5
10 mg/l	25	29	23	23

[a] Adapted from Weis and Weis, 1976b.
[b] The severity of effect is illustrated in Figure 2.

returning the fry to clean water. None of the congenital anomalies caused by insecticides in either Atlantic silversides or in killifish embryos were observed in sheepshead minnows.

Heavy Metal Experiments

Killifish embryos were exposed to either mercuric chloride, cadmium chloride, or lead nitrate at concentrations of 0.01 to 10 mg of the metallic ions per liter of seawater under conditions similar to those in the insecticide studies (Weis and Weis, 1977a). The embryos could apparently tolerate the lead and cadmium to 10 mg/l, but all the embryos exposed to mercury at 1 mg/l died before gastrulation. In lower concentrations, some mercury-exposed embryos failed to complete gastrulation and axis formation. Of those in which morphogenesis did proceed, a significant portion developed forebrain defects which allowed the eye rudiments to converge, sometimes to the point of cyclopia. Some more severe results of incomplete axis formation included anencephaly and poor tail development. Many of the embryos that did not exhibit malformations showed retarded growth. The data are summarized in Table 2.

Examination of fry for skeletal malformations after hatching revealed that lead caused lordosis or, more extremely, inability to uncurl from the embryonic position. Half of the fry exposed to lead at 1 mg/l had a mild form of lordosis and were barely able to swim. All embryos exposed to lead at 10 mg/l had the more extreme condition, but were able to respond to tactile stimulation.

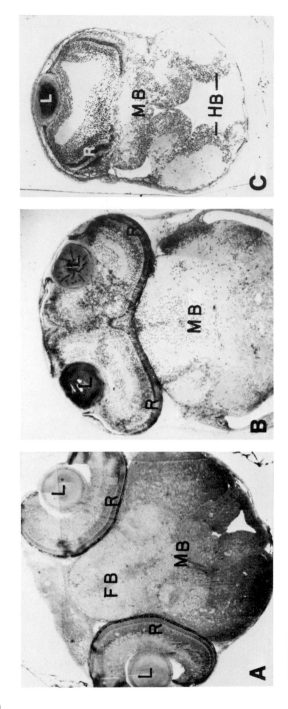

FIGURE 3 Photomicrographs of killifish embryos treated with methyl mercury at 0.04 mg/l, fixed near time of hatching, sectioned coronally, and stained with toluidine blue. A = control embryo; B = synophthalmic embryo; C = cyclopic embryo; L = lens; R = retina; FB = forebrain; MB = midbrain; and HB = hindbrain. (×80.) From Weis and Weis, 1977b. (Reprinted with permission of The Wistar Press.)

TABLE 2 Congenital Abnormalities in Killifish Embryos Exposed to Heavy Metals[a]

	Exposure		Congenital Abnormalities, %			
	Heavy Metal	Dose, mg/l	Axis Formation	Forebrain Defects	General Retardation	Posthatch Lordosis
At Stage 12	Hg^{2+}	0.01	68	0	20	0
(early		0.03	70	22	30	0
blastula)		0.1	39	24	ND[b]	0
		1.0	0	—	—	—
	Cd^{2+}	0.01	77	0	ND	0
		0.10	60	0	ND	0
		1.0	70	2	10	0
	Pb^{2+}	0.1	70	0	ND	0
		1.0	75	0	ND	80
		10.0	79	0	10	100
	Controls	—	83	0	10	0
At Stage 14	Hg^{2+}	0.01	92	0.8	17	0
(late		0.03	95	1.7	25	0
blastula)		0.1	92	22.5	27	0
	Cd^{2+}	10	96	0.8	32	0
	Controls	—	92	0	4	0

[a] Data from Weis and Weis (1977).
[b] ND = not determined.

When these experiments were repeated with methylmercuric chloride, the range of concentrations from no effect to a lethal one was very narrow—0.05 mg/l was embryotoxic and 0.01 mg/l had minimal effect (Weis and Weis, 1977b). At 0.03 to 0.04 mg/l, the forebrain defect with resultant optic anomalies appeared equivalent to that resulting from exposure to inorganic mercury at 0.1 mg/l (Figure 3). In addition, the "tube-heart" syndrome appeared in some embryos as did a general retardation of growth. While embryos with more severe head and cardiovascular abnormalities did not hatch, others, including some with synophthalmia, did hatch and swam normally. Others, which appeared normal *in ovo*, showed the same problem as those exposed to high lead concentrations, i.e., inability to uncurl and swim (Figure 4). The data for methyl mercury exposure are presented in Table 3.

TABLE 3 Abnormalities in Killifish (*Fundulus*) Embryos Exposed to Methyl Mercury[a]

Exposure Time[b]	Experiment No.	Number of Embryos	Methyl Mercury Concentration, mg/l	Early Death, %[c]	Abnormalities, %		
					Heart[d]	Eye[d]	Total[a]
Day 1	1	18	0.01	17	13	7	13
		19	0.03	37	54	38	69
	2	52	0.04	29	26	28	28
Day 2	1	48	0.01	12	0	2	2
		51	0.02	12	19	14	26
		48	0.03	12	63	47	68
		43	0.04	9	89	84	97
	2	51	0.04	0	27	14	33
Day 3	3	53	0.03	9	21	8	21
		51	0.04	8	4	4	4[e]
Controls	1–3	132	—	2	<1	0	<1

[a] Data from Weis and Weis, 1977b.
[b] Day of embryonic development on which exposure was started.
[c] Did not complete axis formation.
[d] % of embryos among those that completed axis formation and that developed abnormality.
[e] Severe growth retardation.

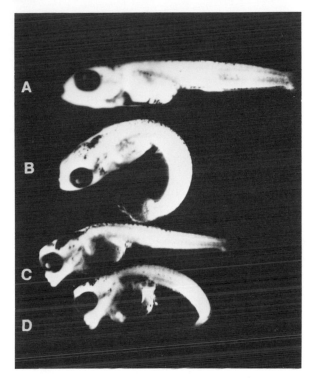

FIGURE 4 One day post-hatch larval killifish. A is a control specimen; B–D were treated with methyl mercury at 0.02 mg/l *in ovo*. The inability to uncurl after hatching (B,D) can occur independently of craniofacial anomalies (C,D). (Approximately ×13.) From Weis and Weis, 1977b. (Reprinted with permission of The Wistar Press.)

Inability to straighten from the embryonic position caused by lead and methyl mercury has also been demonstrated in medaka treated with inorganic mercury (Heisinger and Green, 1975).

When methyl mercury exposure was restricted to 1 day, the second day of development (blastula and gastrula) was the most susceptible. Exposures of 12 h had very little effect, suggesting either a need for longer exposures or a slow passage of methyl mercury across the chorion.

DISCUSSION

We have studied the effects of several insecticides and heavy metals on the development of three native species of estuarine fishes. Of the three species, the killifish proved to be the most useful monitor for evaluating the effects of pollutants at sublethal concentrations. The embryos are readily obtained, easily raised and observed, and respond to a variety of potential teratogens. Since the observations of Stockard (1907, 1910), killifish have been known to develop cyclopia in nonspecific response to several chemicals. In addition to this anomaly, we have found killifish to develop cardiovascular and skeletal anomalies following exposure to certain pollutants. The results of this study indicate that killifish embryos, while responding in several ways to a variety of pollutants, do not seem to be affected significantly by cadmium or malathion. This occurred even at concentrations higher than those that affected embryos of other fish species and killifish adults in studies of fin regeneration (Weis and Weis, 1975, 1976c).

Fish embryos in the natural environment can be exposed to pollutants in three ways: via yolk which is synthesized during oogenesis by exposed females, during the brief but sensitive period between shedding of gametes and elevation of the chorion, and via direct diffusion through the chorionic barrier. Our experiments have involved only the last pathway thereby mimicking a transient influx of chemical pollution into an estuary. Furthermore, in most of our experiments, exposure commenced after the cleavage stages had been completed; this eliminated any naturally abnormal or nonviable eggs from the experimental and control groups. It is likely that if the pollutants had been exposed earlier, there would have been a greater incidence of embryotoxicity and teratogenesis.

The concentrations of pollutants that produced deleterious effects on embryos are in some cases comparable to concentrations actually found in the environment. DDT, for example, has been found in concentrations up to 0.012 ppm (0.012 mg/l) in areas of agricultural runoff (Finley et al., 1970). Carbaryl is generally applied at 0.56 kg/ha. This translates to 0.056 mg/l if applied directly to water which is 1-m deep, or correspondingly more in the shallow water in which many estuarine fishes spawn. Malathion is also used at 0.56 kg/ha, but frequently up to 8.4 kg/ha (Kennedy and Walsh, 1970). Total mercury (organic plus inorganic) concentrations in Newark Bay, N.J., has been reported to be as high as 1.3 μg/l (Cheng, S.-L., N.J. Inst. Tech., personal communication).

We studied the effects of one pollutant at a time. Perhaps more realistic experiments would involve exposure to a combination of chemicals throughout development. In that way, synergistic and/or antagonistic interactions could be observed.

SUMMARY

Selected heavy metals and insecticides have been studied for sublethal effects on embryos of three species of estuarine fish. The insecticide studies indicate that carbaryl and parathion can induce cardiovascular abnormalities and developmental arrest in killifish *(Fundulus hetero-clitus)*, while DDT and malathion had no effect up to 10 ppm. DDT, malathion, and carbaryl caused optic abnormalities in the Atlantic silverside *(Menidia menidia)* at concentrations as low as 0.01 ppm. Sheepshead minnows *(Cyprinodon variegatus)* did not exhibit abnormalities with any insecticide, but malathion caused newly hatched fry to have uncoordinated neuromuscular activity.

Of the heavy metals studied, mercury salts caused craniofacial defects (cyclopia, synophthalmia, and related skeletal anomalies); lead interfered with ability to uncurl after hatching; and cadmium had no observable effect in concentrations below 10 ppm. Methyl mercury was the most potent teratogen, causing anomalies seen in all other experiments with killifish and doing so at lower concentrations.

REFERENCES

Armstrong, P., and J. S. Child. 1965. Stages in the normal development of *Fundulus heteroclitus*. Biol. Bull. 128:143–168.
Costello, D. P., M. E. Davidson, A. Eggers, M. H. Fox, and C. Henley. 1957. Methods for Obtaining and Handling Marine Eggs and Embryos. Marine Biological Laboratory, Woods Hole, Mass. 247 pp.
Finley, M. T., D. E. Ferguson, and J. L. Ludke. 1970. Possible mechanisms in the development of insecticide-resistant fish. Pestic. Monit. J. 3:212–218.
Heisinger, J. F., and W. Green. 1975. Mercuric chloride uptake by eggs of the ricefish and resulting teratogenic effects. Bull. Environ. Contam. Toxicol. 14:665–673.
Kennedy, H. D., and D. F. Walsh. 1970. Effects of Malathion on Two Warmwater Fishes and Aquatic Invertebrates in Ponds. Technical Paper 55. U.S. Bureau of Sport Fisheries and Wildlife, Washington, D.C. 13 pp.
McCann, J. A., and R. L. Jasper. 1972. Vertebral damage to bluegills exposed to acutely toxic levels of pesticides. Trans. Am. Fish. Soc. 101:317–322.
Schreiweis, D. O., and G. J. Murray. 1976. Cardiovascular malformations in *Oryzias latipes* embryos treated with 2,4,5-trichlorophenoxyacetic acid (2,4,5-T). Teratology 14:287–290.
Solomon, H. 1977. The teratogenic effects of the insecticides DDT, carbaryl, malathion, and parathion on developing medaka eggs. Ph.D. Dissertation, Rutgers University, Newark, N.J. 115 pp.
Stockard, C. R. 1907. The artificial production of a single median cyclopean eye in the fish embryo by means of seawater solutions of magnesium chloride. Arch. Entwicklungsmech. Org. 23:249–258.

Stockard, C. R. 1910. The influence of alcohol and other anaesthetics on embryonic development. Am. J. Anat. 10:369–392.

Weis, J. S., and P. Weis. 1975. The effect of insecticides on fin regeneration in *Fundulus*. Trans. Am. Fish. Soc. 104:135–137.

Weis, J. S., and P. Weis. 1976a. Optical malformations induced by insecticides in embryos of the silverside, *Menidia menidia*. Fish. Bull. 74:208–211.

Weis, J. S., and P. Weis. 1977a. Effects of heavy metals on development of the killifish, *Fundulus heteroclitus*. J. Fish Biol. 11:49–54.

Weis, P., and J. S. Weis. 1974. Cardiac malformations and other effects due to insecticides in embryos of the killifish, *Fundulus heteroclitus*. Teratology 10:263–268.

Weis, P., and J. S. Weis. 1976b. Abnormal locomotion associated with skeletal malformations in the sheepshead minnow, *Cyprinodon variegatus*, exposed to malathion. Environ. Res. 12:196–200.

Weis, P., and J. S. Weis. 1976c. Effects of heavy metals on fin regeneration in the killifish, *Fundulus heteroclitus*. Bull. Environ. Contam. Toxicol. 16:197–202.

Weis, P., and J. S. Weis. 1977b. Methyl mercury teratogenesis in the killifish, *Fundulus heteroclitus*. Teratology 16:317–326.

QUESTIONS AND ANSWERS

D. HINTON: What was the incidence of congenital malformations in your control fish?

P. WEIS: They were rare, being approximately 2% of 300 embryos representing all control groups in the methyl mercury experiments—to give an example.

D. HINTON: What types of lesions were encountered?

P. WEIS: We found failure in normal axis formation and tube hearts, but no craniofacial defects. We encountered only one tube-heart syndrome in the controls.

D. HINTON: Do killifish embryos spend any time in the bottom or in the sediments where they might come into contact with much higher concentrations of pollutants?

P. WEIS: The eggs are demersal, and are disseminated at the bottom of estuaries where the sediments are rich in pollutants. We have studied only the effects of pollutants in filtered water—not in sediments. Embryos could receive an additional burden of pollutants in the yolk if females become contaminated prior to laying eggs.

D. HINTON: Do you have any data on amounts of pollutants in the ovary?

P. WEIS: No.

J. MARKOFSKY: Have you studied various temperature effects on these phenomena?

P. WEIS: No, we've worked strictly at 20°C to 22°C.

J. COUCH: We usually think of the organophosphates and the carbamates as potent acetylcholinesterase inhibitors and effectors of neuromuscular lesions. But with kepone we have found that we can induce what we call "broken back syndrome" in fish in parts per trillion range of exposure. It is dose- and time-dependent; in other words, the same effects are produced by reducing the dosage while increasing the exposure time. In contrast to the findings presented above, kepone-exposed fish exhibit tetanic convulsions, and splinter the centra of their vertebrae. Bone splinters intrude the spinal cord. In an attempt at repair, osteoblastic tissue fixes the back in a broken position. The fish do not die immediately, but are paralyzed and fall to the bottom. In nature this has negative survival implications.

P. WEIS: This reminds me also of the work of McCann and Jasper (1972),[1] who found bluegills *(Lepomis macrochirus)* with spinal dislocations and fractures caused by organophosphates. I believe they used malathion but not at the low concentration you used.

J. COUCH: We can induce the broken back syndrome in the sheepshead minnow in a matter of 2 days at about 8 ppb. If we increase the concentration to 10 or 12 ppb, the syndrome develops within 1 day.

[1]McCann, J. A., and R. L. Jasper. 1972. Vertebral damage to bluegills exposed to acutely toxic levels of pesticides. Trans. Am. Fish. Soc. 101:317–322.

Evaluation of Aquatic Pollutants Using Fish and Amphibian Eggs as Bioassay Organisms

W. J. BIRGE, J. A. BLACK, and A. G. WESTERMAN

It is increasingly apparent that the most critical periods for the toxic effects of many environmental trace contaminants are the embryonic and early juvenile stages in the life cycles of many animal species (NAS–NAE, 1973; Birge et al., 1974). The reproductive potential of natural animal populations may be severely restricted or abolished by trace levels of toxicants that are harmless or sublethal to most adult organisms. Accordingly, protective environmental standards and pollution abatement policies, which have been based to a considerable extent on tolerances of adult animals, may not provide adequate protection for embryonic development and reproduction of many species.

Numerous organic toxicants inhibit animal reproduction. For example, Bowes et al. (1973) have found high concentrations of dichloro-diphenyldichloroethylene (DDE) and polychlorinated biphenyls (PCB's) in avian eggs. PCB concentrations have reached 300 ppm wet weight and 3,700 ppm in lipid fractions. These occurrences were correlated with embryonic death and reduced hatchability. Also, PCB compounds adversely affect reproduction of fish and certain invertebrates. Exposure of the water flea (*Daphnia magna*) to Aroclor 1254 administered in a flow-through system at 1.1 ppb resulted in 50% reproductive impairment (Nebeker and Puglisi, 1974). In continuous flow studies by Nebeker *et al.* (1974), Aroclor 1254 at 4.6 ppb produced 100% mortality of eggs of the fathead minnow (*Pimephales promelas*), and 0.52 ppb resulted in 59% to 67% hatchability, compared to 71% to 76% in control populations. Jensen *et al.* (1970) reported that PCB

108

residues of 0.4 to 1.9 ppm in eggs of the Atlantic salmon *(Salmo salar)* resulted in 16% to 100% mortality. As PCB accumulation in fish eggs and tissues may be 75,000 to 200,000 times the concentrations found in water (NAS–NAE, 1973; Hansen *et al.*, 1971), levels as low as 2 to 20 ppt may inhibit egg hatchability. Though generally less toxic than PCB compounds, dichlorodiphenyltrichloroethane (DDT) accumulates in fish eggs and sperm and affects hatchability and survival of fry (Burdick *et al.*, 1964; Cuerrier *et al.*, 1967). Egg concentrations of 0.4 to 4.75 ppm or more produced high embryonic mortality in several species of trout *(Salmo gairdneri, S. clarki, Salvelinus fontinalis, S. namaycush)*. For the coho salmon *(Oncorhynchus kisutch)*, similar results also have been reported for eggs containing DDT residues (NAS–NAE, 1973), and treatment of Atlantic salmon embryos with 50 to 100 ppb DDT retarded behavioral development and impaired balance (Dill and Saunders, 1974). Furthermore, exposure of adult flounder *(Pseudopleuronectes americanus)* to DDT at 2 ppb induced frequent anomalies among offspring (Smith and Cole, 1973). Dieldrin, endrin, 2,4-dichlorophenoxyacetic acid (2,4-D), and other chlorinated hydrocarbons also inhibit piscine reproduction, as reviewed by Hiltibran (1967), Johnson (1968), Johnson and Pecor (1969), and Smith and Cole (1973). It is now evident that the process of reproduction in animals is particularly sensitive to organic pollutants.

Numerous inorganic toxicants also adversely impair animal reproduction, particularly in aquatic species. Eggs, embryos, and early juvenile stages of fish are substantially more susceptible than adults to such metals as cadmium, copper, mercury, and zinc (Skidmore, 1965; Pickering and Vigor, 1965; Brungs, 1969; Hazel and Meith, 1970; McKim *et al.*, 1970; Birge *et al.*, 1974). For example, continuous flow treatment of rainbow trout *(Salmo gairdneri)* eggs with inorganic mercury at 0.1 ppb produced 100% embryonic mortality in 8 days (Birge, 1976). In the same investigation, channel catfish *(Ictalurus punctatus)* and rainbow trout embryos accumulated mercury up to 2,000 times exposure levels. In addition, embryos of the leopard frog *(Rana pipiens)* and domestic fowl *(Gallus domesticus)* were reported to be at least 1,000 times more sensitive to mercury and certain other metals than were adults (Birge and Just, 1975; Birge and Roberts, 1976; Birge *et al.*, 1976). High sensitivity of developmental stages to environmental toxicants was established further in bioassays with boron compounds (Wallen *et al.*, 1957; Birge and Black, 1977b). Continuous flow treatment from fertilization through 4 days posthatching gave boron LC_1 values of 0.001 to 0.1 ppm, 0.2 to 1.4 ppm, and 0.2 to 5.5 ppm for developmental stages of rainbow trout, goldfish *(Carassius au-*

ratus), and channel catfish treated with boric acid or borax at water hardness levels of 50 to 200 ppm calcium carbonate (Birge and Black, 1977b).

The high sensitivity of developmental stages to trace contaminants supports the use of embryo–larval bioassays for establishing water quality criteria. Principal objectives of this investigation included development and use of rapid-scan and continuous flow test procedures for evaluating effects of inorganic and organic toxicants on fish and amphibian embryos and larvae.

MATERIALS AND METHODS

Test species included the narrow-mouthed toad *(Gastrophryne carolinensis)*, largemouth bass *(Micropterus salmoides)*, goldfish, channel catfish, and rainbow trout. Rapid-scan egg bioassays were performed using the technique of Birge and Just (1975), in which test water and toxicants were renewed at regular 12-h intervals. In continuous flow bioassays on fish and amphibian eggs, exposure was maintained from fertilization through 4 days posthatching. Test water was monitored directly for pH, temperature, hardness, dissolved oxygen, conductivity, flow rate, and toxicant concentration, using the procedures of Birge and Black (1977a,b). The pH ranged from 7.0 to 7.8, and dissolved oxygen was maintained at or near saturation. Tests were conducted at hardness levels of 50 to 200 ppm calcium carbonate. Flow rate was set at 200 ml/h for 300-ml exposure chambers. Toxicant and test water were blended in a mixing unit situated ahead of each exposure chamber. Brinkmann peristaltic pumps (model 131900) and Sage syringe pumps (model 355) were used to regulate flow rates and mixing ratios for test water and toxicant. The LC_1 and LC_{50} values were determined by probit analysis of dose–response data for 10 to 14 exposure concentrations (Daum, 1969).

RESULTS

Rapid-scan egg bioassays are particularly useful for broad toxicological surveys. Large numbers of trace contaminants may be efficiently screened and compared for toxicological properties, and alternative animal test species may be evaluated for sensitivity. Data may be used for initial toxicological assessments and as a basis for selecting toxicants or animal species for further study. This screening procedure has been used to investigate a large number of inorganic elements of coal (Birge, 1978). Bioassays for 22 elements have been performed on eggs of the narrow-mouthed toad, goldfish, and rainbow trout. Treatment was initiated at fertilization and maintained through 4 days posthatching, giving exposure periods of 7 days for toad and goldfish and 28 days for trout. Mean water hardness with standard error was 195.0 ± 5.4 ppm calcium carbonate for the goldfish and toad and 104.0 ± 2.0 for the trout.

Elements that proved most toxic to goldfish, in order of decreasing toxicity, were silver, mercury, aluminum, cadmium, arsenic, chromium, cobalt, and lead, with LC_{50} values of 0.03, 0.12, 0.15, 0.17, 0.49, 0.66, 0.81, and 1.66 ppm, respectively. Threshold values (LC_1) for these metals varied from 0.4 and 0.6 ppb for aluminum and silver to 15.0 and 15.5 ppb for cadmium and arsenic. The eight elements that were most lethal to trout embryo–larval stages included mercury, silver, lanthanum, germanium, nickel, copper, cadmium, and vanadium. The LC_{50}'s were 0.005, 0.01, 0.02, 0.05, 0.05, 0.09, 0.13, and 0.16 ppm, respectively. The remaining elements were less toxic; LC_{50} values ranged from 0.17 ppm for thallium to 15.61 ppm for tungsten. Threshold values for the more toxic elements, given in parts per billion, ranged from as low as 0.2 to 0.8 for silver, mercury, germanium, and lanthanum to 6.9 for vanadium. Of 22 elements used to treat eggs of the narrow-mouthed toad, the 8 most lethal were mercury, silver, zinc, chromium, lead, cadmium, copper, and arsenic. The LC_{50} values in ppm ranged from 0.001 and 0.01 for mercury and silver to 0.04 for arsenic, and LC_1's ranged from 0.1 and 0.6 ppb for mercury and silver to 3.2 ppb for lead (Table 1).

Averaging the test data, species sensitivity in increasing order was goldfish, trout, and narrow-mouthed toad. However, the order and degree of differential sensitivity varied significantly for certain elements. Mercury and silver were highly toxic to developmental stages of all species. More selective test responses included high relative toxicity of aluminum to the goldfish, germanium and lanthanum to the trout, and selenium to the toad. While mercury and silver affect a broad cross section of aquatic species, such elements as lanthanum and germanium exert more selective effects on aquatic biota.

The LC_{50} values were averaged for all animal species to provide a working index to the aquatic toxicology of coal elements, permitting initial grouping under three categories (Table 1). The more toxic elements (Group I) had LC_{50} values varying from 0.02 ppm for silver to 0.88 ppm for tin. The LC_{50} ranges for Groups II and III were 1 to 5 ppm and 20 to 47 ppm, respectively. For further reference, coal elements also were ranked using a "most sensitive species index" based on LC_{50} values. The data in Table 1 indicate that coal elements pose a consider-

TABLE 1 Comparative Toxicity of Coal Elements to Fish and Amphibian Embryo–Larval Stages

Mean Toxicity Index[a]		Most Sensitive Species Index			
Element	LC_{50}, ppm	Element	Species	LC_{50}, ppm	LC_1, ppb
Toxicity Group I		Mercury	Toad	0.001	0.1
		Silver	Trout	0.01	0.2
Silver	0.02	Zinc	Toad	0.01	0.6
Mercury	0.04	Lanthanum	Trout	0.02	0.8
Cadmium	0.11	Chromium	Toad	0.03	1.0
Aluminum	0.25	Copper	Toad	0.04	1.0
Cobalt	0.29	Cadmium	Toad	0.04	1.6
Arsenic	0.36	Arsenic	Toad	0.04	1.6
Chromium	0.45	Lead	Toad	0.04	3.2
Lead	0.62	Nickel	Toad	0.05	0.4
Nickel	0.75	Cobalt	Toad	0.05	0.9
Tin	0.88	Germanium	Toad	0.05	1.2
		Aluminum	Toad	0.05	2.3
Toxicity Group II		Tin	Toad	0.09	1.7
		Selenium	Toad	0.09	5.0
Zinc	1.20	Thallium	Toad	0.11	2.4
Vanadium	1.67	Strontium	Toad	0.16	2.4
Copper	1.78	Vanadium	Trout	0.16	6.9
Germanium	1.90	Antimony	Toad	0.30	3.8
Thallium	2.43	Molybdenum	Trout	0.73	22.3
Strontium	2.98	Manganese	Toad	1.42	3.0
Antimony	4.07	Tungsten	Toad	2.90	10.7
Manganese	4.18				
Selenium	4.35				
Toxicity Group III					
Lanthanum	20.25				
Molybdenum	20.56				
Tungsten	47.17				

[a] LC_{50} values at 4 days posthatching averaged for three species, including narrow-mouthed toad, goldfish, and rainbow trout.

able hazard to freshwater ecosystems, depending on the form of coal utilization and the regulations governing waste disposal. Accordingly, continuous flow bioassays and bioaccumulation studies were conducted on the more toxic coal elements (Group I). In continuous flow bioassays, using a water hardness of 101.0 ± 1.8 ppm calcium carbonate, mercury at replicate exposure levels of 0.10 ± 0.03, 0.11 ± 0.01, and 0.14 ± 0.03 ppb consistently produced 100% mortality of trout embryos within 8 days. Trout and channel catfish embryos, which were treated for 8 to 10 days at 0.1 to 2.4 ppb, accumulated mercury in tissues at 782 to 2,000 times exposure levels.

Coal toxicology is further complicated by the fact that trace elements generally are released as complex mixtures. To investigate possible antagonistic, additive, or synergistic interactions, rapid-scan bioassays were used to determine embryopathic effects of mixtures of cadmium, copper, mercury, selenium, and zinc. Two-way metal combinations were tested on eggs of rainbow trout and channel catfish. Without exception, mixtures of nutritionally essential and nonessential metals produced marked synergism. This response to mercury/selenium is summarized in Table 2. Synergism, which was concentration dependent for this mixture, usually was significant at LC_{50} and higher exposure levels. Huckabee and Griffith (1974) also have reported mercury/selenium synergism for the carp *(Cyprinus carpio)*.

In the toxicology of coal elements, test systems must be developed to evaluate complex mixtures of aquatic contaminants. Egg bioassays

TABLE 2   Toxicity of Mercury/Selenium Mixtures to Fish Embryos

Concentration, ppm[a]	Survival at Hatching, Observed (Expected[b]), %			
	Trout	Catfish	Goldfish	Bass
0.001	96(93)	97(100)	—	—
0.002	94(82)	—	—	—
0.005	78(75)	85(96)	—	—
0.007	65(66)	—	—	—
0.010	40(58)	62(92)	99(99)	100(93)
0.025	16(52)	28(80)	87(98)	—
0.050	5(45)	15(62)	73(97)	89(89)
0.075	0(44)	2(57)	60(83)	87(85)
0.100	0	0(43)	49(75)	82(84)
0.250	—	0(33)	42(62)	63(69)
0.500	—	0(22)	38(55)	58(57)
0.750	—	0(12)	26(52)	29(49)
1.000	—	0	0(47)	14(49)
2.500	—	—	—	0
LC_{50}, ppm				
	0.01	0.01	0.16	0.35

[a] Concentrations based on equal proportions of mercury and selenium.
[b] Theoretical survival values for additive effects.

TABLE 3 Effects of Phenol on Developmental Stages of Rainbow Trout (*Salmo gairdneri*)

Water Hardness	Phenol Concentration, ppm		Survival, %[a]		Frequency of Anomalous Survivors at Hatching
	Nominal	Actual (mean ± SE)	At Hatching	4 Days Posthatching	
50 ppm calcium carbonate	0.001	0.0015 ± 0.0003	100	100	0
	0.010	0.0091 ± 0.0007	100	98	2
	0.100	0.068 ± 0.009	92	92	6
	1.000	0.844 ± 0.068	41	40	32
	10.000	8.79 ± 0.40	9	5	73
200 ppm calcium carbonate	0.001	0.0012 ± 0.0003	90	90	0
	0.010	0.0103 ± 0.0009	75	75	1
	0.100	0.070 ± 0.009	58	58	4
	1.000	0.905 ± 0.068	21	20	22
	10.000	9.33 ± 0.91	0	0	—

[a] Percentages were determined as survival in experimental population/control survival.

appear particularly applicable for this purpose. Continuous flow tests are now being used to assess the toxic potential of full suites of contaminants contained in coal-ash effluents. After 522 h of continuous washing of fly ash, leachates produced 100% mortality of leopard frog eggs.

The high sensitivity and simple culture requirements of fish and amphibian eggs also make them especially suitable for tests on organic toxicants, particularly on compounds of high volatility or low solubility, which are difficult to stabilize in conventional aquatic bioassays. A closed bioassay system, devoid of standing air space, was developed to minimize volatility as a test variable. Mechanical homogenization and regulation of flow rate were used to maintain stable suspensions of hydrophobic organics without carrier solvents (Birge *et al.*, 1979). Tests were undertaken for 12 organics, including aromatic amines; organophosphates; aliphatic, aromatic, and chlorinated hydrocarbons; and phthalates. Reproducible exposure levels were maintained down to 1.0 to 0.1 ppb, depending on the toxicant. Initially, sensitivity and reproducibility were demonstrated for this continuous flow bioassay procedure in tests with phenol (Table 3). When administered in hard water, phenol at 0.1 ppm produced mortality or teratogenic impairment in approximately 50% of the test organisms. Control-adjusted survival increased to 90% when exposure was reduced to 1 ppb. Phenol was significantly less toxic when administered in soft water (Table 3). Judging from studies in progress, a wide range of organic compounds appears to be highly toxic to developmental stages of fish and amphibians, producing high frequencies of mortality, teratogenesis, and locomotor impairment.

Developmental stages are subject to a broad spectrum of toxicant-induced responses. Toxic actions may affect biochemical and physiological mechanisms associated with fertilization, genic action, cell proliferation and growth, cellular differentiation, basic metabolism and systemic functions, as well as the hatching process and the initial accommodation to a free-living existence. Consequently, bioassay screening with developmental stages is unlikely to produce "false negative" results, provided that exposure to selected toxicants is maintained continuously from fertilization through 4 to 7 days post-hatching.

SUMMARY

Fish and amphibian embryos and larvae were evaluated as test organisms for aquatic bioassays. Static renewal and continuous flow procedures were developed for testing organic and inorganic toxicants. Static tests were used to compare 22 inorganic elements which occur in coal effluents and emissions. Mercury and silver were the most lethal to eggs and larvae of the rainbow trout, goldfish, and narrow-mouthed toad. The LC_{50} values ranged from 0.005 to 0.12 ppm for mercury and 0.01 to 0.03 ppm for silver. Tungsten was the least toxic, with LC_{50}'s of 2.9 to 120 ppm.

Continuous flow administration of inorganic mercury at 0.1 ppb produced 100% mortality of trout eggs. Test organisms accumulated mercury at 782 to 2,000 times the exposure levels. Mixtures of mercury and selenium were synergistic to fish eggs at LC_{50} or greater concentrations. This interaction was dose-dependent and was not significant at exposure levels that approached threshold. Phenol at 1.0 ppb was toxic to trout eggs; LC_{50} values were 0.54 and 0.08 ppm in soft and hard water, respectively.

ACKNOWLEDGMENTS

This research was supported by National Science Foundation (RANN) grants GI-43623 and AEN 74-08768 A01, Kentucky Institute for Mining and Minerals Research grant 7576-EZ, U.S. Department of the Interior (OWRT) grant A-067-KY, and Environmental Protection Agency (OTS) contract 68-01-4321.

REFERENCES

Birge, W. J. 1976. Effects of Metals on Embryogenesis and Use of Vertebrate Embryos as Sensitive Indicators of Environmental Quality. Final Technical Report, NSF (RANN), Washington, D.C. 140 pp.

Birge, W. J. 1978. Aquatic toxicology of trace elements of coal and fly ash. In J. H. Thorp and J. W. Gibbons, eds. Energy and Environmental Stress in Aquatic Systems, pp. 219–240. DOE Symposium Series (CONF-771114). Government Printing Office, Washington, D.C.

Birge, W. J., and J. A. Black. 1977a. Bioassay Protocol, a Continuous Flow System Using Fish and Amphibian Eggs for Bioassay Determinations on Embryonic Mortality and Teratogenesis. EPA Report No. 560/5-77-002. Environmental Protection Agency, Office of Toxic Substances, Washington, D.C. 59 pp.

Birge, W. J., and J. A. Black. 1977b. Sensitivity of Vertebrate Embryos to Boron Compounds. EPA Report No. 560/1-76-008. Environmental Protection Agency, Office of Toxic Substances, Washington, D.C. 66 pp.

Birge, W. J., and J. J. Just. 1975. Bioassay Procedures Using Developmental Stages as Test Organisms. U.S. Department of the Interior, Research Report #84. NTIS No. PB-240 978. National Technical Information Service, Springfield, Va. 36 pp.

Birge, W. J., and O. W. Roberts. 1976. Toxicity of metals to chick embryos. Bull. Environ. Contam. Toxicol. 16(3):319–324.

Birge, W. J., A. G. Westerman, and O. W. Roberts. 1974. Lethal and teratogenic effects of metallic pollutants on vertebrate embryos. Pp. 316–320 in 2nd Annual NSF–RANN

Trace Contaminants Conference, Asilomar, Calif. National Technical Information Service, Springfield, Va.

Birge, W. J., O. W. Roberts, and J. A. Black. 1976. Toxicity of metal mixtures to chick embryos. Bull. Environ. Contam. Toxicol. 16(3):314–318.

Birge, W. J., J. A. Black, J. E. Hudson, and D. M. Bruser. 1979. Embryo-larval toxicity tests with organic compounds, pp. 131–147. In L. L. Marking and R. A. Kimerle, eds. Aquatic Toxicology. Proceedings of the Second Annual Symposium on Aquatic Toxicology. American Society for Testing and Materials, No. STP-667, Philadelphia, Pa.

Bowes, G. W., B. R. Simoneit, A. L. Burlingame, B. W. deLappe, and R. W. Risebrough. 1973. The search for chlorinated dibenzofurans and chlorinated dibenzodioxins in wildlife populations showing elevated levels of embryonic death. Environ. Health Perspect. 2:191–197.

Brungs, W. A. 1969. Chronic toxicity of zinc to the fathead minnow (*Pimephales promelas* Rafinesque). Trans. Am. Fish. Soc. 98:272–279.

Burdick, G. E., E. J. Harris, H. J. Dean, T. M. Walker, J. Skea, and D. Colby. 1964. The accumulation of DDT in lake trout and its effects on reproduction. Trans. Am. Fish. Soc. 93:127–135.

Cuerrier, J. P., J. A. Keith, and E. Stone. 1967. Problems with DDT in fish culture operations. Nat. Can. 94:315–320.

Daum, R. J. 1969. A revision of two computer programs for probit analysis. Bull. Entom. Soc. Am. 16:10–15.

Dill, P. A., and R. C. Saunders. 1974. Retarded behavioral development and impaired balance in Atlantic Salmon (*Salmo salar*). J. Fish. Res. Board Can. 31:1936–1938.

Hansen, D. J., P. R. Parrish, J. I. Lowe, A. J. Wilson, Jr., and P. D. Wilson. 1971. Chronic toxicity, uptake and retention of Aroclor 1254 in two estuarine fishes. Bull. Environ. Contam. Toxicol. 6(2):113–119.

Hazel, G. R., and S. J. Meith. 1970. Bioassay of king salmon eggs and sac fry in copper solutions. Calif. Fish Game 56:121–124.

Hiltibran, R. C. 1967. Effects of some herbicides on fertilized fish eggs and fry. Trans. Am. Fish. Soc. 96:414–416.

Huckabee, J. W., and N. A. Griffith. 1974. Toxicity of mercury and selenium to the eggs of carp (*Cyprinus carpio*). Trans. Am. Fish. Soc. 103:822–825.

Jensen, S., N. Johansson, and M. Olsson. 1970. PCB-indicators of effects on salmon. Report LFI MEDD of the Swedish Salmon Research Institute. PCB Conference, National Swedish Environment Protection Board, Stockholm.

Johnson, D. W. 1968. Pesticides and fishes—A review of selected literature. Trans. Am. Fish. Soc. 97:398–424.

Johnson, H. E., and C. Pecor. 1969. Coho salmon mortality and DDT in Lake Michigan. Trans. North Am. Wildl. Nat. Resour. Conf. 34:159–166.

McKim, J. M., G. Christensen, and E. Hunt. 1970. Changes in the blood of brook trout (*Salvelinus fontinalis*) after short-term and long-term exposure to copper. J. Fish. Res. Board Can. 27:1883–1889.

NAS–NAE Committee on Water Quality Criteria. 1973. Water Quality Criteria 1972. U.S. Government Printing Office, Washington, D.C. 593 pp.

Nebeker, A. V., and F. A. Puglisi. 1974. Effects of polychlorinated biphenyls (PCB's) on survival and reproduction of *Daphnia, Gammarus,* and the midge *Tanytarsus.* Trans. Am. Fish. Soc. 103:722–728.

Nebeker, A. V., F. A. Puglisi, and D. L. Defoe. 1974. Effect of polychlorinated biphenyl compounds on survival and reproduction of the fathead minnow and flagfish. Trans. Am. Fish. Soc. 103:562–568.

Pickering, Q. H., and W. M. Vigor. 1965. The acute toxicity of zinc to eggs and fry of the fathead minnow. Prog. Fish Cult. 27:153–157.

Skidmore, J. F. 1965. Resistance to zinc sulfate of the zebrafish (*Brachydanio rerio* Hamilton–Buchanan). Ann. Appl. Biol. 56:47–53.

Smith, R. M., and C. F. Cole. 1973. Effects of egg concentrations of DDT and dieldrin on development in white flounder (*Pseudopleuronectes americanus*). J. Fish. Res. Board Can. 30(12, part 1):1894–1898.

Wallen. I. E., W. C. Greer, and R. Lasater. 1957. Toxicity to *Gambusia affinis* of certain pure chemicals in turbid waters. Sewage Ind. Waste 29:695–711.

QUESTIONS AND ANSWERS

P. WEIS: What type of teratomatous lesions did you see in those embryos that did not hatch?

W. BIRGE: We did not analyze anomalies in animals that did not survive the hatching process, since we were concerned principally with delineating environmental thresholds rather than identifying causative mechanisms producing terata. We did find that for most environmental toxicants tested, mortality occurred over the same range of bioassay concentrations that produced terata. This is contrary to evidence that teratogenic effects commonly occur at sublethal exposure levels.

J. WEIS: Did you find any combinations of metals that turned out to be antagonistic to each other?

W. BIRGE: No, we did not find any pronounced antagonism with the animal species or different combinations of metals used thus far. Additive to moderately antagonistic effects were observed for low concentrations of nutritionally essential metals. Approximately the same pattern of response occurred when nonessential metals were combined. This is in contrast to the pairing of essential with nonessential elements that consistently produced greater synergism. However, it should be made clear that this applies to embryopathic effects on trout and catfish eggs, and that synergism occurred at low exposure concentrations (≤ 10 ppb) only in tests with the catfish.

The Kidney of the Quahog (*Mercenaria mercenaria*) as a Pollution Indicator

RICHARD RHEINBERGER, GERALD L. HOFFMAN, and PAUL P. YEVICH

One of the objectives of the Environmental Protection Agency is to develop methods to monitor the marine environment for pollutants and to determine the biological effects of the pollutants on marine ecosystems. In this pursuit, we studied the kidneys of quahogs (a hard-shell clam) as possible indicators of pollution.

Both gross and microscopic histopathological examination revealed that the kidneys of the common quahog (*Mercenaria mercenaria*) collected in polluted areas had larger and more numerous concretions than the kidneys of quahogs obtained from clean areas. To determine if the number, size, and chemical composition of the concretions could be used as an environmental indicator, quahogs were collected from sites along the Rhode Island shore. Their kidneys and the concretions they contained were analyzed histopathologically and chemically.

HISTOPATHOLOGICAL STUDIES

Materials and Methods

Five hundred quahogs were collected from each of nine stations along Narragansett Bay and from a control station (No. 10) at Charlestown Pond, R.I. The bay stations were numbered from No. 1, Sabin Point, to No. 9, Bonnet Shores, in decreasing order of pollution based upon studies by personnel from the University of Rhode Island (Saila *et al.*, 1967; Farrington, 1971; Jeffries, 1972).

119

In the laboratory, all animals were shucked and their kidneys removed. One hundred kidneys from each station were fixed in Helly's fixative for 24 h. They were then cross-sectioned into three or four pieces, dehydrated, and cleared. The cleared sections were examined with a stereomicroscope to determine the number, size, and color of the concretions. These sections were then processed by routine histopathological methods for microscopic examination.

For chemical analysis the concretions were removed from the kidney mass by homogenizing the tissue in a Waring blender and diluting it with water. This mixture was put into a separatory funnel that allowed the concretions to gradually settle to the bottom from which they were drawn off and allowed to dry. A stereomicroscope was used to examine the concretions for unusual characteristics. The concretions were then analyzed chemically and with an electron probe.

Results

Gross Examination: Gross examination of the kidneys showed that there were more and larger concretions in the quahogs collected from the known polluted portions of the Bay than in those from the Charlestown Pond control area (Table 1).

Most of the concretions from the more polluted areas were black and ranged from approximately 390 μm to 485 μm in diameter. They were usually spherical and their outer layer appeared to have a porous texture. The smaller concretions were brownish-red, were more jagged, and lacked the porous appearance. They generally acquired the shape of the

TABLE 1   Concretions in Kidneys

Station Number and Location	Portion of Kidneys with More Than 50 Concretions, %	Animals with Concretions > 300 μm Diam., %	Average Concretion Weights Obtained by Weighing 400 Macerated Kidneys from Each Location, g
1. Sabin Point	57	46	2.8
2. Bullocks Cove	49	34	3.2
3. Conimicut Point	47	30	2.7
4. Potowomut River	31	15	2.4
5. Dyer Island	30	11	2.3
6. Fox Island	28	24	1.7
7. Jamestown bridge	25	15	0.73
8. Rose Island	31	19	1.3
9. Bonnet Shores	21	3	1.2
10. Charlestown Pond	17	1	0.45

FIGURE 1 Concretions separated from the kidney mass. Arrows
point to concretions of various sizes and shapes.

space they occupied. Many of the concretions found in the folds of the
tissue were flat and elongated, while those in the lumen were more spherical (Figure 1).

A few of the concretions found in the kidneys of the control animals
from Charlestown Pond were black, but most were smaller than those
found at the polluted stations. The largest ones ranged from approximately 200 μm to 335 μm in diameter. Most of the concretions were tan
or brown and had the same characteristics as the small concretions from
the other stations. Large white calcium concretions were not found in the
kidneys from the more polluted areas.

Microscopic Examination: The microscopic examination showed that
the concretions had different compositions (Figures 2, 3, and 4). Some
of the larger concretions had centers that varied in size and consisted of

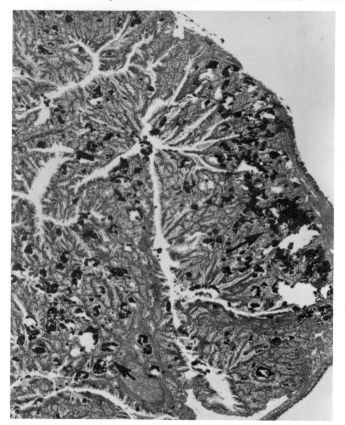

FIGURE 2 Kidney of a quahog collected from Sabin Point. Note the numerous black concretions.

a brownish granular material. This core was surrounded by concentric rings which varied in thickness (Figure 4). Other large concretions consisted of many small, slightly basophilic concretions that were surrounded by a heavier basophilic wall. Smaller concretions, usually brownish in color, appeared to be masses of amebocytes adhering to one another. In some of the quahogs, the cytoplasm of the clear columnar cells of the kidney epithelium was filled with a dense basophilic concretion. In others, there was a very fine granular concentration in the clear cytoplasm.

In kidney tissues of quahogs from Station No. 1 there were extensive histopathological changes when very large concretions were present. A complete loss of the clear columnar epithelium left the concretions lying on the connective tissue of the folds of the kidney. Masses of pink granular amebocytes surrounded many of the concretions and infiltrated the supporting connective tissue. Some of the tissues had the appearance of granulomas with the concretions being the central mass.

Numerous excretory phagocytic brown-staining amebocytes were also seen in the connective tissue of the kidneys and in areas of the renopericardial canal. In many instances these phagocytic amebocytes migrated from the connective tissue into the lumen of the kidney, where they coalesced to form the nucleus around which the concentric layers were deposited. When extensive accumulations of excretory amebocytes were found in the kidney area, there were usually also large numbers of them throughout the connective tissue of the body

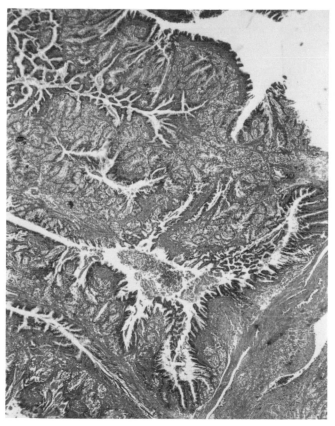

FIGURE 3　Kidney of quahog collected from Charleston Pond. Very few concretions are present.

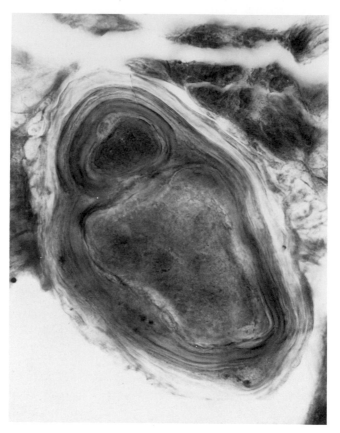

FIGURE 4 Concretion showing two possible centers of formation around which concentric rings have formed.

mass. In some kidneys, foci of hyperplastic cells with a homogenous basophilic cytoplasm were interspersed between the clear columnar epithelial cells of the kidney folds.

The number and size of the concretions and the histopathological changes were smaller in the quahogs from less contaminated stations. The few concretions found in the kidneys of the Charlestown Pond quahogs consisted mostly of fine basophilic granules. When larger concretions were found, they were located in the lumen rather than in the folds of the kidneys. There were no observed histopathological changes in the connective tissue or epithelium of these kidneys.

CHEMICAL ANALYSIS

Materials and Methods

The concretions were separated one at a time by color until several hundred black and brown ones had been obtained for each station except No. 10, where the small number and small size of the brown concretions precluded a good separation. This separation was accomplished with a nylon fiber while observing the colors with a microscope.

The separated concretions were then weighed on a microbalance (Perkin–Elmer Model No. AD-2Z) and transferred to 7-ml (acid-leached) polyethylene vials. A sufficient number of black concretions were separated to allow three or four replicates for each station. One sample weight of brown concretions was obtained from each of the first nine stations.

Typically, 50 black concretions weighed approximately 1 mg, so individual concretions would average approximately 20 μg. However, some large black concretions could weigh as much as 30 to 50 μg. Groups of 100 brown concretions, which were smaller at all the stations, ranged from 0.5 to 3 mg. The average weight was less than 1 mg per 100 or less than 10 μg per concretion.

One hundred microliters of ultra-pure concentrated nitric acid was added to each vial to dissolve the concretions. The vials were capped and allowed to stand at room temperature for 2 days. Five milliliters of deionized water were then added to the acid solutions and mixed thoroughly. Trace metal analysis was performed with a Perkin–Elmer 603 atomic absorption unit for conventional flame analysis. An HGA-2100 was coupled to a Perkin–Elmer 360 atomic absorption unit for all the heated graphite atomization analysis.

Results

In Table 2 the elemental analyses are given as the average percent of the weight of the concretions. No standard deviation is given for the brown concretions, since only one analysis was done for each station. The analysis uncertainty is estimated to be less than 10% for all the data presented in Table 2. The metals listed in the table comprise from 24% to 27.7% of the total mass of the brown and black concretions. These are the only metals that have been quantified and should not be considered as the only inorganic constituents present. For example, electron microprobe analysis of kidney concretions from several sta-

TABLE 2 Average Percent of Metal Concentrations in Black and Brown Concretions Obtained from Quahog Kidneys

Station No. and Concretion Color	No. of Samples	Amount of Metal in Concretions Average % ± SD							Total Concentration
		Calcium	Manganese	Zinc	Iron	Copper	Lead	Cadmium × 10⁻³	
1 Black	3	10.6 ± 0.15	3.88 ± 0.24	5.81 ± 0.10	3.21 ± 0.10	0.46 ± 0.08	0.21 ± 0.02	2.2 ± 0.2	24.2
Brown	1	11.4	4.35	5.54	2.71	0.38	0.14	2.2	24.5
2 Black	3	9.27 ± 0.18	7.40 ± 0.10	6.05 ± 0.01	3.03 ± 0.13	0.29 ± 0.02	0.23 ± 0.02	1.2 ± 0.1	26.3
Brown	1	9.23	8.73	5.51	2.16	0.27	0.16	1.0	26.1
3 Black	4	9.62 ± 0.32	7.14 ± 0.36	5.76 ± 0.20	2.32 ± 0.33	0.32 ± 0.02	0.16 ± 0.02	1.3 ± 0.2	25.3
Brown	1	10.3	7.90	5.39	1.73	0.26	0.12	1.4	25.7
4 Black	3	7.78 ± 0.09	9.62 ± 0.15	5.86 ± 0.13	2.81 ± 0.08	0.20 ± 0.01	0.16 ± 0.02	0.49 ± 0.14	26.4
Brown	1	9.03	12.4	4.38	0.98	0.16	0.085	0.56	27.0
5 Black	3	7.34 ± 0.28	10.8 ± 0.44	5.63 ± 0.98	2.89 ± 0.28	0.16 ± 0.02	0.15 ± 0.01	0.27 ± 0.03	27.0
Brown	1	8.72	12.2	4.83	1.67	0.16	0.088	0.43	27.7
6 Black	3	8.51 ± 0.18	8.86 ± 0.93	5.54 ± 0.18	3.41 ± 0.46	0.15 ± 0.01	0.20 ± 0.01	0.55 ± 0.08	26.7
Brown	1	9.71	10.2	4.79	2.18	0.16	0.11	0.58	27.2
7 Black	3	8.20 ± 0.36	9.0 ± 0.67	5.59 ± 0.34	3.51 ± 0.19	0.13 ± 0.006	0.21 ± 0.02	0.39 ± 0.01	26.6
Brown	1	9.76	10.7	4.70	2.06	0.14	0.11	0.47	27.4
8 Black	4	8.72 ± 0.19	9.20 ± 0.61	5.16 ± 0.34	2.80 ± 0.29	0.14 ± 0.02	0.19 ± 0.02	0.46 ± 0.04	26.2
Brown	1	9.70	11.2	4.06	1.59	0.15	0.092	0.66	26.8
9 Black	3	8.68 ± 0.23	8.48 ± 1.0	5.56 ± 0.02	3.65 ± 0.47	0.12 ± 0.007	0.20 ± 0.04	0.41 ± 0.01	26.7
Brown	1	9.69	11.6	4.13	1.85	0.13	0.094	0.44	27.5
10 Black	3	8.45 ± 0.61	7.44 ± 0.79	3.26 ± 0.18	6.41 ± 0.54	0.027 ± 0.007	0.19 ± 0.02	0.45 ± 0.03	25.8
Brown	0	—	—	—	—	—	—	—	—

tions has indicated the presence of the following elements: gold, platinum, zinc, copper, cobalt, iron, manganese, titanium, cadmium, calcium, potassium, molybdenum, aluminum, magnesium, and strontium. Also, the presence of organic carbon in concentrations ranging from 2% to 4% has been confirmed for some black kidney concretions.

Table 2 presents some interesting comparisons. The percent of the total concentration does not deviate very much from station to station. The lowest percent of manganese content was found at the station that is considered to be the most polluted (Sabin Point). The lowest percent of zinc concentration and the highest percent of iron occurred at the control station. The percent of calcium and manganese concentrations in the brown concretions is always higher than in the black concentrations, but the reverse is true for zinc, iron, and lead. However, in several instances the differences were slight for all five elements. The percent of copper and cadmium concentrations seems to follow a definitive pattern that is related to the number of kidney concretions found at each station.

We speculate that manganese, zinc, iron, copper, lead, and cadmium are present in the concretions as sulfides, and that calcium is either a phosphate or carbonate. We hypothesize that this is the most likely mechanism for forming insoluble compounds of these metals in these quahog kidneys and on the basis of concretion color. Anaerobic conditions should enhance the probability for concretion formation in quahog kidneys.

Reporting metal values as the percent of total weight normalizes all values and does not give absolute differences among locations. For example, quahogs collected at Sabin Point (Station No. 1) had many more kidney concretions than quahogs collected at Charlestown Pond. If the metals are analyzed on an absolute rather than normalized basis, the Sabin Point samples should be many times higher for all metals. Chemical rather than mechanical separation should be used to insure absolute recovery of the concretions from individual kidneys. We are investigating the possibility of analyzing on an absolute basis the metals from the various stations examined in this study.

DISCUSSION

Because of the histopathological and chemical differences between the kidneys of quahogs collected from polluted and those from low pollution sources, we believe that the molluscan kidney with its ability to form solid urine can serve as an indicator of the environmental pollutants in molluscan habitats.

Other than the knowledge we have gained by observation, there is little information about how concretions are formed. We know that they are found in most of the molluscan kidneys and are one of the characteristics that separate the molluscan kidney from the kidneys of vertebrates (Bouillon, 1960; Krahelska, 1910; Potts, 1967; Strohl, 1924; Turchini, 1923). We believe that concretions are formed in two ways. In some instances very fine basophilic granules form in the apical portion of the kidney epithelium and there increase in size until they are extruded into the lumen of the folds of the kidney. There they coalesce with others to form a large mass. At other times, phagocytic amebocytes bearing yellowish, orange, brown, or black material move through the epithelium into the lumen and there coalese into large masses and form concretions. It is possible that the large masses that are seen grossly in kidneys from polluted areas are bound together by various organic compounds, particularly hydrocarbons, as they appear intensely black and (gooey) in the fresh material.

Potts (1967) has stated that the accumulation of concretions does not harm the kidneys. However, our studies showed that there is a loss of epithelium when the accumulations are exceptionally large and that the various chemicals in the concretions, which reflect the chemicals in the environment, may cause an inflammatory response throughout the kidney. We believe that accumulations such as those seen in Figure 3 are not a normal function of the kidney but, rather, that they result from the animals' inability to excrete the masses of concretions. Depuration studies conducted in our laboratory have shown that the kidneys do have the ability to free themselves of these concretions when they are no longer stressed by their environment.

SUMMARY

The kidneys of quahogs collected from 10 variously polluted areas around Rhode Island were analyzed histochemically, chemically, and histopathologically with methods involving light, electron microscopy, and atomic absorption. The quantity, size, and color of the kidney concretions in quahogs collected from the high-pollution areas were different from those collected in areas of low pollution.

Analysis of these concretions revealed a predominance of heavy metals. It was found that interelemental ratios of trace metals depended on the various colors of the concretions. The metals determined were manganese, calcium, zinc, copper, lead, cadmium, and iron. The authors propose that elemental analysis of these concretions and other histopathological features of the molluscan kidney may be a convenient method of tagging the various heavy metal pollutants in the water sediment environment to which the benthic community is exposed.

REFERENCES

Bouillon, J. 1960. Ultrastructure des cellules rénales des mollusques. I. Gasterpodes pulmonés terrestres. Ann. Sci. Nat. Zool. Anim. (12) ser. 2:719–749.

Farrington, J. W. 1971. Benthos lipids of Narragansett Bay—Fatty acids and hydrocarbons. Ph.D. Dissertation. University of Rhode Island. 141 pp.

Jeffries, H. P. 1972. A stress syndrome in the hard clam, *Mercenaria mercenaria*. J. Invert. Pathol. 20(3):242–251.

Krahelska, M. 1910. Über den Einfluss der Winterruhe auf den histologischen Bau einiger Landpulmonaten. Z. Naturwiss. Jena 46:363–444.

Potts, W. T. W. 1967. Excretion in the molluscs. Biol. Rev. Biol. Proc. Cambridge Philos. Soc. 42:1–41.

Saila, S. B., J. M. Flowers, and M. T. Cannario. 1967. Factors affecting the relative abundance of *Mercenaria mercenaria* in the Providence River, Rhode Island. Proc. Nat. Shellfish Assoc. 57:83–89.

Strohl, J. 1924. Pp. 443–607 in H. Winterstein, ed. Handbuch der vergleichenden Physiologie, 2 Bd. Fischer Verlag, Jena, E. Germany.

Turchini, J. 1923. Contribution à l'étude de l'histologie comparée de la cellule rénale. L'excrétion urinaire chez les mollusques. Arch. Morph. Gen. Exp. 18:3–253.

QUESTIONS AND ANSWERS

Z. RUBEN: The convenient thing to do is to analyze chemically the water for the concentrations of the various metals you have examined and to correlate these with the various concretions in the clams. Did you do this?

G. HOFFMAN: The reason for isolating the concretions from the kidney and analyzing them for their metal content is that the kidney concentrates metals from the water. If one were to analyze the entire clam, one would not see differences in the metal levels between the polluted and control areas.

D. JONES: Were all your samples of similar size?

R. RHEINBERGER: Yes.

D. JONES: Did you find larger amounts of concretions in your larger molluscs than in smaller ones?

R. RHEINBERGER: No, the ratios of concretions to kidney mass are approximately the same in large and small animals, although naturally you would find larger amounts of concretions in larger animals.

D. JONES: Approximately how long does it take for these concretions to disappear from the kidneys once you place the animals back in pristine waters?

R. RHEINBERGER: We placed 100 animals in very clean water and sampled them every 2 weeks for 2 months to assess the degree of depuration. Very little change was noticed after 2 months. However, 20 animals left in pristine water for 6 months recovered considerably.

HEAVY
METALS

Comparative Pathology of Experimental Subacute Feline and Canine Methylmercurialism

T. S. DAVIES, S. W. NIELSEN, and B. S. JORTNER

Following the identification of methyl mercury as a hazardous environmental pollutant, investigators established its neurotoxicity (Berglund et al., 1971). Further studies were indicated because the basic mechanism of methyl-mercury-induced injury has not been elucidated and because there is continued risk for humans and wildlife exposed to the compound (Atwood et al., 1976; Spurgeon, 1976). In addition, a dose-dependent variability in the distribution of central nervous system lesions produced by methyl mercury has been observed (Shaw et al., 1975; Wobeser et al., 1976).

The purpose of this study was to compare clinical signs and nervous system lesions produced in subacute feline and canine methylmercurialism. Since reports of serum biochemical changes in experimental methylmercurialism in higher mammals have been limited, serum biochemical profiles were determined.

MATERIALS AND METHODS

Animals and Dosage

Sixteen adult domestic cats were housed in individual cages, and 10 adult, mongrel dogs were housed in box stalls (5 dogs per stall). Fourteen cats (nos. 11–24) were given 0.43 mg of mercury (as methylmercuric hydroxide) [Ventron Corporation, Beverly, Mass.] per kilogram of body weight per day (0.43 mg Hg/kg BW/day). They comprised Group 1. Two cats (Group 2) served as nonexposed controls. Dogs

133

were exposed to 0.43 mg Hg/kg BW/day as methylmercuric hydroxide in Group 3 (nos. 6, 7, 8, and 9) and to 0.64 mg in Group 4 (nos. 10, 11, 12, and 13). Group 5 dogs (nos. 14 and 15) were not exposed to mercury. Doses were adjusted to body weights twice weekly. The compounds were administered in gelatin capsules lubricated [with Lube, Elanco Products, Indianapolis, Ind.] immediately prior to administration. Commercial food [Purina Cat Chow, Ralston Purina Co., St. Louis, Mo.; Joy Dog Food, Best Foods and Farm Supplies, Inc., Oakdale, Pa.] and water were available *ad libitum*. Preexperimental treatment included panleukopenia vaccine [Leukogen TC, Bio-ceutic Lab., St. Joseph, Mo.], canine distemper vaccine [Delcine-H, Dellen Lab., Omaha, Nebr.], and anthelmintics.

Clinical Examination

Following the onset of clinical signs of toxicosis, a daily neurological examination was performed according to the procedure of deLahunta (1971). Terminal serum biochemical profiles of cats 21, 24, and 25, and weekly serum biochemical profiles of dogs, beginning the second week of exposure, were determined at a local hospital, the Windham Community Memorial Hospital in Willimantic, Conn. These profiles were done with a sequential multichannel autoanalyzer (SMA 12/60, Autoanalyzer, Technicon Instruments Corp., Tarrytown, N.Y.), which was not modified for analysis of animal serum samples. Whole blood was obtained by venipuncture, permitted to clot, and spun for 15 min, at 3,600 g in a Sorvall superspeed angle centrifuge (Ivan Sorvall, Newtown, Conn.). Serum was collected and frozen until analysis. Sodium, potassium, calcium, total protein, albumin, total cholesterol, glucose, blood urea nitrogen (BUN), creatinine, total bilirubin, alkaline phosphatase, and glutamic-oxaloacetic transaminase (SGOT) were determined.

Necropsy

Animals died from methyl mercury toxicosis or were killed *in extremis* by injection of barbiturate solution (Tables 1 and 2). Brain, spinal cord, and representative tissue samples of gastrointestinal, respiratory, cardiovascular, endocrine, urogenital, lymphoid, and muscular systems, brachial plexuses, sciatic nerves, and dorsal root ganglia were removed at necropsy and fixed in 10% neutral buffered formalin. Specimens for microscopic examination were embedded in paraffin, sectioned at 6 μm, and stained with hematoxylin and eosin. Lendrum's, cresyl

TABLE 1   Regimen, Exposure Duration, and Dosage for Cats Exposed to Oral Methyl Mercury

Group Number and Dose	Cat Number	Days of Exposure	Clinical Signs Produced at: Total Dose, mg Hg/kg BW[a]	Total Days of Exposure	Total Dose, mg Hg/kg BW	Necropsy, day
1	11	30	12.9	30	12.9	31
(0.43 mg	12	30	12.9	34	14.6	36[b]
Hg/kg BW)	13	30	12.9	34	14.6	36
	14	31	13.3	34	14.6	36
	15	30	12.9	34	14.6	36[b]
	16	30	12.9	34	14.6	37
	17	30	12.9	38	16.3	38[b]
	18	38	16.3	39	16.8	40
	19	38	16.3	40	17.2	40
	20	42	18.1	42	18.1	43[b]
	21	38	16.3	42	18.1	44
	22	42	18.1	42	18.1	44[b]
	23	42	18.1	42	18.1	49[b]
	24	51	21.9	55	23.7	62
Mean ± SE		36 ± 1.8	15.4 ± 0.8	39 ± 1.7	16.6 ± 0.7	41 ± 2.0
2	25	None	None	None	None	67
(Control)	26	None	None	None	None	68

[a] BW = body weight.
[b] Cat found dead; others were sacrificed.

violet, Martius scarlet blue (MSB), Verhoeff's, Holzer's, Bodian's, Cajal's, Weil's, and luxol fast blue stains were used selectively.

RESULTS

Clinical Signs

The mean number of days of exposure until the onset of neurological signs was 36 for cats (Table 1), and 44 (Group 3) or 38 (Group 4) for dogs (Table 2). Clinical signs were similar in both species. Earliest signs of toxicosis were anorexia and symmetrical movements of pelvic limbs ("bunny hopping"). Other signs were related to gait abnormalities. Animals had crouched, hypermetric, ataxic gaits with paretic and abducted pelvic limbs. They frequently collapsed with crossed

TABLE 2   Regimen, Exposure Duration, and Dosage for Dogs
Exposed to Oral Methyl Mercury[a]

Group Number and Dose	Dog Number	Clinical Signs Produced at:		Total Days of Exposure	Total Dose, mg Hg/kg BW	Necropsy, day
		Days of Exposure	Total Dose, mg Hg/kg BW			
3	6	46	19.8	53	22.8	60
(0.43 mg	7	41	17.6	44	18.9	49
Hg/kg BW)	8	41	17.6	45	19.3	46
	9	46	19.8	53	22.8	54
Mean ± SE		44 ± 1.4	18.7 ± 0.6	49 ± 2.5	21.0 ± 1.1	52 ± 3.1
4	10	38	24.3	41	22.2	42
(0.64 mg	11	32	20.5	34	21.8	35
Hg/kg BW)	12	31	19.8	34	21.8	39
	13	49	31.4	63	40.3	69
Mean ± SE		38 ± 4.1	24.0 ± 2.7	43 ± 6.9	26.5 ± 4.6	46 ± 7.7
5	14	None	None	None	None	60
(Control)	15	None	None	None	None	70

[a] From Davies et al., 1977, with permission.

and/or splayed limbs. Righting, wheelbarrowing, hopping, extensor
postural thrust, hemiwalking, and placing reactions became progressively weak and hypermetric. Flexion of limbs onto the dorsum of the
paw indicated proprioceptive impairment.

Animals responded to auditory stimuli but were blind. Menace
response was absent, but pupillary reaction was normal. Abnormal
nystagmus was observed only in cats. Grand mal convulsions occurred, and visceral and somatic motor activity preceded loss of
consciousness. Visceral motor activity included pupillary dilation and
ptyalism; somatic motor signs were clonic spasms, paddling of ex-
tremities, and opisthotonos. The seizures lasted approximately 1 min.
Postictal signs included exhaustion, respiratory distress, mental confu-
sion, and incontinence. Animals were killed when they became terminally recumbent with clonic rigidity and paddling of limbs. Postural
reactions at this time were absent, but patellar, biceps, flexor, and
perineal reflexes were present.

Biochemical Profiles

Significant findings were limited to BUN and SGOT determinations in cats 21 and 24. Values were 81 mg/dl and 115 mU/ml and 85 mg/dl and 100 mU/ml, respectively. Corresponding control (cat 25) values were 27 mg/dl and 55 mU/ml.

Serum total cholesterol concentrations in all exposed dogs increased over the course of the experiment (Figure 1). Values greater than 500 mg/dl (the upper limit of the SMA 12/60 under standard laboratory conditions for human samples) were observed in dogs 7, 8, and 12 (Figure 1). Dilutions of these samples were not analyzed. Alkaline phosphatase and SGOT were elevated in the terminal stages of dogs 7, 8, and 12. Serum alkaline phosphatase concentrations were 285, 175, and 195 mU/ml, respectively, while SGOT values were 210, 270, and 300 mU/ml. Mean control values were 53 mU/ml (alkaline phosphatase) and 30 mU/ml (SGOT).

Microscopic Lesions

In cats, lesions were observed in all major divisions of the brain (Table 3). The primary lesions involved neurons and ranged from acute acidophilic necrosis to complete loss of affected cells with an associated astrocytic reaction. The latter led to formation of gemistocytic astrocytes in affected regions. In regions with severe changes, vacuolation of neuropil and loss of silver-impregnated nerve processes also were noted (Figure 2). More acute neuronal changes were accompanied by microglial infiltrates and neuronophagia.

Lymphocytic perivascular cuffs occasionally containing a few plasma cells and neutrophils were frequently associated with these lesions. Endothelial cell hypertrophy and hyperplasia occurred both in the presence and absence of cuffs. In the depths of a few, random cerebral cortical sulci in most animals, there was a mild lymphocytic leptomeningeal infiltrate often accompanied by histiocytic and endothelial cell hyperplasia.

Lesions were most striking in the cerebral cortex lining the depths of sulci, although here, as in other regions, involvement was not uniform. In the most severe lesions, extensive neuronal loss with associated gemistocytic reactions was observed in the depths of the sulci, with more surviving neurons and less gliosis seen toward the tip of the gyrus. In the cerebellar cortex, necrosis and loss of Purkinje cells and granular cells, and proliferation of Bergmann's astrocytes were

TABLE 3 Distribution of Neuronal Lesions in Cats Exposed to Oral Methyl Mercury

Group Number and Dose	Cat Number	Cerebral Cortex	Basal Nuclei	Diencephalon	Midbrain	Pons	Medulla Oblongata	Cerebellum	Peripheral Nervous System
1 (0.43 mg Hg/kg)	11	+	+	+	+	+	+	+	+
	12	+	+	+	+	+	+	+	
	13	+	+	+	+	+	+	+	
	14	+	+	+	+	+	+	+	
	15	+	+	+	+	+		+	
	16	+	+	+	+			+	
	17	+	+	+		+		+	
	18	+	+	+	+		+	+	
	19	+	+	+	+	+	+	+	
	20	+	+	+	+			+	
	21	+	+	+	+	+	+	+	
	22	+	+	+	+	+	+	+	
	23	+	+	+	+			+	
	24	+	+	+	+	+		+	+
2 (Control)	25								
	26								

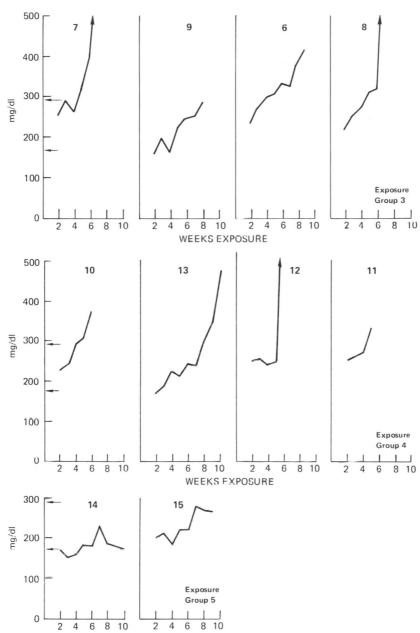

FIGURE 1 Weekly serum cholesterol changes in subacute canine methylmercurialism. Arrows indicate normal range. From Davies *et al.*, 1977, with permission.

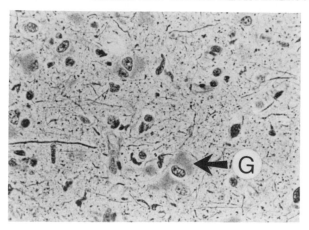

FIGURE 2 Loss of silver-impregnated nerve fibers and proliferation of gemistocytic astrocytes (G) in the cerebral cortex. Cat 20, Group 1. (Bodian's stain. ×63.)

marked. Microglia and necrotic neurons were seen frequently in the
molecular layer (Figure 3).

There were also neuronal necrosis, eosinophilic globules, and glial
reactions in the brain stem and basal ganglia, but of less intensity than
the cortical lesions noted above (Figure 4).

Mild peripheral nervous system lesions, observed in two cats, were
acute neuronal necrosis in dorsal root ganglia (cat 23) and early
demyelination of nerve fibers (cat 11) (Figure 5).

In dogs, the cerebral cortex was the most severely affected portion
of the central nervous system. Cerebral cortical lesions included acute
neuronal necrosis, loss of neurons, gliosis, proliferation of capillaries,
and vacuolation of the neuropil. The reaction was marked by the
presence of hypertrophic astrocytes, including many gemistocytic
forms, proliferation of microglial cells (Figure 6), which were oriented
perpendicular to the pial surface, and neuronophagia. Cerebral cortical
lesions were nonuniform, diffuse, bilateral, and primarily restricted to
the neocortex with infrequent involvement of the pyriform lobes
(Davies *et al.,* 1977).

Perivascular cuffs, consisting of lymphocytes, histiocytes, and, occasionally, plasma cells and neutrophils, occurred in both gray and white
matter. Endothelial cell hypertrophy and hyperplasia were found in the
presence and absence of cuffs.

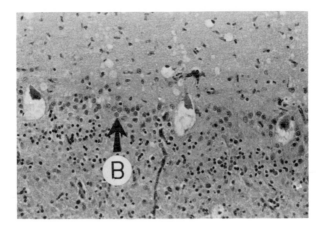

FIGURE 3 Necrosis and loss of Purkinje cells and granular cells, proliferation of Bergmann's (B) glia, and neuronal necrosis with vacuolation of neuropil and microgliosis in the molecular layer of the cerebellar cortex. Cat 15, Group 1. (Hematoxylin and eosin. ×63.)

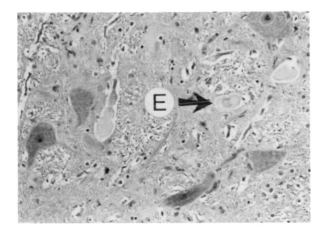

FIGURE 4 Eosinophilic (E) globules in the red nucleus. Cat 22, Group 1. (Hematoxylin and eosin. ×160.)

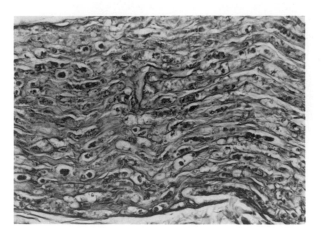

FIGURE 5 Early demyelination of a peripheral nerve. Cat 11, Group 1. (Luxol fast blue. ×63.)

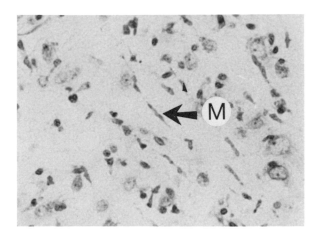

FIGURE 6 Loss of neurons and proliferation of microglial (M) cells in the cerebral cortex. Dog 8, Group 3. (Cresyl violet stain. ×63.) From Davies *et al.*, 1977, with permission.

Leptomeningitis was observed in the depths of random cerebral sulci of all exposed dogs (Figures 7, 8, and 9). Lesions were characterized by proliferation of histiocytes, infiltration by lymphocytes, and, occasionally, plasma cells and neutrophils, endothelial hypertrophy and hyperplasia, fibrin deposition, and hemorrhage. In severe lesions, arterial fibrinoid necrosis was observed. The internal elastic membrane of these vessels had thickened and was fragmented.

Table 4 indicates the distribution of central nervous system lesions beyond the cerebral cortex. Acute neuronal necrosis and astrocytic proliferation were observed in the claustrum of all exposed dogs. Acute neuronal necrosis and/or eosinophilic globules were present in the caudate nucleus, diencephalon, midbrain, pons, medulla oblongata, cerebellar roof nuclei, and gray matter of the spinal cord. Because of staining properties, large globules were interpreted to be necrotic neurons, whereas small globules were considered to be swollen axons. Vacuolation of the neuropil, gliosis, neuronophagia, and perivascular cuffs were associated frequently with these lesions. No significant lesions were found in the cerebellar cortex, dorsal root ganglia, or peripheral nerves.

DISCUSSION

The dogs were more resistant to clinical effects of methyl mercury than the cats. They required larger daily doses to produce clinical signs of

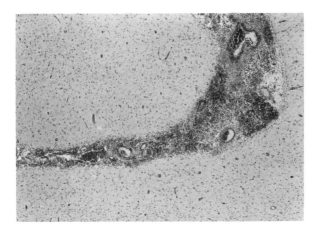

FIGURE 7 Leptomeningitis and hemorrhage in a cerebral sulcus. Dog 8, Group 3. (Hematoxylin and eosin. ×10.) From Davies *et al.*, 1977, with permission.

TABLE 4 Distribution of Neuronal Lesions in Dogs Exposed to Oral Methyl Mercury [a]

Group Number and Dose	Dog Number	Cerebral Cortex	Basal Nuclei	Diencephalon	Midbrain	Pons	Medulla Oblongata	Cerebellum	Spinal Cord
3 (0.43 mg Hg/kg)	6	+	+	+	+	+		+	+
	7	+	+	+		+	+	+	+
	8	+	+	+	+		+	+	+
	9	+	+	+	+	+		+	
4 (0.64 mg Hg/kg)	10	+	+	+	+	+			+
	11	+	+	+	+	+		+	+
	12	+	+	+	+	+	+	+	+
	13	+	+	+	+	+	+	+	+
5 (Control)	14								
	15								

[a] From Davies et al., 1977, with permission.

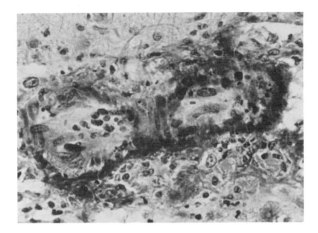

FIGURE 8 Subintimal fibrinoid necrosis in a leptomeningeal artery. Dog 10, Group 4. (Hematoxylin and eosin. ×63.) From Davies *et al.*, 1977, with permission.

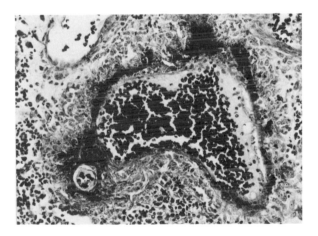

FIGURE 9 Thickened internal elastic membrane of a leptomeningeal artery with a branching vessel. Dog 7, Group 3. (Verhoeff's stain. ×63.)

illness in a similar number of days (Tables 1 and 2). Neurological signs of toxicosis were similar for both species, but abnormal vertical nystagmus was restricted to cats.

Species variations were evident in methyl-mercury-induced serum biochemical changes. Limited data indicated terminal BUN elevation in cats. The most significant change in dogs was a progressive increase in serum cholesterol, possibly reflecting liver dysfunction. Increased feline BUN values occurred in the absence of renal lesions, a finding previously reported by Charbonneau et al. (1974). Elevations in serum alkaline phosphatase and SGOT were thought to be nonspecific.

Microscopic lesions common to both species were neuronal necrosis and loss, gliosis, and perivascular cuffs. Leptomeningitis, accompanied by vascular fibrinoid necrosis in the dog, was much more severe than the mild lymphocytic leptomeningeal infiltrate seen in the cat. Although lesions occurred in the brain stem of both species, they were more pronounced in dogs.

Other important species differences were the lack of lesions in the spinal cord of the cat and the relative resistance of the cerebellar cortex and peripheral nervous system of the dog.

Cats in this study developed clinical signs of illness after 30 to 51 days of exposure to 0.43 mg Hg/kg BW/day as methylmercuric hydroxide. In contrast, cats exposed to approximately 0.45 mg Hg/kg BW/day administered as methyl-mercury-contaminated fish, developed signs in 60 to 83 days (Albanus et al., 1972). A possible explanation for the difference in time of onset of clinical signs was that methyl mercury bound to tissue ligands of the fish homogenate was slightly less toxic than the unbound methylmercuric hydroxide used in this study. Lesions in cats exposed to methyl-mercury-contaminated fish were restricted to the cerebral cortex, the cerebellar cortex, the dorsal root ganglia, and the peripheral nerves (Albanus et al., 1972). Cats exposed to 1.29 and 0.86 mg Hg/kg BW/day as methylmercuric hydroxide developed clinical signs following 15 to 24 days of exposure (Davies and Nielsen, 1977). They had severe lesions in cerebral cortex, brain stem, and cerebellum. The distribution and severity of neuronal lesions in this feline experiment had characteristics in common with the lesions that developed in the experiments of Albanus et al. (1972) and Davies and Nielsen (1977).

SUMMARY

Oral exposure to 0.43 mg (14 cats, 4 dogs) and 0.64 mg (4 dogs) of Hg/kg BW/day as methylmercuric hydroxide produced clinical signs of poisoning in 30 to 51 days. Dogs usually developed clinical signs later than cats receiving an equivalent dose. Significant lesions included neuronal necrosis and loss, gliosis, perivascular cuffing, and leptomeningitis. Affected regions were the cerebral cortex, basal ganglia, brain stem, cerebellum, spinal cord, and peripheral nervous system. Cats were relatively resistant to methyl-mercury-induced lesions of the spinal cord, and dogs to lesions of the cerebellar cortex and peripheral nerves. In cats, blood urea nitrogen was elevated in the terminal stages. Serum total cholesterol increased progressively over the course of exposure in dogs.

ACKNOWLEDGMENTS

This research was supported in part by federal funds made available through the provisions of the Hatch Act and by the Northeastern Research Center for Wildlife Diseases, University of Connecticut, Storrs.

This report is based in part on a Ph.D. thesis by the senior author, which was presented to the Graduate School of the University of Connecticut, Storrs (1978).

REFERENCES

Albanus, L., L. Frankenberg, C. Grant, U. von Haartman, A. Jernelöv, G. Nordberg, M. Rydälv, A. Schütz, and S. Skerfving. 1972. Toxicity for cats of methylmercury in contaminated fish from Swedish lakes and of methylmercury hydroxide added to fish. Environ. Res. 5:425–442.

Atwood, R. G., P. Gates, M. Patnode, R. LacQuaye, T. Clarkson, M. Greenwood, J. C. Smith, E. Nedrow, J. A. Beare, S. Milham, and T. L. Nghiem. 1976. Organic mercury exposure—Washington. Morbidity Mortality Rep. 25:133.

Berglund, F., M. Berlin, G. Birke, R. Cederlof, U. von Euler, L. Friberg, B. Holmstedt, E. Jonsson, K. G. Luning, C. Ramel, S. Skerfving, A. Swensson, and S. Tejning. 1971. Methyl mercury in fish, a toxicologic–epidemiologic evaluation of risks. Report from an Expert Group. Nord. Hyg. Tidskr. Suppl. 4.

Charbonneau, S. M., I. C. Munro, E. A. Nera, R. F. Willes, T. Kuiper-Goodman, F. Iverson, C. A. Moodie, D. R. Stoltz, F. A. J. Armstrong, J. F. Uthe, and H. C. Grice. 1974. Subacute toxicity of methylmercury in the adult cat. Toxicol. Appl. Pharmacol. 27:569–581.

Davies, T. S. 1978. Comparative pathology of canine and feline methyl mercury poisoning. Ph.D. Thesis. University of Connecticut, Storrs.

Davies, T. S., and S. W. Nielsen. 1977. Pathology of subacute methylmercurialism in cats. Am. J. Vet. Res. 38:59–67.

Davies, T. S., S. W. Nielsen, and B. S. Jortner. 1977. Pathology of chronic and subacute canine methylmercurialism. J. Am. Anim. Hosp. Assoc. 13:369–381.

deLahunta, A. 1971. Small animal neurologic examination. Vet. Clin. North Am. 1:191–206.

Shaw, C.-M., N. K. Mottet, R. L. Body, and E. S. Luschei. 1975. Variability of neuropathologic lesions in experimental methylmercurial encephalopathy in primates. Am. J. Pathol. 80:451–469.

Spurgeon, D. 1976. Mercury poisoning in Ontario. Nature (London) 260:476.

Wobeser, G., N. O. Nielsen, and B. Schiefer. 1976. Mercury and mink. II. Experimental methyl mercury intoxication. Can. J. Comp. Med. 40:34–45.

QUESTIONS AND ANSWERS

D. KELLY: Can you comment at all on the possible background to the elevation of plasma cholesterol in dogs? Is it related to any effects on liver or kidney?

T. DAVIES: I suspect that it is an index of liver dysfunction. Elevated cholesterol values also occur in dogs with hypothyroidism but we saw no lesions in the thyroid in these dogs.

D. SCARPELLI: Did you see fatty livers at any stage of your experiments?

T. DAVIES: In none of the animals that I discussed today did I see fatty livers, but in an earlier experiment with cats receiving 1.29 and 0.86 mg of mercury per kilogram of body weight, which are higher levels than those I described today, there was vacuolar degeneration within the liver.

D. SCARPELLI: There was no evidence of disturbances in plasma protein concentration at any time?

T. DAVIES: No.

S. SLEIGHT: Do you have any clue as to why you found species differences in the lesions in the cerebellum? Did autoradiography give you any clue?

T. DAVIES: No, I do not know the reason the cat, more than the dog, reacts to methyl mercury in a fashion similar to pigs. Most reports of methyl mercury poisoning in pigs describe relatively minor damage in the cerebellar cortex. I don't know the reason why. Would you agree, Dr. Tryphonas, that the cerebellar involvement in pigs is relatively minor?

L. TRYPHONAS: That is correct.

L. KOLLER: Have you performed any studies longer than 30 days?

T. DAVIES: We performed a long-term study in dogs, which I did not present today and which was published elsewhere (Davies et al., 1977).[1]

L. KOLLER: Were the changes similar?

T. DAVIES: No, lesions in chronically exposed dogs were restricted to the cerebral cortex. We did not see any brain stem involvement as I described today. In dogs, it would appear that the cerebral cortex is the part of the nervous system most sensitive to the effects of methyl mercury.

L. KOLLER: We fed rabbits 20, 10, and 1 ppm of methyl mercury. The 20-ppm feeding killed most of the rabbits between 3 and 4 weeks. Those on 1 ppm lived. The first death in the 10-ppm group occurred around the 10th week of exposure. All rabbits that received methyl mercury had significantly elevated alkaline phosphatase levels but no significant alteration of other enzymes.

T. DAVIES: Was the alkaline phosphatase elevated terminally or did you see it progressively?

[1] Davies, T. S., S. W. Nielsen, and B. S. Jortner. 1977. Pathology of chronic and subacute canine methylmercurialism. J. Am. Hosp. Assoc. 13:369–381.

L. KOLLER: Progressively.

T. DAVIES: We observed terminal elevations of alkaline phosphatase but considered them to be nonspecific.

L. KOLLER: As a matter of fact, even the rabbits in the 1-ppm group had elevated alkaline phosphatase levels, and we never had deaths in those animals. Also, the pathological changes that occurred in the rabbits were in the cerebellum and not in the cerebrum or in the leptomeninges. The first changes one sees are loss of Purkinje cells and necrosis of the granular and molecular layers. These changes represent quite a remarkable species difference.

S. NIELSEN: Dr. Koller, did you see changes in the serum cholesterol in your rabbits?

L. KOLLER: The cholesterol levels in female rabbits were double those in males, both in the controls and in the methyl-mercury-treated animals.

L. TRYPHONAS: I want to make a comment regarding the cerebellar lesions in pigs. I think that within the dosage range that you used your statement is correct. Once you increase the dosage, the areas in the nervous system exhibiting nerve cell death expand and may include the cerebellum.

R. KIRKPATRICK: Did you measure food consumption in these animals and did you determine weight loss?

T. DAVIES: The animals experienced some weight loss based on twice-weekly body weights. I did not measure food consumption.

R. KIRKPATRICK: Do you think that the elevation in blood urea nitrogen (BUN) might have been due to tissue catabolism in these animals? In other words, did you get a great weight loss or emaciation?

T. DAVIES: In some animals, yes, there was a significant loss.

R. KIRKPATRICK: BUN values will go up when tissue is being catabolized. I think, in many of these studies of contaminants, we need to measure food consumption in order to separate the effects of food restriction from the direct effects of the compound.

T. DAVIES: This is a very interesting observation, because most of the literature I have reviewed indicates that more than 50% of the kidney has to be damaged before there is an increase in BUN values. I was at a loss to explain the increased BUN values that we observed.

D. SCARPELLI: In severe cachexia, where you have a marked decrease in muscle mass due to wasting, BUN levels can be elevated in the absence of renal pathology.

R. ZOOK: Were the cortical lesions in the cerebrum of the cat uniformly distributed? Were any areas affected more than others?

T. DAVIES: The lesions were present in all major lobes without selectivity. The lesions were most severe in the cortex lining the depths of the sulci. No preferential involvement was found in areas such as the occipital cortex as has been reported by many authors.

J. ZINKL: We see probably upwards of 10,000 dogs and cats a year. An elevation in BUN in these animals is usually due to dehydration—not to wasting—because they just are not perfusing their kidneys adequately. I do not agree that the levels you see are elevated. The other thing that I would be concerned about in order to evaluate the kidney function in these animals is whether you saw any change in creatinine, potassium, and phosphorus.

T. DAVIES: I did not see an elevation of creatinine and I believe that there were no significant changes in the two other determinations you mentioned.

J. ZINKL: When were these samples taken? Were the BUN samples taken throughout the experiment?

T. DAVIES: No, the BUN samples were terminal samples when the animals were prostrate. I would like to emphasize that there were a limited number of animals, specifically two exposed cats and one control cat.

J. ZINKL: I would probably attribute the increased BUN to terminal inability to perfuse the kidneys and to excrete urea, and, thus, consider it not significant. I imagine that there was no elevation in urea nitrogen 1, 2, or 3 days before death.

Brain Lesions in Experimental Methyl Mercury Poisoning of Squirrel Monkeys (*Saimiri sciureus*)

B. C. ZOOK, C. R. WILPIZESKI, and
E. N. ALBERT

The contamination of our environment with mercury has led to the poisoning of many aquatic animals and of birds and mammals that eat both seeds and sea creatures (Albanus *et al.*, 1972; Kojima and Fujita, 1973; Nelson *et al.*, 1971; Takeuchi, 1968; Takeuchi *et al.*, 1977; Woheser and Swift, 1976). Over the past 24 years outbreaks of organic mercury poisoning in humans have occurred in Japan, from the ingestion of contaminated seafood, and in Iraq, Pakistan, Guatemala, and the United States, from the ingestion of treated seed grains or of meat from mercury-poisoned animals (Clarkson *et al.*, 1976; Eyl, 1971). There have been efforts to curtail sources of environmental mercury contamination in several countries, but in this hemisphere they are just beginning. Consequently, the problem will be with us for many years. Study of the effects of chronic organic mercury poisoning in animals and humans is therefore needed. This report is one portion of a larger study of the effects of low-level lead or methyl mercury ingestion on the vestibular and auditory systems. Herein are described the neuropathological changes resulting from chronic methyl mercury poisoning of squirrel monkeys.

The purpose of the study from which this material comes was not to study time–dose–lesion relationships, which has been done before. The fundamental purpose was to intoxicate the primates with mercury in such a way that they developed neurologic signs, but did not die or were not too ill to test for loss of hearing or loss of balance.

151

MATERIALS AND METHODS

The purpose of this study was to maintain a state of chronic poisoning to enable repeated clinical evaluations of vestibular and auditory functions over a prolonged course of methyl mercury intoxication. To this end 10 subadult and adult squirrel monkeys (*Saimiri sciureus*) weighing between 349 and 906 g were fed methyl mercury chloride or hydroxide via oropharyngeal tube 3–5 times per week for up to 102 days. Individual doses were adjusted depending on clinical response, in efforts to maintain a chronic state of intoxication; they averaged between 0.12 and 0.24 mg/kg of body weight per day. Total doses per animal ranged from 4.8 to 7.7 mg, averaging about 10.9 mg/kg. The animals were individually housed, observed daily, and fed Purina Monkey Chow *ad libitum* supplemented with fresh fruit. Various tests, e.g., body rotation tests, bilateral caloric stimulations, electronystagmography, computer-averaged auditory evoked responses, etc., were performed to detect vestibular and cochlear changes. Results of these tests will be reported subsequently.

The test monkeys became ill and died, or were euthanized when moribund after 50 to 102 days. The 10 test and 4 controls were necropsied as soon as possible after death. The entire brain was removed and immersed in 10% neutral buffered formalin. The medulla and inferior portions of the cerebellum were fixed intact with the temporal bones. These were decalcified, embedded in celloidin, and serially sectioned at 20 μm. Sections taken every 20 μm were stained with either hematoxylin and eosin or Weil–Weigert stains. Multiple sections of the entire cerebrum and cerebellar cortex were embedded in paraffin, sectioned at 6 μm, and stained with hematoxylin and eosin. Slices from several brains were frozen, sectioned at 25 μm, and stained with gold sublimate for astrocytes.

RESULTS

Clinical Signs

Signs of neurological abnormalities (ataxia, incoordination, and abnormal gait) were first observed 46 to 98 ($\bar{x} = 69$) days after the start of mercury feeding. Amaurosis soon followed. Apparent blindness, clumsiness, and poor hand–eye coordination made ingestion of food difficult. Animals would sometimes peck at food on the floor of their cages much like chickens. Monkeys occasionally rubbed their muzzles against solid objects, possibly because of oral paresthesia. Mean body weights became 94% of the initial weights compared to a mean of 106%

for the four controls. Once neurologic signs were observed the course was progressive, lasting from 4 to 23 ($\bar{x}$ = 12) days. Subtle neurologic changes may have been overlooked, as detailed behavioral studies were not a part of this investigation, and the animals were not observed between 6 P.M. and 8 A.M. or for more than short periods on weekends. Thus, the true course may have been longer. Five of the monkeys died; five were sacrificed when they became moribund. There was a tendency for the primates receiving lower doses to have a somewhat delayed onset and longer course, but there was much individual variation.

Pathological Findings

The cerebral cortex was severely affected in all monkeys. Cortical lesions consisted primarily of necrosis and disappearance of small granular neurons, and of gliosis. In some monkeys perivascular cuffing and rarefaction of the neuropil were observed. The severity of the lesions varied with location in the cortex.

In the cerebral cortex, the most affected areas were (in decreasing order) the striate, prestriate, and temporal cortices of the occipital lobes. The striate cortex was always the most severely involved; the loss of neurons was often associated with cortical atrophy with resultant dilation of the lateral ventricles and widening of the calcarine sulcus (Figures 1 and 2). Somewhat less affected were the temporal, parietal, and insular cortices, followed by the frontal cortex. Lesions became gradually less severe rostrally. Transitions between severely and moderately affected areas were generally gradual, but sometimes abrupt. There were only occasional losses of neurons in the pyriform and cingulate cortices, and then only near the rhinal fissure of the cingulate sulcus. The cerebellar cortex was unaffected.

Necrosis and loss of neurons were most extensive in middle laminae, but all layers were affected and generally disrupted in the visual cortex (Figures 3 and 4). Laminae I and II were least affected. Neuronal necrosis was evidenced by stages of karyolysis, in which many cells had ill-defined nuclear borders, gradations of chromatin loss, and, finally, "ghosts" of previously existing neurons (Figure 5). Karyorrhexis, leaving scattered fragments of nuclear chromatin, was also very common (Figure 6). Neuronal death indicated by shrunken, angular neurons with condensed eosinophilic cytoplasm was more common in the noncalcerine cortex and subcortical nuclei. Satellitosis and neuronophagia were uncommon in cortical lesions, but were prominent in several subcortical nuclei (Figure 7). Many neurons, especially in otherwise unaffected areas, had vesicular nuclei (Figure 8).

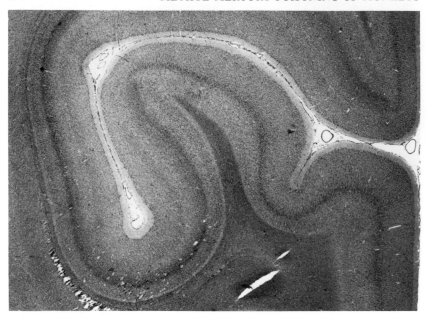

FIGURE 1 Normal striate (calcarine, visual) cortex. Control squirrel monkey. (Hematoxylin and eosin. ×14.)

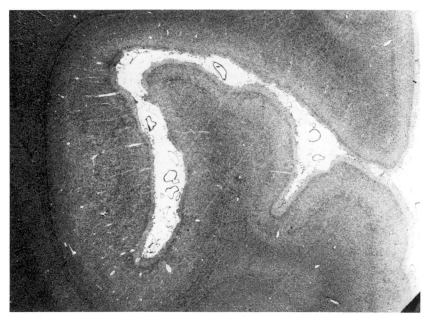

FIGURE 2 Striate cortex, mercury-intoxicated squirrel monkey. There is loss of laminar pattern and atrophy with widening of calcarine fissure. (Hematoxylin and eosin. ×14.)

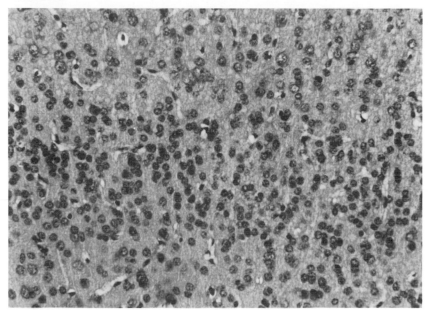

FIGURE 3 Normal middle laminae, striate cortex of control squirrel monkey. (Hematoxylin and eosin. ×260.)

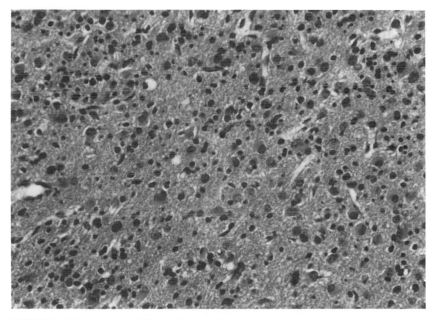

FIGURE 4 Disrupted middle laminae, striate cortex of mercury-poisoned monkey. There is necrosis and loss of neurons and gliosis. (Hematoxylin and eosin. ×260.)

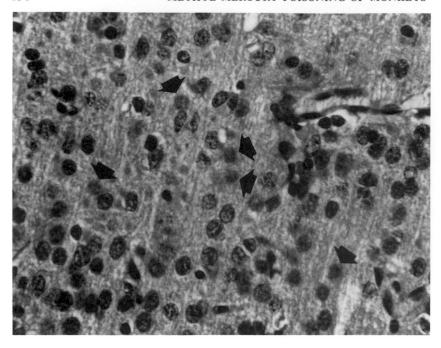

FIGURE 5 Karyolysis of neurons in temporal cortex of occipital lobe in primate with mercury encephalopathy. Affected neurons (arrows) have ill-defined borders and fade leaving "ghosts" behind. (Hematoxylin and eosin. ×520.)

Glial response to cortical injury was intense. Reactive (gemistocytic) astrocytes with abundant eosinophilic cytoplasm were present in considerable numbers (Figure 6). Proliferation of astrocytes and their fibers were particularly obvious on gold sublimate stains (Figures 9 and 10). The reactive astrocytes were most abundant in the cortices of the occipital lobes, were less extensive in the temporal and parietal lobes, and were least numerous in the frontal lobes. The claustrum was the only nucleus that consistently contained reactive astrocytes. The subcortical white matter had a moderate and diffuse astrogliosis. Microgliosis was most prominent in damaged subcortical nuclei; this was especially obvious in the claustrum (Figure 7). Other nuclei were sometimes affected with scattered degenerating neurons with neuronophagia. These included the medial and lateral geniculate bodies, globus pallidus, putamen, and, rarely, other cerebral or cerebellar nuclei. Changes in oligodendroglia were not obvious.

Perivascular infiltration occurred to a moderate degree in the cortices of some monkeys. When present, cuffing was found most often in

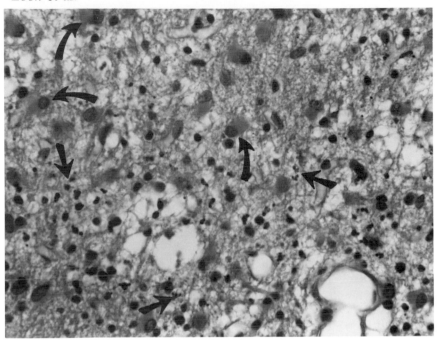

FIGURE 6 Calcarine cortex of intoxicated monkeys. Karyorrhexis (straight arrows) of neurons and gemistocytic astrocytes (curved arrows) are plentiful. There is also vacuolation of the neuropil. (Hematoxylin and eosin. ×416.)

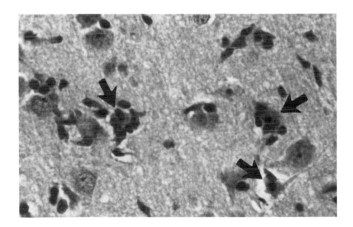

FIGURE 7 Claustrum of methyl mercury poisoned monkey. Many neurons (arrows) have shrunken condensed cytoplasm with pyknotic nuclei, and there is neuronophagia. (Hematoxylin and eosin. ×416.)

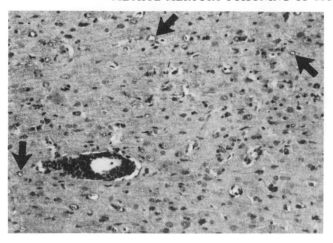

FIGURE 8 Parietal cortex, mercury encephalopathy. There is perivascular mononuclear cell infiltration. Many neurons have vesiculated nuclei (arrows). (Hematoxylin and eosin. ×100.)

the parietal cortex (Figure 8). The cells clustered around vessels were mononuclear leukocytes, mostly macrophages. Vascular changes, other than cuffing and some endothelial proliferation, were not observed. Status spongiosus was inconsistently observed in the calcarine cortex (Figure 6). Loss of myelin was generally minimal. Neither the neuronal, glial, nor vascular changes were obviously related to dose of mercury or length of course.

DISCUSSION

Review of the literature indicates that the neuropathology of chronic organic mercury poisoning varies with age, species, and duration of poisoning. Various salts of ethyl and methyl mercury appear to produce similar lesions. In the adult human, loss of neurons and gliosis in the granular cell layer of the cerebellum is characteristic. Equally important is the loss of small neurons, astrogliosis, and atrophy of the calcarine cortex, and the pre- and postcentral gyri, with less severe changes in other cortical areas. Occasionally, the putamen and, less commonly, the globus pallidus are affected. But generally, mild or no lesions have been reported in other subcortical nuclei, the spinal cord, or peripheral nerves. Vascular lesions and perivascular infiltrations are absent or minimal, but loss of myelin may occur, mostly in cases of

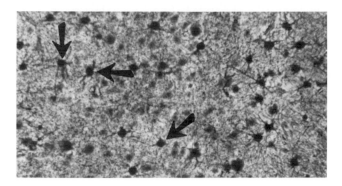

FIGURE 9 Visual cortex, control monkey. Normal fibrous astrocytes are present (bottom) in subcortical white matter. Normal numbers and appearance of protoplasmic astrocytes (arrows) located in deep laminae. (Gold sublimate. ×260.)

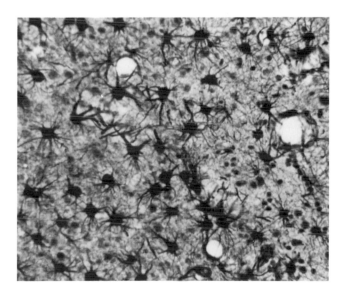

FIGURE 10 Visual cortex, poisoned monkey. Many reactive astrocytes with dense fiber formation in deep laminae. (Gold sublimate. ×260.)

poisonings with prolonged courses (Hay *et al.*, 1963; Hunter *et al.*, 1940; Hunter and Russell, 1954; Schmidt and Harzmann, 1970; Shiraki and Takeuchi, 1971; Takeuchi, 1968).

Natural or experimental alkyl mercury poisoning in fish (Takeuchi, 1968), rabbits (Carmichael *et al.*, 1975), birds (Takeuchi, 1968), rats (Diamond and Sleight, 1972), mink (Wobeser and Swift, 1976; Wobeser *et al.*, 1976), cattle (Jubb and Kennedy, 1963), swine (Davies *et al.*, 1975; Tryphonas and Nielsen, 1973), and cats (Chang *et al.*, 1974; Charbonneau *et al.*, 1976; Davies and Nielsen, 1977; Grant, 1973; Takeuchi, 1968) is reportedly very similar to that in humans, especially with regard to cerebellar cortical changes. Cerebral lesions in these species seem to be less well developed and localized. They tend to involve subcortical nuclei to a greater extent than is common in humans. Spinal cord, peripheral nerves, and nerve roots are also more frequently involved. Vascular lesions have been found in cattle (Jubb and Kennedy, 1963), swine (Tryphonas and Nielsen, 1973), and sometimes in rats (Diamond and Sleight, 1972), cats (Davies and Nielsen, 1977), and dogs (Davies *et al.*, 1977). Cerebellar cortical lesions are generally absent in mice (Berthoud *et al.*, 1976) and dogs (Davies *et al.*, 1977).

Chronic organic mercury poisoning in nonhuman primates, e.g., the rhesus monkey (*Macaca mulatta*), cynomolgus macaque (*M. fascicularis*), stump-tailed macaque (*M. arctoides*), and squirrel monkey (*Saimiri sciureus*), reportedly results in clinical signs and cerebral lesions nearly identical with those in humans (Berlin *et al.*, 1975a; Garman *et al.*, 1975; Grant, 1973; Hunter *et al.*, 1940; Ikeda *et al.*, 1973; Kobayashi *et al.*, 1976; Shaw *et al.*, 1975). Generally, subcortical nuclei are mildly affected, as in humans, except in more acute poisoning where nuclear changes reportedly may predominate (Shaw *et al.*, 1975). Cerebellar cortical lesions, however, are not reported in monkeys despite high mercury deposition in the cerebellum (Berlin *et al.*, 1975b,c; Nordberg *et al.*, 1969; Shaw *et al.*, 1975; Takahashi, 1971).

In the study reported in this paper, squirrel monkeys fed methyl mercury for 50 to 102 days developed signs and cerebral cortical lesions much like those described in humans. The cerebral lesions—consisting of disruption of the cortical laminae with loss of small neurons, astrogliosis, and atrophy—preferentially affected the occipi-.tal cortex, especially the striate area. Subcortical nuclei were similarly affected, except that the claustrum was altered much more than the putamen. The cerebellum was free of the granular cell layer loss generally observed in humans. The unaccountable lack of cerebellar cortical lesions has been noted in several species of nonhuman primates poisoned under somewhat different conditions.

Nonhuman primates seem well suited as models for the study of the cerebral effects of organomercurial poisoning in humans (Minimata disease). Primates seem especially beneficial in studies involving the early detection of subtle neurological signs and behavioral responses (Evans *et al.*, 1975; Hellberg and Nyström, 1972). This study supports the neuropathological findings of others in chronic mercury encephalopathy of monkeys and points out similarities and differences between lesions in humans and those in nonhuman primates and some other species.

SUMMARY

Ten squirrel monkeys fed methyl mercury, doses of 0.12 to 0.24 mg/kg of body weight per day, developed neurological signs including apparent blindness within 46 to 98 days. After a course of 4 to 23 days the monkeys died or were sacrificed *in extremis*. Neuropathological lesions consisted of necrosis and loss of neurons, astrogliosis, and atrophy of the cerebral cortex. The occipital lobes were most affected, especially in the striate cortex. Less affected were the temporal, insular, parietal, and, finally, the frontal cortices. Perivascular cellular infiltration and status spongiosis were present in some monkeys to variable degrees. The claustrum, and sometimes other subcortical nuclei, had neuronal necrosis, neuronophagia, and gliosis. The cerebral lesions of methyl mercury poisoning in squirrel monkeys closely resembled those reported in humans; however, the cerebellar cortex was unaffected, unlike mercury enccphalopathy in humans.

REFERENCES

Albanus, L., L. Frankenberg, C. Grant, U. von Haartman, A. Jernelöv, G. Nordberg, M. Rydälv, A. Schütz, and S. Skerfving. 1972. Toxicity for cats of methylmercury in contaminated fish from Swedish lakes and of methylmercury hydroxide added to fish. Environ. Res. 5:425–442.

Berlin, M., C. A. Grant, J. Hellberg, J. Hellström, and A. Schütz. 1975a. Neurotoxicity of methylmercury in squirrel monkeys. Arch. Environ. Health 30:340–348.

Berlin, M., J. Carlson, T. Norseth. 1975b. Dose-dependence of methylmercury metabolism. Arch. Environ. Health 30:307–313.

Berlin, M., C. Blomstrand, C. A. Grant, A. Hamberger, and J. Trofast. 1975c. Tritiated methylmercury in the brain of squirrel monkeys. Arch. Environ. Health 30:591–597.

Berthoud, H. R., R. H. Garman, and B. Weiss. 1976. Food intake, body weight and brain histopathology in mice following chronic methylmercury treatment. Toxicol. Appl. Pharmacol. 36:19–30.

Carmichael, N., J. B. Cavanough, and R. A. Rodda. 1975. Some effects of methyl mercury salts on the rabbit nervous system. Acta Neuropathol. 32:115–125.

Chang, L. W., S. Yamaguchi, and A. W. Dudley. 1974. Neurological changes in cats following long-term diet of mercury-contaminated tuna. Acta Neuropathol. 27:171–178.

Charbonneau, S. M., I. C. Munro, E. A. Nera, F. A. J. Armstrong, R. F. Willes, F. Bryce, and R. F. Nelson. 1976. Chronic toxicity of methylmercury in the adult cat. Toxicology 5:337–349.

Clarkson, T. W., L. Amin-Zaki, and S. K. Al-Tikriti. 1976. An outbreak of methylmercury poisoning due to consumption of contaminated grain. Fed. Proc. 35:2395–2399.

Davies, T. S., and S. W. Nielsen. 1977. Pathology of subacute methylmercurialism in cats. Am. J. Vet. Res. 38:59–67.

Davies, T. S., S. W. Nielsen, and C. H. Kircher. 1975. The pathology of subacute methylmercurialism in swine. Cornell Vet. 66:33–55.

Davies, T. S., S. W. Nielsen, and B. S. Jortner. 1977. Pathology of chronic and subacute canine methylmercurialism. J. Am. Anim. Hosp. Assoc. 13:369–381.

Diamond, S. S., and S. Sleight. 1972. Acute and subchronic methylmercury toxicosis in the rat. Toxicol. Appl. Pharmacol. 23:197–207.

Evans, H. L., V. G. Laties, and B. Weiss. 1975. Behavioral effects of mercury and methylmercury. Fed. Proc. 34:1858–1867.

Eyl, T. B. 1971. Organic-mercury food poisoning. N. Engl. J. Med. 284:706–709.

Garman, R. H., B. Weiss, and H. L. Evans. 1975. Alkylmercury encephalopathy in the monkey (Saimiri sciureus and Macaca arctoides). Acta Neuropathol. 32:61–74.

Grant, C. A. 1973. Pathology of experimental methylmercury intoxication: Some problems of exposure and response. Pp. 294–312 in H. W. Miller and T. W. Clarkson, eds. Mercury, Mercurials and Mercaptans. Charles C. Thomas, Springfield, Ill.

Hay, W. J., A. G. Rickards, W. H. McMenemey, and J. N. Cumings. 1963. Organic mercurial encephalopathy. J. Neurol. Neurosurg. Psychiatry 26:199–202.

Hellberg, J., and M. Nyström. 1972. The influence of methylmercury exposure on learning–set behavior of squirrel monkeys (Saimiri sciureus). Psychol. Res. Bull. 12:1–9.

Hunter, D., and D. S. Russell. 1954. Focal cerebral and cerebellar atrophy in a human subject due to organic mercury compounds. J. Neurol. Neurosurg. Psychiatry 17:235–241.

Hunter, D., R. R. Bomford, and D. S. Russell. 1940. Poisoning by methylmercury compounds. Q. J. Med. 9:193–213.

Ikeda, Y., M. Tobe, K. Kobayashi, S. Suzuki, Y. Kawasaki, and H. Yonemaru. 1973. Long term toxicity study of methylmercuric chloride in monkeys. Toxicology 1:361–375.

Jubb, K. V. F., and P. C. Kennedy. 1963. Pathology of the Nervous System. Pp. 391–393 in Pathology of Domestic Animals. Vol. 2. Academic Press, New York.

Kobayashi, K., M. Tobe, S. Suzuki, Y. Kawasaki, K. Sekita, K. Matsumoto, K. Yasuhara, Y. Ogawa, K. Hagino, and Y. Ikeda. 1976. Long term toxicity study of methylmercuric chloride in monkeys. Jpn. J. Pharmacol. 26(Suppl.):84.

Kojima, K., and M. Fujita. 1973. Summary of recent studies in Japan on methylmercury poisoning. Toxicology 1:43–62.

Nelson, N., T. C. Byerly, A. C. Kolbye, L. T. Kurland, R. E. Shapiro, S. I. Shibko, W. H. Stickel, J. E. Thompson, L. A. Van Den Berg, and A. Weissler. 1971. Hazards of mercury. Environ. Res. 4:1–69.

Nordberg, G. F., M. H. Berlin, and C. A. Grant. 1969. Methylmercury in the monkey—Autoradiographical distribution and neurotoxicity. Pp. 234–238 in Proceedings of the 16th International Congress on Occupational Health. Tokyo.

Schmidt, H., and R. Harzmann. 1970. Humanpathologische und tierexperimentelle Beobachtungen nach Intoxikation mit einer organischen Quecksilberverbindung ("Fusariol"). Int. Arch. Arbeitsmed. 26:71–83.

Shaw, C.-M., N. K. Mottet, R. L. Body, and E. S. Luschei. 1975. Variability of neuropathologic lesions in experimental methylmercurial encephalopathy in primates. Am. J. Pathol. 80:451–470.

Shiraki, H., and T. Takeuchi. 1971. Minimata disease. Pp. 1651–1665 in J. Minkler, ed. Pathology of the Nervous System. McGraw-Hill, New York.

Takahashi, T., T. Kimura, Y. Sato, H. Shiraki, and T. Ukita. 1971. Time dependent distribution of 203 Hg-mercury compounds in rat and monkey as studied by whole body autoradiography. J. Hyg. Chem. 17:93–107.

Takeuchi, T. 1968. Pathology of Minimata disease. Pp. 141–252 in M. Kutsuna, ed. Minimata Disease. Study Group of Minimata Disease, Kumamoto University, Japan.

Takeuchi, T., F. D'Itri, P. V. Fischer, C. S. Annett, and M. Okabe. 1977. The outbreak of Minimata disease (methylmercury poisoning) in cats on northwestern Ontario Reserves. Environ. Res. 13:215–228.

Tryphonas, L., and N. O. Nielsen. 1973. Pathology of chronic alkylmercurial poisoning in swine. Am. J. Vet. Res. 34:379–392.

Wobeser, G., and M. Swift. 1976. Mercury poisoning in a wild mink. J. Wildl. Dis. 12:335–340.

Wobeser, G., N. O. Nielsen, and B. Schiefer. 1976. Mercury and mink. II. Experimental methylmercury intoxication. Can. J. Comp. Med. 40:34–45.

Yoshino, Y., T. Mozai, and K. Nakao. 1966. Distribution of mercury in the brain and its subcellular units in experimental organic mercury poisoning. J. Neurochem. 13:397–406.

QUESTIONS AND ANSWERS

R. GARMAN: Our studies on two species of macaques have shown histopathological changes that are similar to those in the squirrel monkeys that you studied. Our macaque studies and Dr. Davies' dog studies show comparable cortical changes in that neuronal degeneration is most severe in the depth of the sulci. In contrast, brain stem lesions are not as severe in macaques as they are in dogs; they more closely approximate those in cats and squirrel monkeys. It is difficult to interpret the occasionally observed vesicular changes in the neuronal nuclei, since they were encountered in monkeys dying from conditions unrelated to methyl mercury poisoning.

D. JONES: What is the temporal relationship between behavioral changes and hearing loss and other gross signs of toxicity?

B. ZOOK: Behavioral evaluations were not really a part of this study. Its primary purpose was to evaluate effects of organic mercury on auditory and vestibular systems, including brain, which will be published subsequently.

D. KELLY: What significance do you ascribe to the inflammatory responses described by yourself and Dr. Davies? They strike me as being more severe than those seen in hypoxia or ischemia of the cortex.

B. ZOOK: The distribution of perivascular infiltrates was quite variable, being absent in some brains and quite marked in others. I agree that such lesions are unusual in toxic encephalopathy.

R. GARMAN: I have seen perivascular cuffing and some inflammatory cell infiltrates in the meninges, but only in monkeys acutely poisoned by high dosages of methyl mercury (i.e., those developing prominent neurologic signs within 30 days). Monkeys that received lower dosages and that did not develop neurologic signs for 60, 90, or more days rarely showed any inflammatory reaction (other than a microgliocytosis in areas of neuron degeneration).

D. HASTINGS: Is the amount of mercury in the infected area of the brain increased? And, did you observe dilation of the fourth ventricle in your monkeys?

B. ZOOK: Mercury has been reported to be increased in the brain in areas such as the calcarine area, where the lesions are most severe. Dilation of the fourth ventricle occurred in about one-half of the monkeys and was related directly to severe cell loss in the calcarine area with resultant atrophy of the cortex.

R. GARMAN: There is no correlation between the amount of mercury in the brain and the degree of injury. This is seen characteristically in the lateral geniculate body where there are high levels of mercury and very little neuronal degeneration. This is in contrast to the calcarine cortex where the reverse is true.

D. HASTINGS: In view of the fact that our studies indicate that selenium counteracts methyl mercury toxicity, it would be interesting to know whether or not the enzyme glutathione peroxidase, which breaks down selenium, is involved in the detoxification of mercury.

R. GARMAN: Different areas of the brain have different selenium levels as was demonstrated in 1975 by Höck et al.[1] in the human brain. The lateral geniculate body contains more selenium than does the calcarine cortex. Thus, it may be that the level of selenium in the brain is salient in protecting neurons against damage induced by mercury. It would be interesting to measure the level of selenium in individual neurons and compare that to their susceptibility to mercurial damage.

L. TRYPHONAS: When we exposed pigs to low levels of mercury for long periods, extreme dilation of the laterial ventricles developed due to tissue atrophy. When interpreting the type and severity of neuronal lesions, one should consider not only selenium but also the vascular dynamics of the area.

S. NIELSEN: Did you find any evidence of blindness in these monkeys, and was there any evidence of fibrinoid necrosis of small arteries?

B. ZOOK: Nearly all monkeys appeared to develop blindness, but no lesions were observed in the eyes or visual pathways. Although we searched diligently, we did not identify vascular lesions.

[1] Höck, A., U. Demmel, H. Schicha, K. Kasperek, and L. E. Feinendegen. 1975. Trace element concentration in human brain. Activation analysis of cobalt, iron, rubidium, selenium, zinc, chromium, silver, cesium, antimony and scandium. Brain 98:49–64.

Lead Poisoning in
Mallard Ducks (*Anas platyrhynchos*)

DOUGLAS E. ROSCOE and SVEND W. NIELSEN

Numerous waterfowl mortalities have been attributed to lead poisoning from the ingestion of spent lead shot (Bellrose, 1959). To alleviate this problem the U.S. Fish and Wildlife Service has proposed the substitution of steel for lead shot for hunting waterfowl in areas where lead poisoning poses a threat (U.S. Fish and Wildlife Service, 1976). To help identify such areas, a study of lead poisoning in mallard ducks was initiated at the Northeastern Research Center for Wildlife Diseases at the University of Connecticut in Storrs.

Lead inhibits heme synthetase, the enzyme responsible for the incorporation of iron in protoporphyrin IX (PP) to form heme. In the erythrocytes of lead-poisoned humans, this compound accumulates and bonds with zinc (Lamola and Yamane, 1974). Because zinc protoporphyrin (ZPP) fluoresces red when exposed to blue light, measurement of the intensity of the fluorescence of blood from lead-intoxicated humans can serve as an indicator of lead exposure. A recently developed hematofluorometer (front surface spectrophotometer) available from AVIV Associates, Inc., Lakewood, N.J., is capable of measuring the quantity (μg/dl) of ZPP in one drop of unprocessed blood (Blumberg *et al.*, 1977).

Since erythrocyte fluorescence has been observed in lead-poisoned Canada geese (*Branta canadensis*) and mallard ducks (*Anas platyrhynchos*) (Barrett and Karstad, 1971), adaption of the hematofluorometer for diagnostic testing of lead-poisoned waterfowl seemed feasible.

The use of the hematofluorometer for diagnosis and the pathologic lesions and clinical signs of lead poisoning in mallard ducks are described below.

165

MATERIALS AND METHODS

The study was divided into three experiments: an acute lead-poisoning experiment, a chronic lead-poisoning experiment, and an experiment to correlate PP levels and clinical signs of poisoning.

For the acute lead-poisoning experiment, 30 14-month-old mallard hens were individually caged in 3-hen layer batteries. All birds were assigned wing badge numbers and separated into three treatment groups of 10 each. The birds of groups E and F were given eight and one No. 4 lead shot pellets, respectively, via esophageal catheter. Group G birds served as controls receiving no lead shot. Preweighing of all pellets indicated that the administered shot ranged between 190 and 210 mg. All birds were given washed quartz grit prior to shot administration. In addition, they were provided with continuously flowing water and whole yellow corn *ad libitum*.

Two ducks from each treatment group were sacrificed on days 4, 5, 6, and 8 after shot ingestion. Necropsies were performed on these birds and any others that died spontaneously during the 2-week observation period. Histologic examinations were made of hematoxylin- and eosin-stained sections of brain, liver, kidney, heart, striated muscle, proventriculus, gizzard, and duodenum. Kidney tissues also were subjected to the Ziehl–Neelsen acid-fast stain.

In the chronic lead-poisoning experiment 40 mallard hens were divided into four equal groups receiving eight (group A), two (group B), one (group C), and no (group D) No. 4 lead shot pellets and a commercial duck ration (Duck Growena, Ralston Purina Co., St. Louis, Mo.).

Heparinized blood (0.5 to 1.0 ml) was taken from the brachial vein of each of the chronically poisoned birds prior to shot administration and at 8 h, 2 days, 8 days, and weekly intervals thereafter to the end of the 5-week study. The blood was refrigerated for 48 h prior to testing on the fluorescence spectrophotometer. Exciting and emitting spectra were scanned according to the method established by Lamola *et al.* (1975a). Packed cell volume and hemoglobin concentration in each sample were measured with an AO hemoglobinometer. The mean corpuscular hemoglobin concentration (MCHC) was calculated by dividing the hemoglobin concentration by the packed cell volume. Methanol fixed blood smears were stained with Giemsa's stain and examined microscopically.

An additional 8 male and 8 female mallards were caged outdoors with access to a sand bottom stream. Four ducks were provided with yellow whole corn and administered 18 No. 4 lead shot pellets. The remaining

12 birds were divided into three groups of 4 birds each. These birds were given eight, one, and no No. 4 lead shot pellets and a commercial duck ration.

All birds were checked daily for clinical signs of lead poisoning. Blood samples were collected as each new symptom became evident. Protoporphyrin levels were measured on the hematofluorometer. Blood smears stained with brilliant cresyl blue were examined for reticulocytes.

RESULTS

Five of the group E birds died spontaneously on day 4. Two of these died when handled in preparation for necropsy. Three more birds in group E died on days 5, 6, and 7. One group F mallard died spontaneously on day 11 and another on day 12, whereas none of the group G birds died.

Some of the group E ducks that died spontaneously displayed a bile-stained watery diarrhea, wing droop (radial palsy), and muscular weakness. They supported their heads against the cage, raised and lowered their head and neck feathers, and convulsed prior to death. Emaciation was most pronounced in the two group F mallards that died late in the experiment. These birds had lost 32% and 36% of their initial body weight. Control birds experienced an average weight loss of 3.7%, with a maximum of 7%.

Gross lesions observed in the E and F birds included pectoral muscle atrophy; bile staining of the mucosa of the proventriculus; bile staining, erosion, and desquamation of the gizzard grinding pads; and, in spontaneous mortalities, distention of the gall bladder with dark green bile. Marked erosion and sloughing of the brittle gizzard mucosa were evident in group E birds on all 3 days of necropsy. The gizzard grinding pads of group F birds showed slight evidence of erosion and no sloughing. On day 8 the gizzard of one group F mallard was free of mucosal damage or staining. The gizzard mucosa of control birds was bright yellow and relatively smooth.

Histologic lesions were evident in the kidney, liver, proventriculus, and gizzard. They were generally most pronounced in the mallards that died from the disease. The most severe lesions were found in group E. There were no lesions in the control group, G.

There were acid-fast intranuclear inclusion bodies in the renal proximal convoluted tubules in the kidneys in 70% of the mallards dying of lead poisoning. Two of the three mallards without intranuclear inclusions died on day 4. Inclusions were found in the kidneys of two group

E birds after necropsy on day 6 and in one group F bird on day 6. Inclusions found on day 4 were small (approximately 1 μm in diameter) and few, in contrast to inclusions in ducks necropsied on days 8, 11, and 12, which were large (nearly filling the nucleus) and numerous (Figure 1). Necrosis of the tubular epithelia was also most severe in the birds that died late in the experiment.

In the liver a brown bile pigment had accumulated in Kupffer's cells and hepatocytes, particularly in the vicinity of the hepatic veins of the group E birds.

The proventricular plicae were almost completely absent in the mallards that died spontaneously. In group E mallards they were reduced in height and increased in width with fusion of neighboring plicae.

In gizzards of group E birds, there were lesions of the mucosa. This layer had marked edema and infiltration of the superficial glands and the secreted proteinaceous layer by lymphocytes, histiocytes, and heterophils (Figure 2).

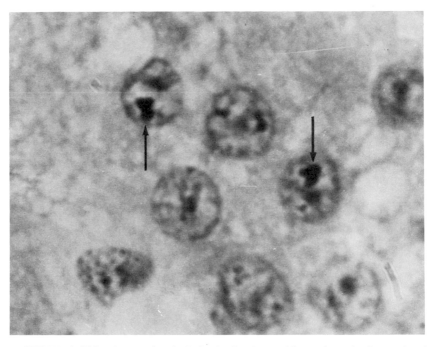

FIGURE 1   Acid-fast intranuclear inclusion bodies (arrows) in renal proximal convoluted tubular epithelium of a lead-poisoned mallard. Ziehl-Neelsen's stain. (×1,000.)

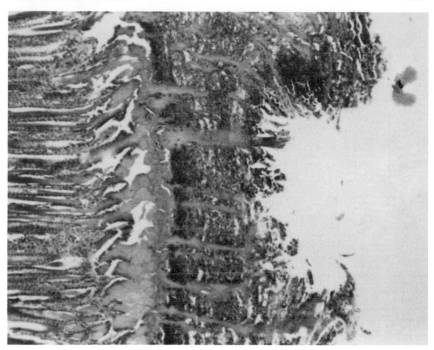

FIGURE 2 Erosion and acute inflammation of the gizzard mucosa of a lead-poisoned mallard. (Hematoxylin and eosin. ×26.)

The fluorescence excitation and emission maxima of blood from lead-treated mallards were 408 ± 1 nm and 635 ± 1 nm, respectively. These maxima are identical to those of free protoporphyrin IX in the detergent solution described by Lamola *et al.* (1975b).

Blood PP levels measured immediately after sampling were similar for group A, B, C, and D ducks. When these blood samples were refrigerated and PP measured at various intervals, samples from the lead-treated birds yielded progressively higher PP levels which stabilized by 48 h and remained constant for at least 9 days. The control PP levels remained low. The highest PP level attained by any of the group D ducks was 36 μg/dl.

Eight hours after shot ingestion, the PP of 40% of the group A ducks was > 36 μg/dl. All lead-treated ducks exceeded 36 μg/dl by the second day. The mean PP levels remained above 36 μg/dl for all treatment groups until the fifteenth day of the experiment. By the thirty-sixth

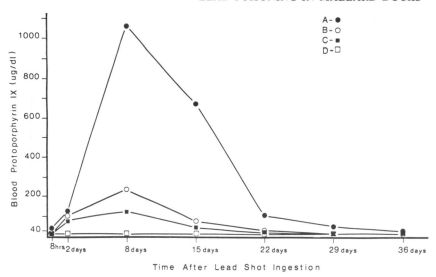

FIGURE 3 Mean blood PP levels of mallard ducks (10 per group) at various intervals after ingesting eight (group A), two (group B), one (group C), and no (group D) number-four lead shot pellets.

day, 92% of the mallards displayed normal PP levels. Figure 3 illustrates the pattern of mean PP levels of group A, B, C, and D ducks at various intervals after shot ingestion.

None of the mallards in the chronic lead poisoning experiment died.

The average and range (in parentheses) of MCHC (g/100 ml) of ducks in groups A, B, C, and D 8 days after shot ingestion were 25 (20–30), 29 (24–32), 30 (27–32), and 31 (26–33), respectively.

Immature and hypochromic erythrocytes were most evident in the smears of peripheral blood from group A ducks on the eighth day of the chronic poisoning experiment (Figure 4). Anisocytosis and poikilocytosis were characteristic findings in these smears. The blood picture returned to normal as the PP levels decreased.

Table 1 summarizes the relationship between PP levels read on a hematofluorometer, modified to read free protoporphyrin IX, and clinical signs of lead-poisoned mallards.

Birds with elevated PP levels exhibited a reticulocytosis in peripheral blood smears. In one bird with a PP level of 975 μg/dl, 49% of the erythrocytes were reticulocytes. Control birds and lead-treated birds displayed reticulocyte counts of 2% to 15% prior to shot ingestion.

DISCUSSION

Since 50% of the ducks fed eight lead shot pellets and the corn ration died within 4 days, and none of the birds fed eight pellets and a commercial ration died within 5 weeks, it was apparent that diet played a significant role in the toxicity of lead. Several authors have observed the enhancement of lead toxicity by a corn diet (Jordan and Bellrose, 1951; Irby *et al.*, 1967; U.S. Fish and Wildlife Service, 1976). The U.S. Fish and Wildlife Service (1976) reported studies by Wendig that suggest that selenium deficiency from a corn diet may contribute to this increased toxicity. Clemens *et al.* (1975) proposed that a diet low in calcium may enhance the severity of clinical signs of lead poisoning and increase the availability of lead to nonosseous tissues of intoxicated mallards. This suggestion was supported by experiments with lead-poisoned rats that showed an increase in lead absorption in rats fed a diet low in calcium and phosphorus (Barltrop and Khoo, 1975).

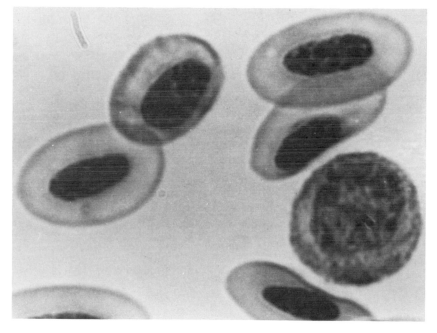

FIGURE 4 Immature, hypochromic erythrocytes and a basophilic erythroblast in a peripheral blood smear from a mallard duck eight days after ingestion of eight number-four lead shot pellets. (Giemsa's stain. ×1,000.)

TABLE 1   Blood Protoporphyrin IX Levels Measured on the
Hematofluorometer and Corresponding Clinical Signs of Lead
Poisoning in Mallard Ducks

Blood Protoporphyrin IX, μg/dl	Number of Ducks	Clinical Signs
801	3	Death Inability to stand, walk, or fly Marked tail and wing droop Loss of voice Green watery diarrhea
501–800	4	Muscular weakness Easily fatigued Unsteady gait Slight tail droop Green watery diarrhea
201–500	7	Hyperexcitability Green watery diarrhea
40–200	11	Green watery diarrhea
0–39	16	No evidence of lead poisoning

The observed increase in the prevalence and size of acid-fast intranuclear inclusion bodies in the renal tubular epithelium with increasing duration and quantity of lead exposure was also reported by Locke
et al. (1966) in their studies with lead-intoxicated mallards. Since
inclusions were not always present in birds dying during our study, it
may be inferred that their absence after acute exposures cannot pre-
clude a diagnosis of lead poisoning. Similarly, Locke *et al.* (1967) did
not consistently observe acid-fast inclusions in the kidneys of spontaneously lead-poisoned Canada geese.

The reduction in height and increase in width of proventricular plicae
have been attributed to the inhibitory effect of lead on rapidly dividing
cells in the crypts (Clemens *et al.*, 1975). The inflammatory lesions and
erosions of the gizzard were due to the corrosive lead salts.

Since increases in blood PP levels were associated with increases in
reticulocytes and other immature erythrocytes, and since these cell
types contain the synthetic machinery necessary to synthesize PP *in
vitro*, it is likely that the PP is synthesized by immature erythrocytes in

lead-poisoned mallards. A. A. Lamola of Bell Laboratories (personal communication, 1977) suggests that since the porphyrin in erythrocytes of lead-poisoned ducks is not chelated to zinc as it is in humans, it readily diffuses from the erythrocytes *in vivo* and is cleared by the liver and kidney. The inverse relationship of the MCHC and PP level suggests that heme synthetase is being inhibited by lead.

Attempts to ascertain the severity of lead poisoning in waterfowling areas has been based partly on shot-ingestion rates determined by examination of gizzards collected during the hunting season. As demonstrated above, this effort is futile without knowledge of the diet.

The hematofluorometer, when modified by substituting the factory installed emission filter with a filter more appropriate for PP measurement, provides an inexpensive, simple method for diagnosis of lead-poisoning in mallard ducks. A single drop of unprocessed blood, previously refrigerated for 2 days, is placed on a glass coverslip and inserted into the instrument. The PP level, in μg/dl, is displayed on a digital meter within 5 seconds.

The hematofluorometer would enable investigators to test live birds during the posthunting season when waterfowl take up winter residence. These birds can serve as barometers of local lead poisoning.

SUMMARY

Eighty-six mallard ducks (*Anas platyrhynchos*) were fed two different diets: a commercial duck pellet in the chronic poisoning experiment, and a whole corn diet in the acute poisoning experiment. In addition the ducks were given various doses of No. 4 lead shot. The diet was a major factor in the toxicity of lead; the corn diet enhanced it. The signs of acute lead poisoning included a bile-stained watery diarrhea, wing droop, muscular weakness, emaciation, head pressing, ruffled head and back feathers, convulsions, and death. Gross lesions of these birds included muscle atrophy, desquamation and bile staining of the gizzard grinding pads, and bile retention with distention of the gall bladder. Histopathological lesions of acutely poisoned mallards were renal tubular necrosis and acid-fast intranuclear inclusions; pigmentation of hepatocytes and Kupffer's cells; proventricular mucosal erosion; and acute inflammation of gizzard mucosa. The birds ingesting lead shot displayed blood fluorescence spectra characteristic of free protoporphyrin IX (PP). PP levels in chronically poisoned birds are proportional to the quantity of lead ingested, and they stabilize in refrigerated blood by 48 h after sampling. The PP is synthesized by immature erythrocytes *in vitro*. Birds with high PP levels had reduced mean corpuscular hemoglobin concentrations. Maximum PP levels were observed by the eighth day after shot ingestion. A simple, fluorometric microtechnique can be used to measure PP levels and to correlate these levels with clinical signs of lead poisoning in mallard ducks.

ACKNOWLEDGMENTS

This study was supported in part by the Northeastern Research Center for Wildlife Diseases, University of Connecticut, and grants from the National Rifle Association, 1600 Rhode Island Avenue, Washington, D.C., and the Shikar-Safari Club International Foundation, 1328 Racine Street, Racine, Wis.

REFERENCES

Barltrop, D., and H. E. Khoo. 1975. The influence of nutritional factors on lead absorption. Postgrad. Med. J. 51:798–799.

Barrett, M. W., and L. H. Karstad. 1971. A fluorescent erythrocyte test for lead poisoning in waterfowl. J. Wildl. Manage. 35:109–119.

Bellrose, F. C. 1959. Lead poisoning as a mortality factor in waterfowl populations. Ill. Nat. Hist. Surv. Bull. 27:235–288.

Blumberg, W., J. Eisinger, A. A. Lamola, and D. Zuckerman. 1977. Zinc protoporphyrin level in blood determined by a portable hematofluorometer; a screening device for lead poisoning. J. Lab. Clin. Med. 89:712–723.

Clemens, E. T., L. Krook, A. L. Aronson, and C. E. Stevens. 1975. Pathogenesis of lead shot poisoning in the mallard duck. Cornell Vet. 65:248–285.

Irby, H. D., L. N. Locke, and G. E. Bagley. 1967. Relative toxicity of lead and selected substitute shot types to game farm mallards. J. Wildl. Manage. 31:253–257.

Jordan, J. S., and F. C. Bellrose. 1951. Lead poisoning in wild waterfowl. Ill. Nat. Hist. Surv. Biol. Notes No. 26–27.

Lamola, A. A., and T. Yamane. 1974. Zinc protoporphyrin in the erythrocytes of patients with lead intoxication in iron deficiency anemia. Science 186:936–938.

Lamola, A. A., M. Joselow, and T. Yamane. 1975a. Zinc protoporphyrin (ZPP): A simple, sensitive, fluorometric screening test for lead poisoning. Clin. Chem. 21:93–97.

Lamola, A. A., S. Piomelli, M. B. Poh-Fitzpatrick, T. Yamane, and L. C. Harber. 1975b. Erythropoietic protoporphyria and lead intoxication: The molecular basis for difference in cutaneous photosensitivity. II. Different binding or erythrocyte protoporphyrin to hemoglobin. J. Clin. Invest. 56:1528–1535.

Locke, L. N., G. E. Bagley, and H. D. Irby. 1966. Acid-fast intranuclear inclusion bodies in the kidneys of mallards fed lead shot. Bull. Wildl. Dis. Assoc. 2:127–131.

Locke, L. N., G. E. Bagley, and L. T. Young. 1967. The ineffectiveness of acid-fast inclusions in diagnosis of lead poisoning in Canada geese. Bull. Wildl. Dis. Assoc. 3:176.

U.S. Fish and Wildlife Service. 1976. Use of Steel Shot for Hunting Waterfowl in the United States. Final Environment Statement. U.S. Dept. of the Interior, Washington, D.C. 276 pp.

QUESTIONS AND ANSWERS

J. ZINKL: Lead poisoning in waterfowl in the United States occurs most often during spring migration, probably because these birds are marginally nourished and therefore are much more susceptible to lead poisoning. An interesting experiment performed at the U.S. Fish and Wildlife Service Laboratory in Denver, Colo., was designed to develop a field method for the detection of sublethal or lethal lead poisoning in ducks. We found mortality in ducks fed cracked corn but not in ducks fed a complete diet of Layena Chow. In birds receiving the corn diet, three No. 9 lead shot pellets were sufficient to kill three out of four birds, and even one No. 9 lead shot pellet was enough to kill one out of four birds. In contrast, as many as 81 No. 9 lead shot pellets did not kill ducks on the Layena diet. This points out the importance of diet in lead poisoning of birds.

The field method that was developed is a relatively simple one based on extraction of 1 cm^3 of blood, with a 1:4 mixture of acetic acid and ethyl acetate which was reextracted with 1 cm^3 of 3 N hydrochloric acid. Exposure to lead is detected by demonstration of protoporphyrin fluorescence by shining a long wave ultraviolet light on the extract. The samples from lead-exposed birds would fluoresce. Unexposed ducks had a very low level of fluorescence. This method is also used to diagnose lead poisoning in cattle (R. Green, 1973).[1] Although the method is rapid and simple, it has a low sensitivity and will be positive only in severely poisoned waterfowl. In poisoned ducks, fluorescence appears approximately 4 days after ingestion of shot and persists for approximately 4 more days.

D. ROSCOE: I would like to make an important point about field testing for lead poisoning. When birds were sampled and tested immediately in the field they had levels below 40 mg/dl, leading one to assume that they were normal. However, when these samples were held for 48 h before testing, the levels went from 40 up to 1,000 where they presumably reached a diffusion equilibrium between the inside of the cell and the surrounding plasma. Usually only about 1% of the protoporphyrin escapes into the plasma.

T. RETTIG: Can you give us some idea as to the availability of the hemato-fluorometer or the use of that instrument?

D. ROSCOE: The instrument is kept at the Northeastern Research Center for Wildlife Diseases. The biologists collect the samples and send them in to us. They are then tested and the results returned.

M. FRIEND: Do you plan to look at the variations present in different species in regard to toxic levels?

D. ROSCOE: Yes, we saw unusually high protoporphyrin levels in the brants (*Branta bernicla*) in Long Island, which were not consistent with the normal

[1] Paper presented by Robert Green in 1973 at the Seminar of the American College of Veterinary Pathologists, San Antonio, Tex.

levels that we were seeing in black ducks (*Anas rubripes*), mallards (*A. platyrhyncos*), blue-winged teal (*A. discors*), and Canada geese (*B. canadensis*). We have done experiments with lead-intoxicated Canada geese. They have normal maximum protoporphyrin levels (~36 to 38 μg/dl) comparable to those of control mallards. In addition, a number of these geese contained shot in their skeletal muscles, apparently as a result of previous encounters with hunters, that did not result in any elevated porphyrin levels. The brant is of particular interest to us. We believe that the elevated protoporphyrin levels may reflect an increased exposure to lead in this species.

B. ZOOK: Is all the lead shot that is given to birds retained in the gizzard? Don't they pass them?

D. ROSCOE: We radiographed the birds serially at weekly intervals and found that they retained the shot for 2 to 3 weeks.

L. TRYPHONAS: So far as inclusion bodies are concerned, we did an experiment with gerbils treated with 0.05% lead acetate in the diet. Up to 2 weeks after treatment we were unable to detect inclusion bodies with the Ziehl–Neelsen acid-fast stain, but were able to detect them using the electron microscope.

Blood Delta-Aminolevulinic Acid Dehydratase (ALAD) to Monitor Lead Contamination in Canvasback Ducks (*Aythya valisineria*)[1]

MICHAEL P. DIETER

Waterfowl may obtain lead through a contaminated food source or by ingesting spent lead shotgun pellets. The lead shot annually kills an estimated 2% to 3% of all waterfowl (Bellrose, 1959). There have been many reports of poisoning from lead shot ingestion by waterfowl (Bagley *et al.*, 1967; Locke and Bagley, 1967a,b,c; Anderson, 1975). Substitutes tested have included either coated, alloyed, water solubilized, or chelated lead shot. Most candidate materials were either ballistically unsatisfactory or as toxic as commercial lead shot, which served only to delay mortality (Longcore *et al.*, 1974).

Since the mid-1950's, canvasback duck (*Aythya valisineria*) populations have not been as abundant as in former decades. As a result, very restrictive hunting regulations have been applied to this species during the past 20 years. The canvasback's inability to maintain population levels can be related to several factors. Deterioration of its nesting, migration, and wintering habitats are three of the most obvious influences. Although the role of lead poisoning in the population decline is clearly difficult to isolate from other factors, lead ingestion is certainly a contributing if not a major factor. An older survey by Bellrose (1959) showed that 12% of 1,216 gizzard samples contained at least one lead shot. Ingestion of one No. 4 commercial lead shot (about 200 mg of lead) will kill mallard ducks (*Anas platyrhynchos*) within 2 weeks, if they are eating a corn diet (Finley and Dieter, 1978). If waterfowl do

[1]A preliminary report of this work was presented at the International Conference on Heavy Metals in the Environment, Toronto, Ontario, Canada, October 1975.

not die outright from lead shot ingestion, there may be delayed mortality from starvation, enhanced natural predation (U.S. Fish and Wildlife Service, 1976), increased disease susceptibility (Hemphill *et al.*, 1971), or other sublethal responses that affect reproduction in subsequent years (Edens *et al.*, 1976).

Canvasbacks have traditionally been a prized species because they forage mainly on wild celery, which makes them especially palatable. Wild celery has largely disappeared and been replaced as a food item by clams, which accumulate lead in shell and tissues in concentrations ranging from 0.5 to 1.0 ppm (unpublished data, Patuxent Wildlife Research Center, 1976). The soft parts of mussels contain from 0.27 to 42 ppm lead on a dry weight basis, depending upon the lead pollution in their local habitats (Chow *et al.*, 1976). Thus both spent shot and food contamination could contribute to the lead in the tissues of canvasback ducks.

We examined the populations at two sites on the canvasback migratory route—the upper Mississippi River pool near LaCrosse, Wis., and the Chesapeake Bay in Maryland. Enzyme and blood residue assays showed that lead contamination in canvasback ducks was identical to that found by Bellrose, whose 1959 survey also revealed a 12% incidence of lead shot in canvasback duck gizzards.

MATERIALS AND METHODS

Canvasback ducks were baited with corn and then captured in welded wire traps. In 1974, they were trapped in the Chesapeake Bay at Persimmon Point, Va., and in the Mississippi River at pool 7 north of LaCrosse, Wis. In 1975, we trapped in the Chesapeake Bay at Cove Point, Md., and again at pool 7 in Wisconsin (Table 1). The ducks were sexed and aged by plumage characteristics (Carney, 1964) and cloacal examination (Hochbaum, 1942). Blood samples were drawn from the alar vein and divided into aliquots for enzyme and lead determinations. The ducks were then weighed, banded, and released. Delta-aminolevulinic acid dehydratase (ALAD) was assayed at Patuxent Wildlife Research Center; blood lead residues were assayed at Environmental Trace Substances Research Center, Columbia, Mo.; and tissue lead residues were assayed at WARF (Wisconsin Alumni Research Foundation) Institute, Inc., Madison, Wis. Burch and Siegel (1971) have reported detailed methods for the determination of enzyme activity; Finley *et al.* (1976c) have outlined the methods for lead. Units of ALAD enzyme activity are defined as 0.100 increase in absorbance per ml erythrocytes per hour (38°C) at 555 nm with a 1.0-cm light path.

TABLE 1 Blood Sample Collections from Trapped and Released Canvasback Ducks

| | | Number of Samples | | |
		Enzyme Assay	Lead Assay	Both Assays
Sampling	Location			
Jan.–Feb. 1974	Chesapeake Bay	95	48	48
Nov. 1974	Mississippi River	71	49	49
Jan.–Feb. 1975	Chesapeake Bay	106	51	51
Nov. 1975	Mississippi River	[a]	50	0

[a] One hundred samples were taken, but enzyme measurements were inaccurate and were discarded because of mechanical failure.

RESULTS

During four successive winters blood samples were obtained from 272 canvasback ducks. The ducks were trapped, sampled, and released on the upper Mississippi River and the Chesapeake Bay. Blood ALAD enzyme activity was measured during three samplings (N = 272) and blood lead during all four (N = 198) (Table 1).

Figure 1 presents data from both sexes collected in both locations. It shows a highly significant negative correlation between ALAD enzyme activity (1 to 272 ALAD units) and lead concentration in the blood (0.02 to 1.83 ppm).

Based on previous lead dosing experiments with mallard ducks, blood samples with less than 50 ALAD enzyme units were regarded as abnormal and were expected to contain elevated lead residues greater than 0.200 ppm. During each of the four samplings, approximately 50 blood samples, including all those with abnormal ALAD enzyme levels, were assayed for lead residues. Thirty-two or 11.8% of the 272 samples exhibited ALAD enzyme inhibition, and 22 or 11.2% of 198 samples contained greater than 0.200 ppm lead. Blood samples with less than the 50 ALAD units without elevated lead concentrations or the inverse were identified. They provided some measure of variation for our estimate of lead contamination. Of the samples with less than 50 ALAD enzyme units, 10 had only borderline concentrations of lead (between 0.100 and 0.200 ppm); 9 samples with greater than 0.200 ppm of lead

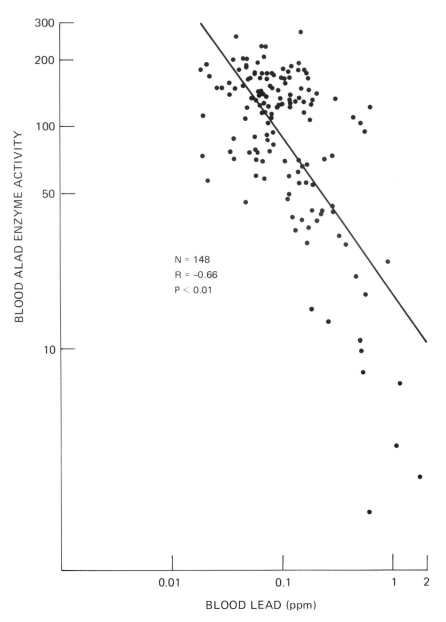

FIGURE 1 Correlation between lead concentration and δ-aminolevulinic acid dehy-dratase (ALAD) activity in the blood of canvasback ducks. The line of fit is described by the power regression: 1n blood ALAD activity = 1n 18.00–0.71 1n blood lead.

failed to be detected in the ALAD enzyme bioassay. If these corrections were considered, the incidence of lead contamination would become 41/272 or 15.1% based on ALAD, 12/198 or 6.1% based on lead, and would average 10.6% based on both criteria, suggesting that the reliability of the bioassay was quite acceptable.

Table 2 compares normal and abnormal values for enzyme activity and lead concentration at three samplings. The 1974 Chesapeake Bay data were part of a previous report on environmental pollutants (Dieter *et al.*, 1976) and were included here to provide the best estimate of lead pollution in canvasback ducks. The 1975 Mississippi River data were not included in Table 2 because the enzyme measurements were faulty. In the overall survey, about 12% of the samples averaged less than half of the normal blood ALAD enzyme activity and more than 0.500 ppm blood lead. The 1974 data in Table 2 suggest that further north in the migratory route (Mississippi River) fewer ducks carried elevated blood lead levels, and there was definitely a much greater difference in ALAD enzyme activity between normal and abnormal values.

The data in Table 3 show that neither age nor sex affected ALAD enzyme activity nor blood lead concentrations. Nor did it matter if the samples were divided into normal and abnormal values before comparison.

The distribution of tissue lead was examined in more detail in 25 canvasback ducks collected on the Chesapeake Bay in 1975. Ducks with a range of blood ALAD enzyme activity were purposely selected for tissue analyses. Table 4 shows that at least twice as much lead was found in blood, brain, and liver of ducks exhibiting abnormal (<50 units) as in those with normal (>50 units) ALAD enzyme activity. There was of course a significant correlation between blood lead and blood ALAD enzyme activity (r = −0.83). These values comprise a subset of the values in Figure 1. In addition, there were significant correlations between blood ALAD enzyme activity and lead in the liver (r = −0.88) and brain (r = −0.65). There is a relationship between lead in the blood and that in the brain and liver (Figure 2). Blood and tissue lead were significantly correlated in both organs (N = 25, r = 0.75, P < 0.01), indicating that circulating lead accurately reflected the concentrations in these organs.

There were three unique ducks in the 25-bird sample. One, which probably had been chronically exposed to low lead levels, had less than 0.100 ppm lead in the blood or liver; however, this duck had the highest bone lead of all ducks—36.4 ppm. The other two ducks, which probably had been acutely exposed to high lead levels, had greater than 0.500 ppm blood lead and greater than 1.30 ppm lead in the liver, but had less bone lead (8.4 and 12.3 ppm) than the other 12 ducks.

TABLE 2 Incidence and Degree of Lead Contamination in the Canvasback Duck Population. Means ± SE, Sample Numbers in Parentheses

Sample Origin	Blood ALAD, units		Blood Lead, ppm[b]	
	Normal Values[a]	Abnormal Values	Normal Values	Abnormal Values
Chesapeake Bay, 1974	105 + 5 (79)	29 ± 4 (16)	0.062 ± 0.005 (32)	0.263 ± 0.048 (16)
Mississippi River, 1974	163 ± 5 (66)	14 ± 5 (5)	0.085 ± 0.008 (44)	0.890 ± 0.270 (5)
Chesapeake Bay, 1975	113 ± 4 (95)	27 ± 5 (11)	0.175 ± 0.022 (40)	0.580 ± 0.130 (11)
3-Year Average	127 (240)	23 (32)	0.107 (116)	0.578 (32)

[a] Normal values include samples with greater than 50 units ALAD enzyme activity, and abnormal values, those with less.
[b] Each year only a portion of the samples were analyzed for lead.

TABLE 3 ALAD Enzyme Activity and Lead
Concentration in the Blood of 148 Canvasback Ducks
of Different Ages and Both Sexes

Number of Ducks	ALAD Units		Lead Concentrations, ppm	
Males	*Mean*	*Range*	*Mean*	*Range*
Mature 97	108	(1–207)	0.182	(0.02–1.210)
Immature 16	122	(7–182)	0.176	(0.019–1.200)
Females				
Mature 30	118	(3–272)	0.171	(0.020–1.830)
Immature 5	123	(2–233)	0.210	(0.038–0.625)

DISCUSSION

For over 50 years lead has been recognized as the cause of death of
thousands of waterfowl (Wetmore, 1919). In addition to mortality, lead
also causes abnormal brain development (Brown, 1975; Maker *et al.*,
1975; Reiter *et al.*, 1975; Bouldin *et al.*, 1975), nervous disorders
(Silbergeld and Chisolm, 1975; Silbergeld and Goldberg, 1975), and an
array of biochemical changes (Vallee and Ulmer, 1972; Goyer and
Rhyne, 1973; Goyer and Moore, 1974). One of the most sensitive
biochemical changes in response to lead is the inhibition of

TABLE 4 Comparison of Tissue Lead Concentrations in 25
Canvasback Ducks with Normal and Abnormal ALAD Enzyme
Activities. Means ± SE

	Normal Values[a] (N = 16)	Abnormal Values (N = 9)
	ALAD Units	ALAD Units
Average ALAD enzyme activity	119	26
Range of ALAD enzyme activity	55–181	1–44
Tissue lead	ppm	ppm
Blood	0.20 ± 0.04	0.58 ± 0.13
Brain	0.23 ± 0.04	0.43 ± 0.11
Liver	0.27 ± 0.03	0.86 ± 0.25
Bone	12 ± 2	18 ± 4

[a] Normal values include samples with greater than 50 units ALAD enzyme activity and abnormal values
those with less.

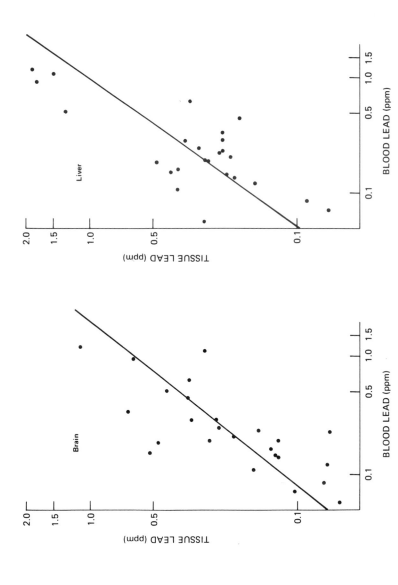

FIGURE 2 Correlation between lead concentration in blood and tissues of canvasback ducks. The lines of fit are described by these power regressions: ln brain lead = ln 0.61 + 0.72 ln blood lead; ln liver lead = ln 0.99 + 0.78 ln blood lead. These relationships were statistically significant (N = 25, r = 0.75, P < 0.01).

δ-aminolevulinic acid dehydratase (ALAD) enzyme activity in erythro-cytes (Tola *et al.*, 1973).

The ALAD enzyme bioassay is ideally suited for studies of lead contamination in migrating waterfowl populations. It has superb quantitative characteristics because the test is specific for lead (Gibson *et al.*, 1955), requires only a small blood sample (Sassa *et al.*, 1975; Chakrabarti *et al.*, 1975), shows an unusual degree of reproducibility (Dieter *et al.*, 1976; Finley *et al.*, 1976a,b; Dieter and Finley, 1978a), is extremely sensitive (Tola *et al.*, 1973; Finley *et al.*, 1976a), is responsive over a broad range of concentrations (Finley *et al.*, 1976b), and is relatively simple to perform (Burch and Siegel, 1971). The biological aspects of the bioassay have proven no less valuable than the quantitative aspects since ALAD activity was unaffected by age or sex (Dieter *et al.*, 1976; Finley *et al.*, 1976b; Table 3), or breeding condition of waterfowl (Finley *et al.*, 1976c). The duration of enzyme inhibition after lead shot ingestion was 3 months (Dieter and Finley, 1978a), which is long enough to detect lead poisoning during any phase of the migratory season, but not so persistent to cause interference from previous years.

The ALAD enzyme bioassay has also been used extensively to survey lead in humans, other mammals, birds, and fishes (Hernberg *et al.*, 1970; Mouw *et al.*, 1975; Ohi *et al.*, 1974; Dieter *et al.*, 1976; Hodson, 1976). In our initial canvasback survey (Dieter *et al.*, 1976), ducks with less than 50 ALAD enzyme units had more than 0.200 ppm blood lead. Successive surveys and studies in mallard ducks (Finley *et al.*, 1976b) supported the premise that less than 50 ALAD units and more than 0.200 ppm blood lead constituted abnormally high levels. A month after dosage with one No. 4 commercial lead shot, ALAD enzyme inhibition was significantly correlated with blood lead concentration ($P < 0.01$, $N = 60$):

ln ALAD activity = ln 18.22–0.64 ln blood lead concentration.

This shows that 51 units of ALAD enzyme activity equals 0.200 ppm lead, verifying that the two criteria selected for lead contamination were meaningfully related to lead shot ingestion.

Results indicate that the incidence of lead contamination in canvas-backs was 11% to 12%, whether based on ALAD enzyme inhibition (32/272 or 11.8%) or on lead levels (22/198 or 11.2%). Also, there was an indication of possible geographical differences in lead contamination. Obviously this would be difficult to verify since waterfowl are highly mobile and omnivorous. They intermingle among flocks and may even vary their migratory routes. More revealing, perhaps, was the sharp delineation between normal and abnormal values in ducks from the Mississippi River. In 1974 the abnormal ones had extremely low ALAD enzyme activity and high lead compared to normals.

By comparison the values from the Chesapeake Bay tended to be more normally distributed and the abnormal values were more difficult to recognize. This suggests that sources other than lead shot have contributed to their high lead contamination. When one considers the volume of lead contained in one lead pellet compared to an equivalent amount consumed in various food items, the ingestion of pellets must be the dominant source. However, the data indicate that lead contamination of food items could be significant at some locations. Likely candidates are the mussels and clams, which concentrate lead in their shells and tissues and constitute a major food item for canvasbacks at Chesapeake Bay (Perry and Uhler, 1976; Chow et al., 1976). In chronic feeding studies with mallards, 25 ppm of lead nitrate produced an average blood lead concentration of 0.150 ppm and depressed ALAD enzyme activity by 40% after 3 weeks. This persisted for 12 weeks, and by then the ducks fed 5 ppm lead showed the same magnitude of response (Finley et al., 1976a). It is not unreasonable, then, to suspect that some of the canvasback blood samples with concentrations between 0.100 and 0.200 ppm lead and intermediate degrees of ALAD enzyme inhibition might have been chronically exposed to lead through the food chain.

It has been argued that inhibition of blood ALAD enzyme activity posed no serious consequences because of large enzyme reserves (Waldron, 1974), but tissue enzyme reserves are not as easily replaced. We measured brain and liver ALAD enzyme inhibition in mallards 1 month after feeding them one No. 4 commercial lead shot pellet (Dieter and Finley, 1978b). The inhibition in these critical organs was as severe as in the blood. This could have serious consequences, because the ALAD enzyme is involved in the biosynthesis of coenzymes, and of protoporphyrins for incorporation into cytochromes. These products are in turn utilized to support critical cellular processes such as detoxification by the liver and nerve transmission by the brain. It was therefore alarming to find in this survey that ducks with elevated blood lead also contained proportionally elevated levels of lead in the brain and liver, because ALAD enzyme activity in the liver of humans (Secchi et al., 1974) and in the kidney and liver of rats (Hammond, 1973; Mouw et al., 1975; Suketa et al., 1975; Buchet et al., 1976) are also inhibited by low lead concentrations. In addition, enzymes other than ALAD in these tissues, e.g., brain adenylcyclase (Nathanson and Bloom, 1975), coenzyme oxidation in the brain (Bull *et al.*, 1975), hepatic mixed-function oxidases (Scoppa *et al.*, 1973), and hepatic gluconeogenic processes (Cornell and Filkins, 1974), are inhibited by low lead concentrations. Results of the samplings emphasize that the sublethal consequences of lead poisoning are particularly significant because they have gone largely undetected by conventional sampling techniques. Their effects are subtle and gradual and do not occur as noticeable changes in annual rates of mortality or reproduction.

SUMMARY

Waterfowl obtain some environmental lead through the food chain, but most of it by ingesting spent shotgun pellets. The lead shot ingestion results in severe inhibition of a lead-specific enzyme in the erythrocytes, δ-aminolevulinic acid dehydratase (ALAD). Our studies with mallard ducks have shown that the ALAD blood enzyme test is an ideal bioassay. It is specific for lead, sensitive, and accurate over a wide range of blood lead concentrations. In addition, it persists for 3 months after lead shot ingestion. Over a 4-year period the incidence and degree of lead contamination was monitored in about 400 ducks sampled and released from the upper Mississippi River and the Chesapeake Bay. Twelve percent of the ducks exhibited less than half of the normal blood ALAD enzyme activity and averaged greater than 0.5 ppm blood lead. In addition, 25 ducks were killed and tissue lead concentrations measured. At least twice as much lead was found in the blood, brains, and livers of ducks exhibiting abnormal ALAD enzyme activity as in those with normal activity.

ACKNOWLEDGMENTS

I want to thank the personnel at the Migratory Bird and Habitat Research Laboratory (MBHRL) and the Northern Prairie Wildlife Research Center (NPWRC) for their cooperation and for sharing their scientific expertise. This survey could not have been completed without the special efforts of Matthew C. Perry of MBHRL, who permitted me to obtain blood samples throughout this canvasback research program. I am also indebted to J. R. Serie of the NPWRC for his help and to the personnel at the Upper Mississippi River Wildlife and Fish Refuge who let me use their facilities.

REFERENCES

Anderson, W L. 1975. Lead poisoning in waterfowl at Rice Lake, Illinois. J. Wildl. Manage. 39(2):264–270.

Bagley, G. E., L. N. Locke, and L. N. Young. 1967. Lead poisoning in Canada geese in Delaware. Avian Dis. 11(3):601–608.

Bellrose, F. C. 1959. Lead poisoning as a mortality factor in waterfowl populations. Ill. Nat. Hist. Surv. Bull. 27(3):235–288.

Bouldin, T. W., P. Mushak, L. A. O'Tuama, and M. R. Krigman. 1975. Blood–brain barrier dysfunction in acute lead encephalopathy: A reappraisal. Environ. Health Perspect. 12:81–88.

Brown, D. R. 1975. Neonatal lead exposure in the rat: Decreased learning as a function of age and blood lead concentrations. Toxicol. Appl. Pharmacol. 32:628–637.

Buchet, J. P., H. Roels, G. Hubermont, and R. Lauwerys. 1976. Effect of lead on some parameters of the heme biosynthetic pathway in rat tissues in vivo. Toxicology 6:21–34.

Bull, R. J., P. M. Stanaszek, J. J. O'Neill, and S. D. Lutkenhoff. 1975. Specificity of the effects of lead on brain energy metabolism for substrates donating a cytoplasmic reducing equivalent. Environ. Health Perspect. 12:89–95.

Burch, H. B., and A. L. Siegel. 1971. Improved method for measurement of delta-aminolevulinic acid dehydratase activity of human erythrocytes. Clin. Chem. Winston-Salem 17:1038–1041.

Carney, S. M. 1964. Preliminary keys to waterfowl age and sex identification by means of wing plumage. U.S. Fish Wildl. Serv. Spec. Rep. Wildl. No. 82.

Chakrabarti, S. K., J. Brodeur, and R. Tardif. 1975. Fluorometric determination of δ-aminolaevulinate dehydratase activity in human erythrocytes as an index to lead exposure. Clin. Chem. Winston-Salem 21(12):1783–1787.

Chow, T. J., H. G. Snyder, and C. B. Snyder. 1976. Mussels (Mytilus sp.) as an indicator of lead pollution. Sci. Total Environ. 6:55–63.

Cornell, R. P., and J. P. Filkins. 1974. Depression of hepatic gluconeogenesis by acute lead administration. Proc. Soc. Exp. Biol. Med. 147:373–376.

Dieter, M. P., and M. T. Finley. 1978a. Erythrocyte δ-aminolevulinic acid dehydratase activity in mallard ducks: Duration of inhibition after lead shot dosage. J. Wildl. Manage. 42(3):621–625.

Dieter, M. P., and M. T. Finley. 1978b. δ-Aminolevulinic acid dehydratase enzyme activity in blood, brain, and liver of lead-dosed ducks. Unpublished.

Dieter, M. P., M. C. Perry, and B. M. Mulhern. 1976. Lead and PCB's in canvasback ducks: Relationship between enzyme levels and residues in blood. Arch. Environ. Contam. Toxicol. 5:1–13.

Edens, F. W., E. Benton, S. J. Bursian, and G. W. Morgan. 1976. Effect of dietary lead on reproductive performance in Japanese quail, Coturnix coturnix japonica. Toxicol. Appl. Pharmacol. 38:307–314.

Finley, M. T., and M. P. Dieter. 1978. Toxicity of an experimental lead–iron shot versus commercial lead shot in mallards. J. Wildl. Manage. 42(1):32–39.

Finley, M. T., M. P. Dieter, and L. N. Locke. 1976a. Sublethal effects of chronic lead ingestion in mallard ducks. J. Toxicol. Environ. Health 1:929–937.

Finley, M. T., M. P. Dieter, and L. N. Locke. 1976b. δ-Aminolevulinic acid dehydratase: Inhibition in ducks dosed with lead shot. Environ. Res. 12:243–249.

Finley, M. T., M. P. Dieter, and L. N. Locke. 1976c. Lead in tissues of mallard ducks dosed with two types of lead shot. Bull. Environ. Contam. Toxicol. 16(3):261–269.

Gibson, K. D., A. Neuberger, and J. J. Scott. 1955. The purification and properties of delta-aminolevulinic acid dehydratase. Biochem. J. 61:618.

Goyer, R. A., and J. F. Moore. 1974. Cellular effects of lead. Adv. Exp. Med. Biol. 48:447–462.

Goyer, R. A., and B. C. Rhyne. 1973. Pathological effects of lead. Int. Rev. Exp. Pathol. 12:1–77.

Hammond, P. B. 1973. The relationship between inhibition of δ-aminolevulinic acid dehydratase by lead and lead mobilization by ethylenediamine tetraacetate (EDTA). Toxicol. Appl. Pharmacol. 26:466–475.

Hemphill, F. E., M. L. Kaeberle, and W. B. Buck. 1971. Lead suppression of mouse resistance to *Salmonella typhimurium*. Science 172:1031–1032.

Hernberg, S., J. Nikkanen, G. Mellin, and H. Lilius. 1970. δ-Aminolevulinic acid dehydratase as a measure of lead exposure. Arch. Environ. Health 21:140–145.

Hochbaum, H. A. 1942. Sex and age determination of waterfowl by cloacal examination. Pp. 209–307 in Transactions of the North American Wildlife Conference. American Wildlife Institute, Washington, D.C.

Hodson, P. V. 1976. δ-Aminolevulinic acid dehydratase activity of fish blood as an indicator of a harmful exposure to lead. J. Fish. Res. Board Can. 33:268–271.

Locke, L. N., and G. E. Bagley. 1967a. Lead poisoning in a black duck. Bull. Wildl. Dis. Assoc. 3:37.

Locke, L. N., and G. E. Bagley. 1967b. Case report: Coccidiosis and lead poisoning in Canada geese. Chesapeake Sci. 8(1):66–69.

Locke, L. N., and G. E. Bagley. 1967c. Lead poisoning in a sample of Maryland mourning doves. J. Wildl. Manage. 31(3):515–518.

Longcore, J. R., R. Andrews, L. N. Locke, G. E. Bagley, and L. T. Young. 1974. Toxicity of lead and proposed substitute shot to mallards. U.S. Fish Wildl. Serv. Spec. Sci. Rep. Wildl. No. 183.

Maker, H. S., G. M. Lehrer, and D. J. Silides. 1975. The effect of lead on mouse brain development. Environ. Res. 10:76–91.

Mouw, D., K. Kalitis, M. Anver, J. Schwartz, A. Constan, R. Hartung, B. Cohen, and D. Ringler. 1975. Lead: Possible toxicity in urban vs. rural rats. Arch. Environ. Health 30:276–280.

Nathanson, J. A., and F. E. Bloom. 1975. Lead-induced inhibition of brain adenyl cyclase. Nature 255:419–420.

Ohi, G., H. Seki, K. Akiyama, and H. Yagyu. 1974. The pigeon, a sensor of lead pollution. Bull. Environ. Contam. Toxicol. 12(1):92–98.

Perry, M. C., and F. M. Uhler. 1976. Availability and utilization of food organisms in canvasbacks. Paper presented at the Annual Meeting of the Atlantic Estuarine Research Society, Rehoboth Beach, Delaware, May, 1976.

Reiter, L. W., G. E. Anderson, J. W. Laskey, and D. F. Cahill. 1975. Developmental and behavioral changes in the rat during chronic exposure to lead. Environ. Health Perspect. 12:119–123.

Sassa, S., S. Granick, and A. Kappas. 1975. Effect of lead and genetic factors on heme biosynthesis in the human red cell. Ann. N.Y. Acad. Sci. 244:419–440.

Scoppa, P., M. Roumengous, and W. Penning. 1973. Hepatic drug metabolizing activity in lead-poisoned rats. Experientia 29:970–972.

Secchi, G. C., L. Erba, and G. Cambiaghi. 1974. Delta-aminolevulinic acid dehydratase activity of erythrocytes and liver tissue in man. Arch. Environ. Health 28:130–132.

Silbergeld, E. K., and J. J. Chisolm, Jr. 1975. Lead poisoning. Altered urinary catecholamine metabolites as indicators of intoxication in mice and children. Science 192:153–155.

Silbergeld, E. K., and A. M. Goldberg. 1975. Pharmacological and neurochemical investigations of lead-induced hyperactivity. Neuropharmacology 14:431–444.

Suketa, Y., M. Aoki, and T. Yamamoto. 1975. Changes in hepatic δ-aminolevulinic acid in lead-intoxicated rats. J. Toxicol. Environ. Health 1:127–132.

Tola, S., S. Hernberg, S. Asp, and J. Nikkanen. 1973. Parameters indicative of absorption and biological effect in new lead exposure: A prospective study. Br. J. Ind. Med. 30:134–141.

U.S. Fish and Wildlife Service. 1976. Steel: Final Environmental Statement. Proposed Use of Steel Shot for Hunting Waterfowl in the United States. U.S. Government Printing Office, Washington, D.C. 276 pp.

Vallee, B. L., and D. D. Ulmer. 1972. Biochemical effects of mercury, cadmium, and lead. Ann. Rev. Biochem. 41:91–128.

Waldron, H. A. 1974. The blood lead threshold. Arch. Environ. Health 29:271–273.

Wetmore, A. 1919. Lead Poisoning in Waterfowl. U.S. Dept. Agr. Bull. 792. 12 pp.

QUESTIONS AND ANSWERS

G. CHOULES: How were the corroborative lead assays done?

M. DIETER: They were determined by atomic absorption spectrophotometry.

J. ZINKL: I wonder about the possibility of secondary poisoning of raptors after eating lead-poisoned birds.

M. FRIEND: We have a number of unconfirmed cases of lead poisoning in bald eagles (*Haliaeetus leucocephalus*). In many of these cases we have both direct evidence and circumstantial evidence that they were poisoned by consuming lead-poisoned waterfowl. The stomachs of some of these birds contained not only waterfowl feathers but also lead shot. Whether or not lead poisoning in these instances was due to the ingested shot is still an open question. Eagles regurgitate bones of ingested animals in a cast which also contains large numbers of lead shot. It is more probable that the source of lead is from the organs of ingested prey.

M. DIETER: It is important to emphasize that some rather severe biochemical effects accrue from lead poisoning long before death occurs. A case in point is that 0.2 ppm of lead in the blood of waterfowl leads to a 75% inhibition of ALAD enzyme activity.

J. PAYNE: Is there interference in the ALAD assay from pesticides and has anyone studied heme synthetase to see if this enzyme is increased in poisoned birds?

M. DIETER: I am not aware of any literature on interference of the assay from pesticides. I do not know if anyone has studied heme synthetase levels. However, a study of erythrocyte protoporphyrin levels could be done. This would give some direct information on the activity of heme synthetase.

B. ZOOK: Is 0.8 ppm of lead in blood diagnostic for poisoning in waterfowl?

M. DIETER: I said that in the canvasback duck there is a cutoff point at 50 ALAD units, which corresponds to 0.2 ppm blood lead. The relationship between blood lead and liver lead levels is close to 1:1. The point that I would like to make is that these canvasbacks carry more than normal background levels of lead. There are papers stating that 3 to 10 ppm of lead in the liver is diagnostic, but I believe these are higher than those causing sublethal harmful effects.

M. FRIEND: We use 8 ppm as the higher level of normal. Anything above that in waterfowl is considered a clear case of lead poisoning. Values between 6 and 8 ppm are presumptive evidence of poisoning and our diagnosis is then based on a combination of clinical signs, gross pathology, and liver lead levels. Both recent avian deaths attributed to aspergillosis and a recent outbreak of avian cholera in South Dakota have apparently occurred in lead-poisoned birds. These are again examples of how chronic lead poisoning in birds can lower resistance to infectious agents.

J. NEWMAN: What is the significance of the 12% incidence of abnormal ALAD blood activity in the canvasback duck? Is there a relationship to geography?

M. DIETER: I cannot give you any firm answer on that. The incidence of low ALAD blood levels matches well with the incidence of lead shot found by Bellrose (1959) in the gizzards of canvasbacks, although he collected his sample population from a different geographic location. It is not clear whether lead poisoning is responsible for the decline in the canvasback population which has been occurring over the past 20 years.

W. BIRGE: Do you have any data on biological half-life for lead in the various tissues of waterfowl?

M. DIETER: I do not have any specific data that I can quote. However I can state that lead appears in the blood and then is transferred to various internal organs from which it enters bone where it is stored. No one knows what happens to lead after it enters bone. It has been demonstrated that laying mallard hens can accumulate very high concentrations of lead in bone.[1] Whether or not such lead can subsequently be mobilized, thereby becoming toxic, remains to be shown.

W. BIRGE: What about the concentration in eggs and any possible deleterious effects on reproduction?

M. DIETER: There is a recent paper on quail by Edens *et al.*[2] where sublethal levels of lead in the diet were shown to decrease reproductive success, including the number of eggs laid.

[1] Finley, M. T., and M. P. Dieter. 1978. Influence of laying on lead accumulation in bones of mallard ducks. J. Toxicol. Environ. Health 4:123–129.

[2] Edens, F. W., E. Benton, S. J. Bursian, and G. W. Morgan. 1976. Effect of dietary lead on reproductive performance in Japanese quail, *Coturnix coturnix japonica*. Toxicol. Appl. Pharmacol. 38:307–314.

Accumulations of Lead in Rodents from Two Old Orchard Sites in New York

WANDA M. HASCHEK, DONALD J. LISK, and
ROBERT A. STEHN

Poisoning by lead, an environmental pollutant, is recognized as a common and important cause of illness, especially in children, young dogs, and cattle. In contrast, wildlife is seldom poisoned by lead, with the exception of waterfowl and some other game birds which ingest spent lead shot (Zook et al., 1972). Lead interferes with many enzymatic processes in the nervous system, hematopoietic system, and kidneys (Goyer and Rhyne, 1973) and causes structural damage to mitochondria in the kidney (Goyer, 1968). The diagnostic morphological feature of lead intoxication is the presence of characteristic acid-fast intranuclear inclusion bodies predominently in the proximal tubular epithelial cells of the kidney (Choie and Richter, 1972).

Several major sources of environmental contamination by lead are exhaust products of leaded gasoline, metal smelting plants, lead arsenate pesticides, phosphate fertilizers, lead-based paints, and spent lead shot from hunting activities (Goldsmith et al., 1976). For many years compounds such as lead arsenate or calcium arsenate have been used for insect control in apple orchards in this country, while compounds such as phenylmercuric acetate have been used for fungal control. Lead, arsenic, and mercury remain concentrated in the upper few centimeters of soil in contact with shallow-rooted plants (Jones and Hatch, 1937; McLean et al., 1944), but generally in forms that cannot be absorbed by plants (Lisk, 1972).

To determine plant absorption of these compounds, a number of vegetable crops and millet were grown on the site of an uprooted apple orchard to which lead, arsenic, and mercury compounds had been applied for many years. Appreciable absorption of lead and, to a lesser

192

extent, arsenic was observed only in carrots and millet. Because the elemental content of tissues from indigenous rodents might better reflect the lead concentrations in the soil, an investigation of lead accumulation in rodents inhabiting old orchards was undertaken.

MATERIALS AND METHODS

Two old orchards were investigated in this study. One, belonging to the Pomology Department at Cornell University, Ithaca, N.Y., was established in 1927. It had received an average of four applications of lead arsenate (0.9–1.35 kg/380 l of water, 2,850 l/ha) annually from the early 1930's until 1969. Phenylmercury fungicides (0.12 to 0.71 l/1,900 l of water, 2,850 l/ha) were applied 4 to 12 times per year between 1950 and 1970. The second orchard, located at New Paltz, N.Y., had been treated with lead arsenate (~0.9 kg/380 l; 950.5 l/ha) 2 to 4 times per year from about 1920 to 1971.

Meadow voles (*Microtus pennsylvanicus*) were trapped in grassy runways under apple trees in these orchards and in a control, nonorchard area. Pine voles (*Pitymys pinetorum*) and white-footed mice (*Peromyscus leucopus*) were trapped in the New Paltz orchard, although these species were not found in either the Ithaca orchard or in control areas. All animals were taken from areas located well over 180 m from the nearest paved road to obviate lead contamination from vehicle exhausts (Williamson and Evans, 1972).

Representative soft tissue sections and femurs were taken at necropsy and fixed in 10% buffered formalin. The femurs were demineralized under water pump vacuum in 10% formic acid buffered to pH 4.5 with sodium citrate. Tissue sections were then embedded in paraffin, sectioned at 4 to 6 μm, and stained with hematoxylin and eosin. The kidney was also stained with the modified Ziehl–Neelsen acid-fast stain (Lillie, 1954). Sections of kidney were also fixed in 6% glutaraldehyde in phosphate buffer at pH 7.3 for electron microscopic examination. Lead determination was performed on soil, kidney, liver, and skeletal samples. Tissues were dry-ashed to 475°C and analyzed by conventional anodic stripping voltammetry using a Princeton Applied Research Corp. Model 174 Polarographic Analyzer (Gajan and Larry, 1972).

RESULTS AND DISCUSSION

Concentrations of lead found in the soils and animal tissues at the varying sampling locations are shown in Tables 1 and 2. Lead concen-

TABLE 1   Concentrations of Lead in Tissues of Meadow Voles
(*Microtus pennsylvanicus*) Trapped in Old Orchards

Meadow Vole		Location of Capture	Concentration of Lead, ppm, dry weight			
Sex	Maturity		Soil[a]	Kidney	Liver	Bone
M	Adult	Orchard, Ithaca	1,342.0	18.7	3.0	111.0
M	Adult	Orchard, Ithaca	1,342.0	14.2	4.1	73.0
M	Subadult	Orchard, Ithaca	1,342.0	24.0	4.9	103.0
F	Adult	Orchard, Ithaca	1,342.0	23.0	4.8	232.0
F	Subadult	Orchard, Ithaca	1,342.0	21.0	3.4	108.0
M	Adult	Orchard, New Paltz	6,326.0	34.0	10.5	182.0
F	Adult	Orchard, New Paltz	6,326.0	41.0	9.0	303.0
M	Adult	Control area, Ithaca	14.2	2.0	1.3	18.7
M	Subadult	Control area, Ithaca	14.2	3.3	1.6	32.6
M	Subadult	Control area, Ithaca	14.2	7.2	2.0	8.6
F	Adult	Control area, Ithaca	14.2	3.9	1.7	13.5

[a] The arsenic concentrations (ppm, dry wt.) of these soils were: Ithaca orchard, 298; New Paltz, 290; and Ithaca control, 4.4.

trations were higher in the tissues of voles trapped in orchards as compared to control animals. This reflects the high concentrations of lead in the orchard soils.

There were no histological lesions in meadow voles from the control area, but lesions were present in both meadow and pine voles inhabiting the orchards. Renal lesions were confined mainly to the proximal convoluted tubules and involved the tubular epithelial cells. Many of the cells were enlarged, some had irregular luminal borders, and a few had degenerated with subsequent necrosis and sloughing into the lumen. Varying degrees of karyomegaly were observed in affected epithelial cells; the chromatin appeared to have decreased and was marginated at the periphery of the nuclei.

Prominent intranuclear inclusion bodies of varying sizes were observed in the enlarged nuclei of affected epithelial cells (Figure 1). The inclusion bodies were generally solitary, spherical to oval in shape, nonrefractile, and acidophilic in the hematoxylin and eosin stain. Some of them consisted of a dense central core surrounded by a layer of lower density. The inclusions were acid-fast with the Ziehl–Neelsen stain.

Electron microscopy of the kidney showed intranuclear inclusions consisting of a dense central core of amorphous material with a peripheral fibrillar meshwork. This has been described in other reports of lead poisoning (Goyer, 1968; Goyer *et al.*, 1970; Choie and Richter, 1972). There was similar fibrillar material within many tubular lumena. Occasionally, small accumulations were present in the cytoplasm of the tubular epithelial cells. The mitochondria of the proximal renal tubules were swollen, oval to rounded in shape, and had shortened and marginated cristae. At times, these mitochondria obliterated other cytoplasmic details.

Mild karyomegaly with margination of chromatin and occasional small intranuclear inclusion bodies were found in the hepatocytes of several pine voles and in the brain of one pine vole. These inclusions were present in astrocytes, large glial cells, and endothelial cells in the outer cerebral cortical gray matter. No inflammatory response was associated with the above lesions, which are regarded as diagnostic of lead poisoning (Choie and Richter, 1972; Zook *et al.*, 1972).

The femurs of immature meadow and pine voles were examined by light microscropy. The following observations were made in both species. The epiphyseal plate was thin due to lack of cartilaginous differentiation and maturation. The primary spongiosa was absent. The secondary spongiosa of the metaphysis consisted of many short, thick

TABLE 2 Concentrations of Lead in Tissues of Adult Female Pine Voles (*Pitymys pinetorum*) and White-Footed Mice (*Peromyscus leucopus*) Trapped in the New Paltz Orchard

| | Concentration of Lead, ppm dry weight | | |
Species	Kidney	Liver	Bone
Pine vole	30.0	5.5	401.0
Pine vole	27.0	8.3	306.0
White-footed mouse	14.6	5.7	81.0
White-footed mouse	9.7	2.1	28.0

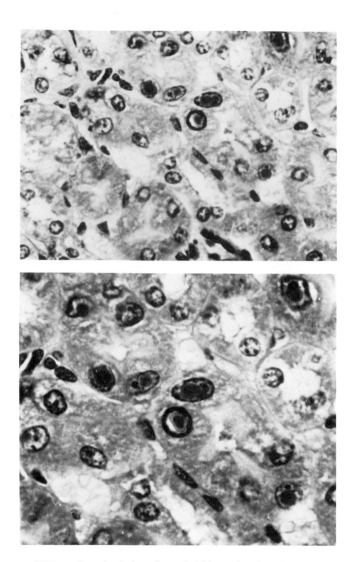

FIGURE 1 Renal tubules of a vole kidney showing characteristic intranuclear inclusion bodies. Some inclusion bodies consist of a dense central core surrounded by an outer layer of lower density. (Hematoxylin and eosin. Top × 1,250; Bottom × 1,950.)

trabeculae with a wide, mineralized chondroid core such as that described by Eisenstein and Kawanone (1975). This is indicative of arrested osteocytic chondrolysis (Whalen *et al.*, 1973). There were cementing lines present in both the epiphyseal trabeculae and the cortical bone. These findings are indicative of arrested osteocytic osteolysis (Bélanger *et al.*, 1968).

The rodents appeared to accumulate lead in the following order: pine voles > meadow voles > white-footed mice. This correlates with their degree of subsurface feeding and movement. Although all of the dietary habits of these animals are not known, the fossorial pine voles spend most of their time beneath the soil surface consuming root stocks, stems of grasses and herbs, roots of trees, and, undoubtedly, some adhering soil particles containing high lead concentrations. Meadow voles, while also frequenting subsurface runways, spend more time above ground consuming the upper parts of plants. White-footed mice generally remain above ground and are, in part, arboreal foragers which consume fruit and seeds of plants. All of the animals tend to occupy a relatively small area throughout their lives. Pine voles occupy an average of 90 m² (Gettle, 1975), while the area occupied by meadow voles and white-footed mice in orchards extends from 0.1 to 0.4 ha.

This study indicates that heavy metal pesticides, particularly lead compounds, are not rapidly dissipated in orchards. Residues have been observed in the soil and in tissues of indigenous rodents. Similarly, elevated levels of lead have been found in vegetation, invertebrates, shrews, voles, and mice inhabiting areas near highways (Williamson and Evans, 1972). The expected high intake of lead by predator species, such as hawks, owls, foxes, or weasels, which feed exclusively on animals from such areas, is presently being investigated.

SUMMARY

Rodents inhabiting old orchards, which had received heavy applications of lead arsenate, were examined for lead levels and histological evidence of lead toxicity. Lead, as well as arsenic, remains concentrated in the upper few centimeters of soil. Lead concentrations in kidney, liver, and bone tissues of orchard-trapped rodents were markedly higher than in tissues from control animals (up to 400 ppm compared to an average of 20 ppm in bone). Lead accumulation correlated well with the degree of subsurface feeding and movement of the rodents: pine voles > meadow voles > white-footed mice.

Intranuclear inclusion bodies, diagnostic of lead poisoning, were found in renal epithelial cells of the proximal convoluted tubules in voles from orchards that had been treated with lead arsenate. Inclusions were also present in the hepatocytes, and in astrocytes and endothelial cells of the outer cerebral cortex of some voles. There was evidence of arrested osteocytic osteolysis in the long bones.

ACKNOWLEDGMENT

The authors wish to thank Donald C. Elfving and Carl A. Bache who contributed significantly to this study.

REFERENCES

Bélanger, L. F., L. Jarry, and H. F. Uhthoff. 1968. Osteocytic osteolysis in Paget's Disease. Rev. Can. Biol. 27:37–44.

Choie, D. D., and G. W. Richter. 1972. Lead poisoning: Rapid formation of intranuclear inclusions. Science 177:1194–1195.

Eisenstein, R., and S. Kawanone. 1975. The lead line in bone—A lesion apparently due to chondroclastic indigestion. Am. J. Pathol. 80:309–316.

Gajan, R. J., and D. Larry. 1972. Determination of lead in fish by atomic absorption spectrophotometry and by polarography. I. Development of the methods. J. Assoc. Off. Anal. Chem. 55:727–732.

Gettle, A. S. 1975. Densities, movements and activities of pine voles (*Microtus pinetorum*) in Pennsylvania. Masters thesis. Pennsylvania State University, State College. 66 pp.

Goldsmith, C. D., Jr., P. F. Scanlon, and W. R. Pirie. 1976. Lead concentrations in soil and vegetation associated with highways of different traffic densities. Bull. Environ. Contam. Toxicol. 16:66–70.

Goyer, R. A. 1968. The renal tubule in lead poisoning. I. Mitochondrial swelling and aminoaciduria. Lab. Invest. 19:71–77.

Goyer, R. A., and B. C. Rhyne. 1973. Pathological effects of lead. Pp. 1–77 in G. W. Richter and M. A. Epstein, eds. International Review of Experimental Pathology. Vol. 12. Academic Press, New York.

Goyer, R. A., P. May, M. M. Cates, and M. R. Krigman. 1970. Lead and protein content of isolated intranuclear inclusion bodies from kidneys of lead-poisoned rats. Lab. Invest. 22:245–251.

Jones, J. S., and M. B. Hatch. 1937. The significance of inorganic spray residue accumulations in orchard soils. Soil Sci. 44: 37–62.

Lillie, R. D. 1954. In Histopathologic Technic and Practical Histochemistry. The Blakiston Company, Inc., New York. 598 pp.

Lisk, D. J. 1972. Trace metals in soils, plants and animals. Adv. Agron. 24:257–325.

McLean, H. C., A. L. Weber, and J. S. Joffe. 1944. Arsenic content of vegetables grown in soils treated with lead arsenate. J. Econ. Entomol. 37:315–316.

Whalen, J. P., N. O'Donohue, L. Krook, and E. A. Nunez. 1973. Pathogenesis of abnormal remodeling of bones: Effects of yellow phosphorus in the growing rat. Anat. Rec. 177:15–22.

Williamson, P., and P. R. Evans. 1972. Lead: Levels in roadside invertebrates and small mammals. Bull. Environ. Contam. Toxicol. 8:280–288.

Zook, B. C., R. M. Sauer, and F. M. Garner. 1972. Lead poisoning in captive wild animals. J. Wildl. Dis. 8:264–272.

QUESTIONS AND ANSWERS

D. HINTON: Did you see any evidence of renal tubular hyperplasia?

W. HASCHEK: No.

G. MIGAKI: What has been your experience, or that of others, with animals exposed to high levels of lead as a result of smoke pollution from city dumps?

W. HASCHEK: There have been several reports concerning rats collected from sewer areas and city dumps. In these cases quite high levels of lead, as well as intranuclear inclusion bodies, were found in tissues.

J. NEWMAN: What are the ecological effects of heavy metal exposure in various animal populations?

R. STEHN: Little is known because the data are limited. For example, most of the information on the effects of lead on pine voles is from orchard populations only. We have very little control information on similar animals from natural woodland habitats.

J. NEWMAN: I have been interested in the relationship between the threshold level of a contaminant in food and its ultimate bioaccumulation by the host. Is there a certain threshold level for a contaminant below which biocontamination does not occur and above which it does occur?

W. HASCHEK: I am not familiar with this phenomenon. I would suspect that any amount of lead in mammals would be reflected in some degree of bioaccumulation.

B. ZOOK: Were any of the voles trapped in nonorchard areas?

W. HASCHEK: Yes, the controls were from nonorchard areas.

B. ZOOK: What was the lead level in the livers of control animals?

W. HASCHEK: It was between 1 and 3 ppm on a dry weight basis.

B. ZOOK: Did you compare your dry weight measurements with those for wet weights?

W. HASCHEK: No. Several reports state that any values over 3 ppm are indicative of lead poisoning. However, they do not indicate whether their calculations were made on a wet or dry weight basis.

L. TRYPHONAS: It is interesting that carrots and millet accumulated more lead than the other plants. Did you actually analyze the carrots?

W. HASCHEK: Yes, all the edible portions of the crops were analyzed. They were thoroughly washed before analysis.

L. TRYPHONAS: So the carrot absorbs more lead from soil than other plants?

W. HASCHEK: Yes, and corn even more so.

Lead Contamination of Mammals and Invertebrates Near Highways with Different Traffic Volumes

PATRICK F. SCANLON

The road and highway system of the United States is extensive and the adjacent verdant areas incorporated in their rights-of-way provide much habitat for wild animals. Food and forage crops are grown close to highways and large numbers of people live and/or work in their immediate vicinity.

Smith (1976) has estimated the size of the roadside ecosystem in the United States to be 3.04×10^7 ha. This ecosystem is subjected to contamination by materials exhausted by vehicular traffic. A major contaminant is lead from alkyl lead compounds that have long been used as antiknock additives in gasoline. Smith (1976) reviewed estimates of lead quantities used and exhausted by cars. The average lead content of leaded gasoline is approximately 0.66 g/l. Yearly consumption of lead in gasoline increased from 90×10^6 kg in the mid-1940's to more than 220×10^6 kg in 1970. An estimate of the lead released in automobile exhausts in 1970 was 136×10^6 kg. Automobiles are the primary source of aerosol lead in the atmosphere contributing 98.4% (Hicks, 1972).

As lead is a major contaminant of roadside ecosystems, the effects on animal forms living there and/or consuming vegetation or animals grown in such ecosystems are of interest. This report presents results of two studies conducted on the lead levels in animals, plants, and soil near highways of different traffic volumes. Difficulties in interpreting impacts of lead pollution on mammals are discussed.

200

MATERIALS AND METHODS

The first study was conducted in fall 1974 in Montgomery, Craig, and Giles Counties, Va. Soil, plants, insects, earthworms, small rodents, and shrews were collected from three areas associated with highways of different traffic volume and from three remote forested areas (controls). Traffic volumes on the roads were 21,040, 8,120, and 1,085 vehicles per day on highways U.S. 460, Va. 114, and Va. 421, respectively. All control areas were farther than 500 m from any road. Soil, plants, and earthworms were sampled at 3, 6, 12, and 18 m from roadways. Mammals and insects were trapped wtihin 20 m of roads. Lead levels were determined using a model 290 Perkin Elmer atomic absorption spectrophotometer and were calculated in terms of micrograms per grams of dry matter.

The second study, which was initiated in August 1976, was part of a project to determine heavy metal (lead, cadmium, nickel, and zinc) contamination of roadside environments. Four study areas associated with highways were chosen in Fairfax, Montgomery, and Craig Counties, Va. One had a much higher traffic flow than any area in the first study. Daily traffic counts in vehicles per day were approximately 100,000 on I. 95, 24,000 on U.S. 460, 7,500 on Va. 114, and 500 on Va. 42. Two control areas, each in grassland and more than 500 m from any road, were also sampled. Soils and vegetation were sampled at 3, 6, 12, 24, and 48 m from the edge of the paved surface. Mammals were trapped within 20 m of the roadways. Samples were collected in August/ September 1976 and again in November/December 1976. Samples were analyzed for lead using an Instrumentation Laboratory Model 351 atomic absorption spectrophotometer. Lead concentrations were calculated as micrograms per gram of dry matter.

Comparisons were made using the Kruskal—Wallis or Wilcoxon Rank Sum nonparametric tests. Dunn's test for multiple comparisons was used to determine order of significance when significant differences were indicated by the Kruskal–Wallis Test (Hollander and Wolfe, 1973).

RESULTS

Data from the first study indicated that lead levels in soils decreased significantly ($P < 0.05$) as distance from highways increased. The lead content of all soil samples from roadside areas exceeded those from control areas. Lead concentrations in soils from highway areas ranged from 109.7 μg/g to 19.9 μg/g; soil from control areas had a mean lead content of 7.8 μg/g. Goldsmith *et al.* (1976) provide a fuller description of the results. Lead concentrations in vegetation showed a general decreasing trend as distance from the highway increased and as traffic volume declined.

Lead concentrations in grasshoppers from all roadside areas did not exceed 4 μg/g. Much higher levels of lead were found in earthworms (Table 1). Lead concentrations were considerably higher in the high traffic volume area and tended to decline as distance from the highway increased.

Lead concentrations also were determined for 10 species of mammals, which were trapped in roadside and control areas. These data are presented in Table 2.

Significant (P < 0.05) differences among areas were found in lead concentrations in least shrews (*Cryptotis parva*), meadow voles (*Microtus pennsylvanicus*), and white-footed mice (*Peromyscus leucopus*). In all instances the higher lead levels occurred in animals from the higher traffic density areas. Significant (P < 0.05) differences were not found among areas in lead levels of short-tailed shrews (*Blarina brevicauda*) though differences among traffic areas (P < 0.09) and between pooled traffic vs. pooled control areas (P < 0.14) approached significance.

The lead concentrations in white-footed mice from the heaviest traffic area were significantly different (P < 0.05) from those from control areas. In the area of lowest traffic, there was a significant (P < 0.05) difference between the concentrations in short-tailed shrews and those in meadow voles. In all cases the higher lead concentrations were found in short-tailed shrews. In the area of lowest traffic the difference

TABLE 1   Lead Concentration in Earthworms Recovered at Various Distances from Roadways with Heavy and Light Traffic[a]

Traffic, Vehicles/Day	Lead Concentrations, μg/g dry wt.		
	6 m from Highway	12 m from Highway	18 m from Highway
21,040	51.01	50.31	32.10
1,085	—[b]	8.51	11.65

[a] Adapted from Goldsmith and Scanlon (1977).

[b] Not measured.

in lead concentrations between short-tailed shrews and least shrews approached significance (P < 0.06).

The small mammals that were recovered from the study areas in numbers adequate for statistical tests represented two trophic levels. There were species that were primarily herbivorous—white-footed mice, golden mice (*Ochrotomys nuttalli*), and meadow voles—and those which were insectivorous—the short-tailed shrews, least shrews, and masked shrews (*Sorex cinereus*). In general, higher or significantly higher levels of lead were found in shrews than in the white-footed mice, golden mice, or in the meadow voles. This was true for both traffic areas and control areas.

In the second study lead levels declined significantly (P < 0.05) as distance from the roadway increased and as traffic volume decreased. Near the roadway with 100,000 vehicles per day, lead concentrations in soil ranged from 1,483 μg/g at 6 m to 188 μg/g at 48 m in August/September. Soil from control areas had 20 μg/g of lead. In the area with the lowest traffic volume, lead concentrations in soil 3 m from the roadway did not exceed 30 μg/g, while control levels were 10 μg/g. Lead in vegetation declined significantly as distance from the roadway increased and as traffic volume decreased. Vegetation samples collected in November/December had a significantly (P < 0.05) higher lead content than samples collected in August/September. There was considerable variation in lead concentration among species of vegetation.

Three mammalian species were recovered from all areas in the second study. These were two herbivorous species, meadow voles and white-footed mice; and one insectivorous species, the short-tailed shrew. Lead concentrations in meadow voles ranged from 22.4 μg/g in the 100,000 vehicles per day area to 1.1 μg/g in a control area. Lead concentrations in white-footed mice ranged from 16.3 μg/g in the 100,000 vehicles per day area to 0.9 μg/g in a control area. In short-tailed shrews, lead concentrations were 87.3 μg/g in the 100,000 vehicles per day area and 1.7 μg/g in the control area. Lead levels increased as traffic volume increased. Within areas insectivorous shrew species tended to have higher lead concentrations than herbivorous rodents.

DISCUSSION

Interpretation of the significance of lead contamination of wildlife inhabitants of roadside habitats is extremely difficult. Much existing data on lead concentrations in soil, vegetation, and animals have been

TABLE 2 Lead Concentrations in Mammals Collected from Three Roadsides and Three Control Areas

Species	Mean Lead Concentrations, μg/g dry wt. $\pm$ SE (Number of Animals in Parentheses)					
	21,040 Vehicles/Day	8,120 Vehicles/Day	1,085 Vehicles/Day	Control	Control	Control
Masked shrew (Sorex cinereus)	(0)	13.7 ± 1.1(4)	(0)	(0)	(0)	16.6 ± 2.6(8)[x]
Short-tailed shrew (Blarina brevicauda)	34.8 ± 9.5(3)[x]	15.8 ± 3.3(2)	11.6 ± 0.6(5)[x]	14.1 ± 3.8(3)[x]	14.6 ± 0.6(2)[x]	13.9 ± 1.2(8)[x]
Least shrew (Cryptotis parva)	(0)	10.4 ± 0.4(8)[a]	6.5 ± 1.4(2)[b,x,y,z]	(0)	(0)	(0)
White-footed mouse (Peromyscus leucopus)	15.6 ± 1.2(10)[c,u]	9.7 ± 0.9(5)[b]	(0)	5.0 ± 0.6(4)[e,u]	7.6 ± 0.4(16)[c,u]	6.4 ± 0.2(30)[d,u]
Meadow vole (Microtus pennsylvanicus)	(0)	12.1 ± 0.2(2)[a]	6.9 ± 0.6(5)[b,u]	(0)	(0)	(0)

[a,b,c,d,e] Means in the same *row* with different superscripts are significantly (P < 0.05) different. The order of significance being $a > b > c > d > e$. Means with none, the same, or multiple superscripts for comparisons in one direction (e.g., xy) are not significantly different.

[x,u] Means in the same *column* with different superscripts are significantly (P < 0.05) different, the order of significance being $x > y$.

[z] The difference between concentrations in *B. brevicauda* and those in *C. parva* in the study area with 1,085 vehicles per day approached significance (P < 0.06).

SOURCE: Adapted from Goldsmith and Scanlon, 1977.

gathered without at least some of the following considerations: sampling at different traffic volumes, sampling at multiple distances from roadways, sampling at various soil depths, multiple sampling of the same species of plant or animal, multiple sampling over different seasons, and sampling with different extraction procedures. Accordingly, comparison of results is difficult. Smith (1976) and Goldsmith (1975) have reviewed much of the available data on lead concentrations in soils, vegetation, and animals. Of interest here is the finding of significant differences between season of collection and lead concentrations in vegetation.

Little information is available on what proportion of lead contamination of vegetation results from uptake of soil lead. Smith (1976) discusses this. In one sense herbivores are exposed to the same amount of lead regardless of whether it is in or on the plant. However, there may be effects on palatability of contaminated vegetation from either absorbed lead or surface contamination that may influence rates of ingestion. The whole question of palatability of vegetation contaminated with exhausted lead compounds is largely unstudied.

In high traffic areas, herbivores are confronted with extremely high levels of lead in vegetation, in the air they breathe, and in the soil they contact. Insects have low lead contents. Yet herbivorous animals have lower lead levels than insectivores (Quarles *et al.*, 1974; Goldsmith and Scanlon, 1977). A number of possibilities arise. Herbivorous species may select vegetative parts with the least amount of surface contamination, they may be less tolerant of high levels thereby dying without accumulating high levels, or they may be more efficient in voiding ingested or inhaled lead. While much is known about absorption of lead and sites of lead storage in human and domestic animal tissues (Chisolm, 1971; Ammerman *et al.*, 1977), little if any work has been done on sites of lead storage and absorption of lead in small wild mammals.

A probable cause of high levels was found in earthworms. Quarles *et al.* (1974) discussed lead levels in shrews in relation to published information on shrew diets, most of which are confined to insects. Because lead concentrations in insects are low, it is difficult to conceive of shrews acquiring high lead levels without consuming a high proportion of earthworms. More information is needed on food habits of shrews, especially on a year-round basis, as well as on the vulnerability of lead-contaminated earthworms to predation by shrews. Information on toxicity of lead to shrews and small mammals is also lacking.

Toxicity of lead in roadside environments is probably also influenced by contamination from other materials, including such heavy metals as cadmium and zinc. Hiller *et al*. (1977) and Blair *et al*. (1977) reported that both cadmium and zinc increased with traffic volume and with nearness to the highways. The whole problem of multiple contaminants and whether each enhances toxicity of the other has scarcely been addressed at this point for any wild species.

SUMMARY

Insects, earthworms, small rodents, and shrews were collected from the immediate vicinity of highways of different traffic density and from control areas more than 500 m from any roadway. Samples of soil and vegetation were assayed for lead using atomic absorption spectrophotometry. Soil lead levels increased as traffic volume increased, and decreased significantly as distance from the highway increased. Greater lead concentrations in vegetation were observed in heavy traffic areas and in areas closest to the highways. Lead levels were extremely low in insects from all areas. They were highest in earthworms recovered close to highways and from those recovered near roadways with the highest traffic volumes. Shrews had higher lead levels than plant-consuming rodents. Problems in interpreting data on lead contamination of wildlife are discussed.

ACKNOWLEDGMENTS

I wish to acknowledge the contributions of C. D. Goldsmith, Jr., C. W. Blair, A. L. Hiller, and N. L. Schauer to this report. Funding for a portion of the work reported herein was provided by Department of Transportation Contract No. 60226.

REFERENCES

Ammerman, C. B., S. M. Miller, K. R. Fick, and S. L. Hansard II. 1977. Contaminating elements in mineral supplements and their potential toxicity: A review. J. Anim. Sci. 44(3):485–508.

Blair, C. W., A. L. Hiller, and P. F. Scanlon. 1977. Abstr. Heavy metal concentrations in mammals associated with highways of different traffic densities. Va. J. Sci. 28(2):61.

Chisolm, J. J., Jr. 1971. Lead poisoning. Sci. Am. 224(2):15–23.

Goldsmith, C. D., Jr. 1975. Lead levels in soil, vegetation and animals associated with highways of various traffic densities. M.S. thesis, Virginia Polytechnic Institute and State University, Blacksburg. 66 pp.

Goldsmith, C. D., Jr., and P. F. Scanlon. 1977. Lead levels in small mammals and selected invertebrates associated with highways of different traffic densities. Bull. Environ. Contam. Toxicol. 17(3):311–316.

Goldsmith, C. D., Jr., P. F. Scanlon, and W. R. Pirie. 1976. Lead concentrations in soil and vegetation associated with highways of different traffic densities. Bull. Environ. Contam. Toxicol. 16(1):66–70.

Hicks, R. M. 1972. Airborne lead as an environmental toxin. Chem. Biol. Interact. 5(6):361–390.

Hiller, A. L., C. W. Blair, and P. F. Scanlon. 1977. Abstr. Heavy metal concentrations in soils and vegetation associated with highways of different traffic densities. Va. J. Sci. 28(2):64.

Hollander, M., and D. A. Wolfe. 1973. Nonparametric statistical methods. John Wiley & Sons. New York. 503 pp.

Quarles, H. D., R. B. Hanawalt, and W. E. Odum. 1974. Lead in small mammals, plants and soil at varying distances from a highway. J. Appl. Ecol. 11(3):937–969.

Smith, W. H. 1976. Lead contamination of the roadside ecosystem. J. Air Pollut. Control Assoc. 26(8):753–766.

QUESTIONS AND ANSWERS

W. HASCHEK: Did you examine a composite of the whole animal or did you examine individual tissues?

P. SCANLON: Lead levels were determined using carcasses that were freeze-dried and ashed after the stomach contents were removed to allow for subsequent analysis of food habits and to prevent any dilution by trap bait. Lead levels were calculated on a dry weight basis. I cannot provide information as yet on individual organs.

J. NEWMAN: One possible explanation for the high lead levels in shrews: Although there are low levels in the insects that constitute their food, shrews also feed on dead rodents—in particular, on their livers and kidneys.

P. SCANLON: I expect that many animals are more selective in their feeding habits than is currently assumed, which is why studies of food habits are so important. The question I have is how selective an animal is while feeding in contaminated areas. Is he going to eat a piece of plant that happens to have 1,400 ppm of lead on it? There is a good possibility he doesn't.

P. WEIS: I would suggest that the vegetarian mammals you studied ingest lead rather than inhale it because they are nocturnal animals, and probably sleep in their burrows during rush hour traffic.

P. SCANLON: Your point is a good one; however, one must take the settling patterns of lead around highways into consideration. From what my engineering colleague tells me, it appears that exhausts remain airborne as long as there are cars moving. Settling of lead occurs during periods of low traffic movement, which coincides in part with the nocturnal activity period of these animals.

J. NEWMAN: It is quite probable that the high levels of lead in rodents and shrews provide a source of lead for raptors and other predators and scavengers that range along the highway.

P. SCANLON: We are quite conscious of this possibility. In fact, we have noted that raptors, particularly American kestrels (*Falco sparvarius*), use the roadways extensively. Shrikes (*Lanius* sp.) also use roadways extensively as can be seen in Florida during the winter. However, it is much more difficult to collect sufficient data on birds because capturing large numbers of them is so difficult. We have considered trapping them and analyzing feathers for lead content, but since feathers are grown during a restricted period and possibly in a distant area, it is unclear how useful feathers would be for monitoring lead.

M. MIX: Are you assuming that all of the lead present is derived from automobile exhaust? Or is there a possibility that some of the lead in your study area may be derived from decay of long-lived parent elements of uranium, thorium, etc.?

P. SCANLON: The latter possibility always exists, but since we are taking specimens from different distances from the roadways, that possibility is negligible. The reassuring thing about our method is that when you get 50 m from the roadway, the lead levels approach the levels of the control areas. There is also the possibility of local pockets of high concentrations such as old home sites. However, we have scrupulously tried to avoid these areas.

Immune Response Altered by Lead in CBA/J Mice

LOREN D. KOLLER

Environmental contaminants such as lead (Hemphill *et al.*, 1971), cadmium (Cook *et al.*, 1975), sulfur dioxide (Fairchild *et al.*, 1972), cobalt sulfate (Gainer, 1972), arsenicals (Gainer and Pry, 1972), methyl mercury (Koller, 1975), and polychlorinated biphenyls (PCB's) (Friend and Trainer, 1970) are synergistic with infectious agents in animals. Some of them are apparently immunosuppressive. Circulating antibody titers to infectious agents from animals exposed to lead, cadmium, mercury (Koller, 1973), methyl mercury, dichlorodiphenyltrichloroethane (DDT) (Wassermann *et al.*, 1971), and PCB's (Koller and Thigpen, 1973) were significantly lower than those from control animals (Table 1). More recently it was reported that chronic exposure to lead (Koller and Kovacic, 1974), cadmium (Koller *et al.*, 1975), and methyl mercury (Koller *et al.*, 1977a) produced a significant decrease in antibody synthesis (Table 2). It is obvious from these studies that many environmental contaminants interfere with the immune system.

Because it has been confirmed that numerous environmental contaminants, often at subclinical doses, inhibit the immune response, their action on lymphocytes, plasma cells, accessory cells (e.g., the macrophage) or a combination thereof must be examined. Antibody responses to many antigens require cooperation between at least two types of lymphocytes for optimal expression. One cell type is thymus-derived (T cells); another is bone marrow-derived (B cells). T cells may function as cytotoxic cells or may amplify, help, or suppress B cells. T cells do not produce antibody. B cells mature independently of thymic influence. They differentiate into antibody-producing cells and are

209

TABLE 1   Mean Serum Antibody Titers Against Pseudorabies or Influenza Virus in Rabbits Chronically Exposed to Lead, Cadmium, Mercury, Methyl Mercury, or Polychlorinated Biphenyls (PCB's). Virus Inoculations Were Administered 7 Days Apart

Substance	Dose, ppm	Primary Response[a]	Secondary Response[b]
Lead	2,500	230[c]	308[c]
Cadmium	300	490[c]	705[c]
Mercury	10	641[c]	813[d]
Control	0	2,074	1,667
Methyl Mercury	10	80	160
	1	160	320
Control	0	320	320
PCB			
1221	75	28	69[c]
1242	75	13[d]	56[c]
1254	75	15	67[c]
Control	0	29	138

[a] 7 days after first virus inoculation.
[b] 10 days after the third virus inoculation.
[c] Highly significant (P < 0.01).
[d] Significant (P < 0.05).

often influenced by T cells. A third cell type, the macrophage, is an important accessory cell. It cooperates with T cells in aiding the response of B cells to antigens.

METHODS AND RESULTS

To examine the effect of lead and cadmium on B cells (Koller and Brauner, 1977), 28-day-old CBA/J mice were given 1,300, 130, or 13 ppm lead as lead acetate, or 300, 30, or 3 ppm cadmium as cadmium chloride in deionized water for 70 days. The controls were given deionized water. There were 25 mice in each group. All mice were fed a diet (Oregon State University [OSU] Rodent Chow) containing less than 1.2 ppm lead and 0.05 ppm cadmium by analysis.

Spleens from each group of mice were pooled into five samples of five spleens each and homogenized in glass tissue grinders. The spleen cells were washed three times with cold Hanks' Balanced Salt Solution (HBSS), layered on a Ficoll–Hypaque gradient, and centrifuged at room temperature for 40 min at 400 × g. The band of mononuclear cells was collected and washed three times with cold HBSS. The cells were counted and the viability was determined by the trypan blue exclusion test. Cell viabilities were never less than 90%.

Sheep red blood cells were collected in Alsever's solution and stored at 4°C for no longer than 7 days. Prior to use, the cells were washed three times in phosphate-buffered saline (PBS) and resuspended to a final concentration of 5% in PBS. Rabbit antisheep red-blood-cell serum was heat inactivated at 56°C for 30 min and stored at –20°C. The hemagglutination titer was 1:400. Autologous mouse complement serum was obtained from blood collected by cardiac puncture and stored at –70°C.

Sheep erythrocytes (E) were sensitized with an equal volume of a 1:800 dilution of rabbit antisheep red-blood-cell serum (A) for 30 min at 37°C. The EA complex was subsequently washed in PBS and resuspended in Veronal®-buffered saline (VBS) to the initial volume of sheep red blood cells. The EA complex was sensitized with an equal volume of a 1:10 dilution of mouse complement serum (C) for 30 min at 37°C. After two washes in VBS, the EAC complex was resuspended in HBSS eight times the initial volume of sheep red blood cells.

The spleen lymphoid cells (1.5×10^6 in 0.5 ml HBSS) were incubated with 0.5 ml EAC at 37°C for 5 min. After centrifugation at $200 \times g$ for 5

TABLE 2   Mean Number Antibody-Forming Cells per Million Spleen Cells in Mice Exposed to Lead, Cadmium, or Methyl Mercury

Substance	Dose, ppm	Plaques per Million Spleen Cells	
		Primary Response	Secondary Response
Lead	0	500	500
	13	350[a]	150[a]
	130	400	120[a]
	1,300	300[a]	50[a]
Cadmium	0	1,800	1,600
	3	1,200[b]	750[a]
	300	1,600	650[a]
Methyl Mercury	0	2,200	1,100
	1	1,700	650
	5	1,600	1,300
	10	700[a]	825

[a] Highly significant ($P < 0.01$).
[b] Significant ($P < 0.05$).

min, the cells were incubated on ice for 1 to 2 h without removing the supernatant. After gentle resuspension, an aliquot of cells was mixed with an equal volume of crystal violet solution (1 mg/ml in minimal essential media [MEM] with 10% fetal calf serum). Two hundred cells were examined microscopically in a hemacytometer. Only stained cells with three or more bound erythrocytes were counted as EAC rosettes.

The percent of EAC rosette formation by spleen cells from mice exposed to lead and cadmium for 10 weeks was generally less than rosette formation by control animals (Table 3). The number of EAC rosettes was significantly less in animals that received 130 or 1,300 ppm lead or 30 ppm cadmium. The impaired rosetting was not due to toxicity of lead or cadmium since the percent viability was similar for treated and control groups. Therefore, the direct effect of these compounds on B cells could account in part for their suppression of the humoral immune response.

Koller and Roan (1977) have also determined that lead and cadmium actually activated macrophages instead of suppressing their activity.

Eighteen-day-old Swiss Webster and CBA/J mice were given 3, 30, or 300 ppm cadmium as cadmium chloride, or 13, 130, or 1,300 ppm lead as lead acetate orally in deionized water for 70 days. The controls were given deionized water. The diet (OSU rodent chow) fed to all mice contained less than 1.2 ppm lead and 0.05 ppm cadmium.

All mice were inoculated intraperitonially (IP) with 3 ml mineral oil 5 days before termination. Peritoneal exudate cells (PEC) were obtained at death by injecting 7 ml of cold Hanks' Balanced Salt Solution (HBSS) containing 10 U heparin IP using a 20 g, 1.5-in disposable needle.

TABLE 3   Percent EAC Rosettes Formed by Spleen Cells from Groups of 25 Mice Exposed to Lead or Cadmium for 10 Weeks

Dose, ppm		EAC, %	
Lead	Cadmium	Lead	Cadmium
0	0	53.5	45.1
13	3	46.5	42.9
130	30	$41.5^{a,b}$	$30.9^{a,c}$
1,300	300	41.3^a	42.7

[a] Significant at $P < 0.05$ (one-way analysis of variance).
[b] LSD − 9.64.
[c] LSD − 13.63.

Leaving the needle in place, the cavity was gently massaged and the fluid withdrawn. The PEC were washed three times in cold HBSS and centrifuged at 900 revolutions per minute (rpm) to remove the oil. The PEC were transferred to a clean siliconized tube (one tube per mouse sample) and diluted to 4 to 6 ml in culture medium (medium 199, 15% fetal calf serum, 100 U/ml penicillin, 100 g/ml streptomycin, Hepes buffer pH 7.2), depending on the size of the pellet. Viability was tested using the trypan blue dye exclusion test. Two ml of the diluted cell solution was added to Leighton tubes that contained coverslips (Wheaton no. 1, 9 × 50 mm). The cells were cultured for 2 h at 37°C. To obtain optimal phagocytosis, a 10% solution of washed sheep red blood cells (SRBC) was incubated for 1 h with an equal volume of diluted opsonin (rabbit anti-SRBC, Microbiological Associates). Preliminary testing showed the optimal opsonin dilution to be 1:1600 hemagglutination units in microtiter plates. A PEC monolayer was washed once with warm phosphate buffered saline (PBS) followed by addition of 2 ml of the opsonized washed SRBC (diluted to 10% in culture medium) and incubated for 1 h at 37°C. The coverslips were carefully removed, rinsed in two solutions of PBS, fixed in absolute methanol, stained with Giemsa, and mounted onto slides. The cells were then counted. A macrophage was considered positive for phagocytosis if two or more SRBC were engulfed. The cells with three or more SRBC around the macrophage in the form of a rosette were counted separately. Two slides of macrophages were prepared from each mouse and 200 cells were counted per slide.

Acid phosphatase concentrations in macrophages were determined using a Coleman U.V. Digital Spectrophotometer. A 0.2-ml solution containing 1×10^6 macrophages was added to 1.0 ml of buffered substrate (Boehringer Mannheim Corporation, Acid Phosphatase Test Kit, Catalog No. 15988), which contained 50 mmol citrate buffer (pH 4.8), 550 μmol p-nitrophenyl-phosphate, and 12.8 mol sodium chloride. This solution was kept at 37°C for 30 min after which the reaction was terminated by adding 10 ml of 0.02 N sodium hydroxide. Color change was determined with the spectrophotometer at a wavelength of 405 nm by comparing treatment samples to a reagent blank.

These two compounds, lead and cadmium, stimulated phagocytosis (Table 4) and increased acid phosphatase levels (Table 5) in macrophages. It was concluded that the macrophage does not contribute to the immunosuppressive activity produced by lead and cadmium.

A study recently completed in our laboratory was mitogen stimulation of lymphocytes from lead- and cadmium-exposed mice. Splenic

TABLE 4   Phagocytosis by Peritoneal Macrophages from Mice Exposed to Lead or Cadmium for 10 Weeks

| Dose, ppm | | Phagocytosis, % | | | |
| | | CBA | | SW | |
Lead	Cadmium	Lead	Cadmium	Lead	Cadmium
0	0	56	54	43	43
13	3	67	56	49	38
130	30	54	56	57[a]	45
1,300	300	72[a]	63[a]	62[b]	59[a]

[a] Significant at $P < 0.05$.
[b] Significant at $P < 0.01$ (t test).

lymphocytes were separated by nylon wool column filtration. Mitogens used for stimulation were lipopolysaccharide *E. coli* (LPS), 055:B5 and 0111:B4, phytohemagglutinin P (PHA), and concanavalin A (Con A). The stimulation index (SI) was calculated by dividing the CPM of spleen cells with mitogen by CPM of spleen cells with culture medium only.

Lead inhibited the SI of Con A and PHA in T lymphocyte cultures (Table 6). This effect was remarkable at all three lead exposure levels in animals that had previously been inoculated with Bacillus Camette Guérin (BCG) (Table 6). The lower dosages of cadmium resulted in the

TABLE 5   Mean Concentration of Acid Phosphatase in Peritoneal Macrophages from Mice Exposed to Lead or Cadmium for 10 Weeks

| Dose, ppm | | Concentration Acid Phosphatase, Stimulation Index Units (SI) | |
Lead	Cadmium	Lead	Cadmium
0	0	18.37	21.72
13	3	12.81	28.12
130	30	57.63[a]	34.80[b]
1,300	300	24.50	28.40

[a] Significant at $P < 0.01$ (t test).
[b] Significant at $P < 0.05$.

TABLE 6   Lymphocyte Stimulation Induced by Mitogens (LPS, Con A, PHA, and PPD) from Mice Exposed to Lead and Cadmium for 10 Weeks. One Group was Inoculated with BCG at 4 (1 g) and 9 (0.25 mg) Weeks

		Stimulation Index				
		T Lymphocytes		T Lymphocytes-BCG		
Substance	Dose, ppm	Con A	PHA	Con A	PHA	PPD
Lead	0	133	20	253	23	14
	13	108	12	76	13	36
	130	125	20	65	14	13
	1,300	61	11	78	9	5
Cadmium	0	133	20	253	23	14
	3	25	7	85	11	6
	30	85	10	137	14	6
	300	155	36	109	22	6

lowest SI for T lymphocytes. As in the lead-exposed animals, those mice inoculated with BCG and exposed to cadmium had the lowest SI. The SI for PPD was reduced by both lead and cadmium. Therefore, it was concluded that lead and cadmium affect T cells, which is indicated by the inhibition of cell division by mitogens. The inhibition of T cells suggests impairment of cell mediated immunity.

The immune response in aged mice chronically exposed to lead has also been investigated (Koller *et al.*, 1977b). The immunological assays examined were mitogen (LPS, Con A, PHA) stimulation of lymphocytes; erythrocyte–antibody (EA), erthrocyte–antibody–complex (EAC), and phagocytosis of macrophages; and FAC of splenic lymphocytes. The results were interpreted by comparing data obtained from aged mice to data collected from young adult mice (Table 7). It was apparent by this comparison that the aged mice were naturally immunosuppressed. Therefore, the results obtained from lead-exposed mice were unpredictable.

DISCUSSION

Many environmental contaminants are directly toxic, causing overt clinical signs and occasionally death. However, at subclinical concen-

TABLE 7   Immunological Assay Values of Young Adult Control Mice Compared to Those of Aged Control Mice

Immunological Assay	14-Week-Old (Young Adults) Mice	Aged Mice
Phagocytosis by macrophages, %	56	40
EAC lymphocyte rosettes, %	54	43
Lymphocyte stimulation index		
LPS 055:B5	27.5	6.9
LPS 0111:B4	17.9	4.6
Con A	32.6	9.3
PHA-P	3.2	1.8

trations many of these compounds suppress the immune system thereby rendering the host more susceptible to infectious agents. It is apparent from the above studies that lead and cadmium are two pollutants that affect the immune response of animals.

The direct effect of chronic exposure to lead or cadmium on B and T cells could account in part for their suppression of humoral and cell-mediated immune responses. Subclinical doses of at least some environmental contaminants are indirectly detrimental to health.

SUMMARY

The effects of lead and cadmium on the immune system of mice were investigated. The action of these metals on lymphocytes and macrophages was examined. After mice were exposed to the two metals, their B cells had inhibited EAC-rosette formation. The results from this B cell specific test indicate that lead and cadmium directly suppress the B cell. Lead and cadmium actually activated macrophages instead of suppressing their activity demonstrated by stimulated phagocytosis and increased acid phosphatase levels in macrophages. It was concluded that the macrophage did not contribute to the immunosuppressive activity produced by lead and cadmium. Mitogen stimulation of lymphocytes from lead- and cadmium-exposed mice and the effect of lead and cadmium on aged mice were also studied.

REFERENCES

Cook, J. A., E. O. Hoffmann, and N. R. Diluzio. 1975. Influence of lead and cadmium on the susceptibility of rats to bacterial challenge. Proc. Soc. Exp. Biol. Med. 150:741–747.

Fairchild, G. S., J. Roan, and J. McCarroll. 1972. Atmospheric pollutants and the pathogenesis of viral respiratory infection. Sulfur dioxide and influenza infection in mice. Arch. Environ. Health 25:174–182.

Friend, M., and D. P. Trainer. 1970. Polychlorinated biphenyl: Interaction with duck hepatitis virus. Science 170:1314–1316.

Gainer, J. H. 1972. Increased mortality in encephalomyocarditis virus-infected mice consuming cobalt sulfate: Tissue concentrations of cobalt. Am. J. Vet. Res. 33:2067–2073.

Gainer, J. H., and T. W. Pry, 1972. Effects of arsenicals on viral infections in mice. Am. J. Vet. Res. 33:2299–2307.

Hemphill, F. E., M. L. Kaeberle, and W. B. Buck. 1971. Lead suppression of mouse resistance to Salmonella typhimurium. Science 172:1031–1032.

Koller, L. D. 1973. Immunosuppression produced by lead, cadmium and mercury. Am. J. Vet. Res. 34:1457–1458.

Koller, L. D. 1975. Methylmercury: Effect of oncogenic and nononcogenic viruses in mice. Am. J. Vet. Res. 36:1501–1504.

Koller, L. D., and J. A. Brauner. 1977. Decreased B lymphocyte response after exposure to lead and cadmium. Toxicol. Appl. Pharmacol. 42:621–624.

Koller, L. D., and S. Kovacic. 1974. Decreased antibody formation in mice exposed to lead. Nature 250:148–149.

Koller, L. D., and J. G. Roan. 1977. Effects of lead and cadmium on mouse peritoneal macrophages. J. Reticuloendothel. Soc. 21:7–12.

Koller, L. D., and J. E. Thigpen. 1973. Reduction of antibody to pseudo-rabies virus in polychlorinated biphenyl exposed rabbits. Am. J. Vet. Res. 34:1605–1606.

Koller, L. D., J. H. Exon, and J. G. Roan. 1975. Antibody suppression by cadmium. Arch. Environ. Health 30:598–601.

Koller, L. D., J. H. Exon, and J. A. Brauner. 1977a. Methylmercury: Decreased antibody formation in mice. Proc. Soc. Exp. Biol. Med. 155:602–604.

Koller, L. D., J. G. Roan, J. A. Brauner, and J. H. Exon. 1977b. Immune response in aged mice exposed to lead. J. Toxicol. Environ. Health 3:535–543.

Wassermann, M., D. Wassermann, E. Kedar, and M. Djauaherian. 1971. Immunological and detoxication interaction in P, P,-DDT fed rabbits. Bull. Environ. Contam. Toxicol. 6:426–435.

QUESTIONS AND ANSWERS

R. KIRKPATRICK: Do you have any evidence that suggests that these compounds may have an effect on the adrenal gland rather than a direct effect on the immune system?

L. KOLLER: Histologically, there is no morphologic alteration in the adrenal gland.

R. KIRKPATRICK: Did you look at circulating corticoid levels?

L. KOLLER: No, most of these studies were performed *in vitro*, so we have not measured corticosteroid levels in intact animals.

R. KIRKPATRICK: One of the reasons I ask is that we have studied PCB in several species of mice and found an increase in corticosteroid levels, even at very low concentrations of PCB ingestion.

J. REIF: Did you also measure erythrocyte rosette forming cells and would you comment on your method of purifying lymphocytes? Are you sure that there are no monocytes present which might be masking this decrease in the EAC-forming cells?

L. KOLLER: Mice do not form erythrocyte rosettes like humans, so we cannot consider them as models in this regard. I think perhaps only 2% or so will form erythrocyte rosettes. Therefore, we separate the cells on Ficoll–Hypaque gradient, pass them through nylon wool columns, and then test the system by plating out the macrophages.

J. REIF: Does this remove the macrophages?

L. KOLLER: There is a small amount of contamination by them.

H. CASEY: Can you detect any histological differences in these mice compared to controls?

L. KOLLER: No.

J. ZINKL: Vos (1973)[1] has shown that TCDD (2, 3, 7, 8-tetrachlorodibenzo-*p*-dioxin) affects lymphoid tissues following exposure at low levels. He also demonstrated that the tuberculin response in guinea pigs exposed to TCDD was depressed. These results implicate that TCDD affects T cells.

L. KOLLER: That is correct. However, one must be careful about attaching a significance to morphologic changes in the lymphoid system. For example, any infectious agent will affect the immune system and secondarily will lead to marked stimulation of lymphoid follicles. Furthermore, certain parts of the lymphoid system such as the thymus undergo natural involution with aging. I feel much more comfortable utilizing specific immunological assays to a specific antigen to measure alterations of the lymphoid system rather than to depend on morphologic changes.

J. ZINKL: I also believe that Vos demonstrated that there was a difference in the circulating corticosteroid level in guinea pigs treated with TCDD as compared to control animals. He was unable to explain whether the immunosuppressive effects were adrenal gland mediated or whether TCDD exerted a direct effect on T cells.

R. MASAKE: Did you do any tests on B cells and T cells from animals treated with lead?

L. KOLLER: We have been working on our *in vitro* system of antigen stimulation for the past year and we now feel it is ready to use in such experiments. Our next set of studies will involve interchanging T and B cells and macrophages from normal animals with those of lead-treated animals. By creating several different mixtures of cells we will be able to see what effect these cells have on one another. In this way we hope to be able to test T helper cell function, suppressor cell effects, and so on.

R. MASAKE: One possibility is that the T cell is probably affected at the plasma cell level. Also, what about B cells?

L. KOLLER: Remember, a T cell does not differentiate into a plasma cell; a B cell transforms into a plasma cell. We inhibited antibody formation by rosetting so that there were less cells producing antibodies. When we

---

[1]Vos, J. G., J. A. Moore, and J. Zinkl. 1973. Effects of TCDD on the immune system of laboratory animals. Environ. Health Perspect. 5:149–162.

performed our EAC tests, we were not testing antibody secretion. There are immunoglobulin receptors on the surface of B cells for the Fc fragment and complement. We examined only one receptor complement and did demonstrate inhibition of the complement receptor. What that has to do with antibody secretion, I am not sure. Some T cells on the other hand do have Fc receptors that are activated, but T cells also have a soluble factor such as lymphokine or they may come directly in contact with B cells, which then will result in stimulating B cells to produce more antibodies.

D. HASTINGS: Did you have these animals on an adequate selenium diet?

L. KOLLER: Yes, we have been using selenium diets, both excess and deficient ones, and these have not shown too many changes in mice. We have a study in progress with different combinations of methyl mercury and selenium. The findings from this study will be of interest.

H. TRYPHONAS: Have any studies been done on the effect of lead or cadmium on complement levels in monkeys or in man?

L. KOLLER: I am not sure.

AIR
POLLUTION

The Effects of Air Pollution on Wildlife and Their Use as Biological Indicators

JAMES R. NEWMAN

Air pollution is a major human health hazard. But with the exception of studies on the effect of acid rain on aquatic animals (Dochinger and Seliga, 1976) little attention has been given to the harmful effects of air pollution on wildlife. The neglect of such studies does not reflect the importance of the issue. The health of wildlife populations is an important measure of environmental quality. There are numerous reports relating air pollutants to widespread sickness and, in many cases, death in large numbers of domestic animals (Lillie, 1970; Newman, 1975). These episodes often occurred in rural areas where it was likely that large numbers of wildlife were also exposed to pollutants. This paper considers the impact of air pollution on terrestrial wildlife, in particular the sensitivity of the house martin (*Delichon urbica*) to air pollution, and the use of wildlife as biological indicators of air pollution. Information is based on a literature review and on the results of a study in Czechoslovakia.

HARMFUL EFFECTS ON WILDLIFE: MAMMALS

Air pollution has increased markedly since the Industrial Revolution in the 1850's. Reports of sickness and death of domestic animals were published as early as the 1870's and appeared frequently through the turn of the century (Lillie, 1970; Newman, 1975). In England, the problem was so severe that in 1878 a royal commission was called to investigate the livestock losses attributable to air pollution (Royal Commission, 1878).

223

Reports of incidents involving free-living mammals also appeared early but not as frequently (Table 1). One of the earliest wildlife reports involved the death of fallow deer (*Dama dama*) in 1887, which resulted from arsenic emissions from a silver foundry in Germany (Tendron, 1964).

Another early but more detailed report on the harmful effects of air pollution is also from Germany. In 1936, widespread killing of game mammals by arsenic emissions occurred in Germany's Tharandt forest. Between 60% and 70% of the red deer (*Cervus elaphus*), roe deer (*Capreolus capreolus*), and wild rabbits (*Oryctolagus cuniculus*) died. The deer exhibited defective hair growth and antler formation, cirrhosis, splenic fibrosis, and emaciation (Prell, 1936).

Along with the greater environmental awareness of the 1960's came an increase in the number of reported incidents of injury and death to wildlife from air pollution. Harmful effects to mammals have been observed for a variety of pollutants ranging from heavy metals to gaseous pollutants. Mammals from North America, Europe, and Africa have been affected. The reported effects of exposure to air pollutants have included death, population declines, debilitating industry-related diseases, bioaccumulation of air pollutants, and physiological changes associated with stress.

Reports of industry-related debilitating diseases are widespread. Hais and Masek (1969) reported the loss of balance and hair in red deer and roe deer caused by exposure to arsenic emissions in Czechoslovakia. These debilitating effects contributed to the freezing deaths of deer during the winter. Industrial fluorosis exists in deer populations of

TABLE 1   Early Air Pollution Incidents Involving Wildlife[a]

Date	Location	Species	Pollutant(s)	Effects
1887	Germany	Fallow deer	Arsenic	Death of fallow deer
1900	England	Moths	Sulfur dioxides and particulates	Melanism
1927	Texas, U.S.A.	Small birds	Hydrogen sulfide	Death of many birds
1936	Germany	Rabbits, red and roe deer	Arsenic	Death of 60% to 70% of game population

[a] Adapted from Newman (1975).

Canada (Karstad, 1967) and the United States (Robinette *et al.*, 1957; Kay *et al.*, 1975; Newman and Yu, 1976). In the state of Washington, the teeth of both young and old blacktailed deer (*Odocoileus hemionus columbianus*) have severely deteriorated by pitting, chipping, and excessive wear (Newman and Yu, 1976). Asbestosis has occurred in free-living baboons (*Papio* sp.) and rodents, e.g., the Namaqua rock rat (*Rattus namaquensis*), in the vicinity of asbestos mines in South Africa (Webster, 1963).

The harmful effects of air pollution have been observed at great distances from the pollutant source. An example is the high incidence of total and partial blindness that was found in a herd of bighorn sheep (*Ovis canadensis*) in the San Bernardino Mountains of California. This region has a high concentration of oxidants, which are known to be eye irritants. The source of these oxidants is more than 160 km away in Los Angeles (Light, 1973).

Air pollutants concentrate in the tissue of wildlife. For example, fluoride has been found in high concentrations not only in the bones of deer (Karstad, 1967; Kay *et al.*, 1975; Newman and Yu, 1976), but also of cottontail rabbits (*Sylvilagus floridanus*), hares (*Lepus* sp.), muskrats (*Ondatra zibethica*), ground squirrels (*Citellus columbianus*), woodchuck (*Marmota monax*), deer mice (*Peromyscus* sp.), and voles (*Microtus* sp.) (Gordon, 1969; Karstad, 1970). High concentrations have also been observed in carnivores such as the red fox (*Vulpes fulva*) (Karstad, 1970) and the barn owl (*Tyto alba*) (J. R. Newman, unpublished data, 1977). The effects of these accumulations in wildlife are not generally known.

Some of the most extensive studies on the physiological and ecological effects of air pollution on wildlife have been conducted in Czechoslovakia. Western Czechoslovakia is a region of heavy industry with high emissions of sulfur dioxide, oxides of nitrogen, fly ash, and cement dust. It also has intensively managed game populations. Wild hares (*Lepus europaeus*) have been studied in detail by Nováková and her co-workers at the Institute of Landscape Ecology in Prague. They reported significant changes in the urine of hares and attributed them to air pollution. Hares from areas with high concentrations of sulfur dioxide and fly ash have a more acid urine compared to hares from pollution-free areas. Animals from areas with heavy cement dust have a more basic urine (Nováková, 1970). Changes in the calcium:phosphorus ratio in the blood also occur. In areas of heavy sulfur dioxide and fly ash, the calcium deficiency is close to a hypocalcemic condition. This difference is greatest in young and adult female hares. Levels of phosphorus in the blood are higher in adult hares, especially males (Nováková and Roubal, 1971). Changes in the erythrograms of these hares are associated with changes in calcium and phosphorus concentrations in the blood. Hypoproteinosis, as well as declines in albumins, α-globulins, and β-globulins occurs in hares from areas of heavy air pollution. These changes in erythrograms are comparable to those observed in animals with infections or allergic reactions (Nováková et al., 1973). Changes in the age structure of these populations have also been observed. In regions with sulfur dioxide and fly ash, the ratio of 1-year-old hares to adult hares is 35% below that observed in pollution-free areas. In areas with heavy cement dust, the ratio of 1-year-old hares to adult hares is 35%—greater than in the control areas (Nováková, 1969).

HARMFUL EFFECTS ON WILDLIFE: BIRDS

Reports on the sensitivity of birds to air pollution are common and widespread. In 1971, a heavy fog and high sulfur emissions from a pulp mill in British Columbia, Canada, killed between 200 and 500 songbirds. There were inflammations of the intestinal tract and hemorrhages in the brain and trachea of the birds (Harris, 1971). Cadmium from industrial sources has been found to accumulate in house sparrows in Japan. Nishino et al. (1973) have reported a high mortality in these populations.

Wellings (1970) observed respiratory damage in house sparrows from urban areas in California. General declines in passerine bird populations have been associated with high fluoride emissions from an aluminum plant in Czechoslovakia (Feriáncová-Masárová and Kalivodová, 1965). In London, during the past 70 years, there has been a decline and subsequent return of bird populations to the inner city. This population change has been attributed to the reduction of the once high concentrations of sulfur dioxide and other pollutants in central London (Cramp and Gooders, 1967; Gooders, 1968).

The sensitivity of birds to air pollutants is not unexpected. Birds have high respiration rates and spend most of their time in the air, the transporting medium of such pollutants. The observed sensitivities and the potential use of birds as bioindicators of air pollution led us to investigate in 1976 the relation of various air pollutants to the distribution of certain bird species in Czechoslovakia. We paid particular attention to the house martin, which Feriáncová-Masárová and Kalivodová (1965) and Cramp and Gooders (1967) reported to be adversely affected by different air pollutants. This species is an insectivorous migratory bird that nests on the overhangs of houses in the villages and cities of Europe and Britain. A census of active nests in villages and towns in regions of known pollution showed the house martin to be rare or absent in areas heavily polluted with fluoride, sulfur dioxide, fly ash, cement dust, or nitrogen oxides (Table 2). House martin populations were 60% lower in moderately polluted areas than in control areas, indicating that this migratory species has preferred pollution-free areas over potentially suitable nesting sites in heavily polluted areas. The reasons for this avoidance of high air pollution areas are under investigation. Reduced aerial insect food supply may be a cause.

DISCUSSION

From these studies of wildlife and supporting studies on the effects of air pollution on domestic and laboratory animals (Newman, 1975), five biological indicator groups can be identified:

Sentinels: highly sensitive species found in the environment, whose responses act as early warning signals to a particular air pollutant.

Bioassay Monitors: species with known life histories and known characteristic responses to a given air pollutant.

Detectors: species found in an environment that show a reasonably characteristic response to a given air pollutant.

Thrivers: species whose presence or abundance indicates the occurrence of a particular pollutant or environmental condition associated with that pollutant.

Accumulators: species that collect and accumulate an air pollutant in reasonably large quantities, thereby showing trophic level accumulation.

Wildlife species may be used as one or more of these biological indicators. For example, there have been several biological indicators suggested for fluoride. Bees, which are highly sensitive to fluoride (Marier and Rose, 1971), could act as sentinel species. Game populations, especially managed populations (e.g., deer), whose general health, ecology, and response to a given pollutant are known, would make suitable bioassay monitors. The house martin, which responds not only to fluoride but to other air pollutants as well, is an example of a potential detector species. Deer or other large herbivores would also be suitable detector species for fluoride. Numerous studies (Wentzel and Ohnesorge, 1961; Pfeffer, 1965; Sierpinski, 1972) have shown that infestations of trees by insects are associated with various air pollu-

TABLE 2   Relationship Between Nesting Density of the House Martin (*Delichon urbica*) and Air Pollution

Primary Pollutant	Mean Nest Density, i.e., Number of Nests per Side of Building by Relative Pollutant Concentration		
	Light	Moderate	Heavy
Cement dust	0.39(147)[a]	0.14(210)[b]	0.00(53)[b]
Fluoride	—	0.43(189)	0.01(76)[b]
Sulfur dioxide and fly ash	—	0.32(326)	0.00(125)[b]
Sulfur dioxide and nitrogen oxides	0.52(77)	0.05(214)[b]	0.00(70)[b]
Other	—	—	0.10(42)[b]
Mean nest density for combined pollutants (total sides of buildings sampled)	0.43(224)	0.23(939)[b]	0.01(366)[b]

[a] Numbers in parentheses are sample sizes.
[b] $P < 0.05$, student t-test, when compared to control areas that had a mean of 0.50(615); 0.50 nests per side of building in a sample totaling 615 sides of buildings.

tants. Industrial fluoride contributes to defoliating insect infestations in lodgepole pine (*Pinus Contorta* var. *latifolia*) in Montana (Carlson *et al.*, 1977). These species are examples of the thriver group. Finally, numerous wildlife species (e.g., rodents), which accumulate large amounts of fluoride, can act as accumulator species for fluoride. When used as biological indicators of air pollution, various wildlife species are extremely valuable for several reasons: they reveal the biological changes and ecological trends in the environment due to the pollutant; they indicate pathways and points of accumulation of air pollutants in the environment; they show an integration of many environmental factors, including the effect of pollutants on the total environment; and they correlate physical and chemical measures of pollutants with biological effects. This kind of information enables environmental managers to develop more meaningful environmental assessments and air quality standards.

SUMMARY

This paper reviews the effects of air pollution on wildlife and presents the preliminary results of a recent study of the effects of air pollution on birds in Czechoslovakia. The house martin (*Delichon urbica*) shows differential sensitivity in nesting to various levels of air pollution (sulfur dioxide, nitrogen oxides, particulates, and fluoride). This species may serve as a useful biological indicator of air pollution. Information on the effects of air pollution on game populations (hares, deer, and pheasants) in Czechoslovakia is also presented. Wildlife is a valuable biological indicator of air pollution.

ACKNOWLEDGMENTS

I would like to thank the U.S. Environmental Protection Agency, the U.S. National Academy of Sciences, and the Czechoslovak Academy of Science for their partial support of this study. I wish especially to thank the Institute of Landscape Ecology in Prague, particularly Drs. E. Hadác and E. Nováková, for their hospitality and discussions on biological indicators of air pollution. Finally, I wish to thank Dr. Robert L. Rudd for his influence and encouragement to study wildlife problems. Acknowledgment is also given to the Bureau for Faculty Research, Western Washington State University, for the typing of this paper.

REFERENCES

Carlson, C. E., W. E. Bousfield, and M. D. McGregor. 1977. The relationship of an insect infestation on lodgepole pine to fluorides emitted from a nearby aluminum plant in Montana. Fluoride 10:14–21.

Cramp, S., and J. Gooders. 1967. The return of the house martin. Lond. Bird Rep. 31:93–98.

Dochinger, L. S., and T. A. Seliga, eds. 1976. Proceedings of the First International Symposium on Acid Precipitation and the Forest Ecosystem. Northeast Forest Experiment Station, USDA Forest Service Gen. Tech. Rep. NE-23, Upper Darby, Pa. 1,074 pp.

Feriáncová-Masárová, Z., and E. Kalïvodová. 1965. Nickolko poznamok vplyve fluorovych exhalátov v okoli Hlinikarne v Ziari nad Hronom na kvantitu hniezdiacich vtakov. Biologia (Bratislava) 20:341–346.

Gooders, J. 1968. The swift in central London. Lond. Bird Rep. 32:93–98.

Gordon, C. C. 1969. Cominco American Report II. Department of Environmental Studies, University of Montana, Missoula. 15 pp.

Hais, K., and J. Masek. 1969. Effects of some exhalations on agricultural animals. EPA translation of Ucinky nekterych exhalaci na hospodarska zvirata. Ochr. Övzdusi 3:122–125.

Harris, R. D. 1971. Birds collected (die off) at Prince Rupert, B. C., September 1971. Canadian Wildlife Service, unpublished report, 1 October 1971.

Karstad, L. 1967. Fluorosis in deer (*Odocoileus virginianus*). Bull. Wildl. Dis. Assoc. 3:42–46.

Karstad, L. H. A. 1970. Wildlife in changing environment. Pp. 73–80 in D. E. Elrick, ed. Environmental Change: Focus on Ontario. Simon & Schuster, New York.

Kay, C. E., P. C. Tourangeau, and C. C. Gordon. 1975. Industrial fluorosis in wild mule and whitetail deer from western Montana. Fluoride 8:182–191.

Light, J. T. 1973. Effects of oxidant air pollution on forest ecosystems of the San Bernardino Mountains. Pp. 1–12 in O. C. Taylor, ed. Oxidant Air Pollution Effects on a Western Coniferous Forest Ecosystem, Task B Report, part B. Air Pollution Research Center, University of California, Riverside.

Lillie, R. J. 1970. Air Pollutants Affecting the Performance of Domestic Animals. Agric. Handbk. 380. U.S. Department of Agriculture, Washington, D.C. 109 pp.

Marier, J. R., and D. Rose. 1971. Environmental Fluoride. NRC Publ. No. 12,226. National Research Council of Canada, Ottawa. 32 pp.

Newman, J. R. 1975. Animal Indicators of Air Pollution: A Review and Recommendations. Corvallis Environmental Research Lab. Rep. CERL-006. U.S. Environmental Protection Agency, Corvallis, Oreg. 192 pp.

Newman, J. R., and M. Yu. 1976. Fluorosis in black-tailed deer. J. Wildl. Dis. 12:39–41.

Nishino, O., M. Arari, I. Senda, and K. Kuboto. 1973. Influence of environmental pollution of the sparrow. (Kankyo osen no suzume ni oyobosu eikyo.) Jpn. J. Public Health (Nihon Koshu Eisei Zasshi) 20:1.

Nováková, E. 1969. Influence des pollutions industrielles sur les communautés animales et l'utilisation des animaux comme bioindicateurs. Pp. 41–48 in Air Pollution, Proceedings of the First European Congress on the Influence of Air Pollution on Plants and Animals, April 22–27, 1968, Wageningen. Centre for Agricultural Publishing and Documentation, Wageningen, the Netherlands.

Nováková, E. 1970. Influence of industrial air pollution on urine reaction in horses. Pp. 712–715 in Nutrition, Proceedings of the Eighth International Congress, Prague, 1969. Excerpta Medica, Amsterdam.

Nováková, E., and Z. Roubal. 1971. Taux de calcium et de phosphore dans le sérum sanguin des lièvres exposés aux pollutions de l'air. Pp. 529–536 in Actes du X Congrès Union Internationale des Biologistes du Gibier, Paris, 3–7 mai 1971, L'Office National de la Chasse, Paris.

Nováková, E., A. Finková, and Z. Sová. 1973. Etudes préliminaires des protéines sanguines chez le lièvre commun exposé aux pollution industrielles. Pp. 69–84 in Nemzetközi Vadászati Tudományos Konferencia Elöadásai. Lectures of the International Scientific Hunting Conference. Section II. Apróvadgazdálkodás (small game farming), Sopron, Budapest, 1971.

Pfeffer, A. 1965. The Effect of Air Pollution with SO_2 on the Countryside. Pp. 171–183 in Preprints of the Czechoslovak Reports. International Symposium on the Control and Utilization of SO_2 and Fly Ash from Flue Gases of Large Thermal Power Plants. House of Scientific Workers, Liblice.

Prell, H. 1936. Die Schädigung der Tierwelt durch die Fernwirkungen von Industrieabgasen. Arch. Gewerbepathol. Gewerbehyg. 7:656–670.

Robinette, W. L., D. A. Jones, G. Rogers, and J. S. Gashwiler. 1957. Notes on tooth development and wear for Rocky Mountain mule deer. J. Wildl. Manage. 21:134–152.

Royal Commission. 1878. Noxious vapors. Royal Commission Report 614. Her Majesty's Stationary Office, London. 72 pp.

Sierpinski, Z. 1972. Die Bedeutung der sekundären Kiefernschädlinge in Gebieten chronischer Einwirkung industrieller Luftrerunreinigungen. (The significance of secondary pine parasites in areas of chronic exposure to industrial pollution.) Mitt. Forstl. Bundesversuchsanst. Wien. 97:609–615.

Tendron, G. 1964. Effects of air pollution on animals and plants. No. A87.389 in Council of Europe Report, Vol. 4. European Conference on Air Pollution, June 24 to July 1, 1964, Strasbourg. Council of Europe, Strasbourg.

Webster, I. 1963. Asbestosis in non-experimental animals in South Africa. Nature (London) 197:506.

Wellings, S. R. 1970. Respiratory damage due to atmospheric pollutants in the English sparrow, *Passer domesticus*. In Project Clean Air. Research Project S-25. Department of Pathology, University of California, Davis.

Wentzel, K. F., and B. Ohnesorge. 1961. Zum Auftreten von Schadinsekten bei Luftverunreinigung. (Occurrence of insect pests with air pollution.) Forstarchiv. 32:177–186.

QUESTIONS AND ANSWERS

P. SCANLON: Did you attempt to count the insects in your sampling areas?

J. NEWMAN: No, not yet. The numbers of swallows (*Hirundo rustica*), another species of bird that is closely related to the house martin, did not seem to be affected by air pollution. Both of these species are insectivorous birds.

M. FRIEND: The Fish and Wildlife Service is in the process of banning lead shot for waterfowl hunting. One of the criticisms of this ban is that the lead, found in wing bones, is not due to ingestion of lead shot, but to air pollution emanating from automobile exhausts. Do you have plans for sampling environmental lead in the bird population in Czechoslovakia?

J. NEWMAN: No, but the Institute of Landscape Ecology in Prague is doing a general survey of heavy metals in wildlife populations. At the end of the molting season, they collect the feathers of some species of ducks and analyze them for lead, cadmium, and other heavy metals.

M. FRIEND: In examining a number of American black brant (*Branta bernicula*) from the die-off on the East Coast this past winter, we noticed lead levels that were too high to be background but certainly not indicative of ingested lead shot. The question arose whether this was lead contamination from vegetation eaten by these birds. Automobile exhaust was considered as another possible source of contamination, because the lead levels were above what are considered normal background levels.

H. CASEY: Has anybody studied urban pigeons in this country from the perspective of monitoring environmental pollutants?

J. NEWMAN: A paper by Tansey and Roth[1] suggested the use of pigeons as biological indicators. Pigeons would be suitable monitors because they are sedentary, do not migrate, and are found in areas with high concentrations of various air pollutants.

[1] Tansey, M. F., and R. P. Roth. 1970. Pigeons, a new role in air pollution. J. Air Pollut. Control Assoc. 20:306–307.

P. SCANLON: I have a comment relative to the last question. Ohi et al.[2] used pigeons (*Columba livia*) to monitor lead pollution and indicated their potential value as indicators. Pigeons are city dwellers even though their grain diet originates in a rural, uncontaminated origin. Thus, the only source of lead contamination of pigeons is air pollution or ingested grit, which, in all likelihood, receives lead contamination from air pollution.

J. NEWMAN: I might make one additional comment. We tend to think that air pollution has adverse effects; it does have some beneficial effects and acts like an antiseptic. Nováková[3] and others have found that wild hares and pheasants from regions of heavy air pollution in Czechoslovakia have a reduced incidence, actually reduced to zero, of protozoan and nematode infections. In particular, pheasants, if they were living in regions of high sulfur dioxide ($\sim$300 mg/m^3) and fly ash (500 metric tons/km/yr), did not exhibit gapeworm (*Syngamus trachea*) disease, whereas in control regions this disease is quite common.[4]

D. SCARPELLI: Continued exposure to sulfur dioxide and nitrogen oxides has a deleterious effect on the lungs and produces rather severe diseases of the terminal bronchioles in humans. Could you tell us whether this happens in birds?

J. NEWMAN: I have not looked at this and do not know of any specific information. I think there are some studies in several of the zoos in the United States describing the incidence of lung cancer in ducks.[5]

J. LAST: The smog in Czechoslovakia is probably a reducing smog. Sulfur dioxide and nitrogen oxides would not be very deleterious to the terminal bronchioles. It's the oxidants in American smog that cause the damage.

J. NEWMAN: Yes, Czechoslovakia does not have problems with photochemical oxidants as yet but, because of the increasing number of automobiles in Prague, the city is beginning to have problems with atmospheric oxidants.

[2] Ohi, G., H. Seki, K. Akiyama, and H. Yagyu. 1974. The pigeon: A sensor of lead pollution. Bull. Environ. Contam. Toxicol. 12:92–98.

[3] Nováková, E. 1969. Influence des pollutions industrielles sur les communautés animales et l'utilisation des animaux comme bioindicateurs. Pp. 41–48 in Air Pollution, Proceedings of the First European Congress on the Influence of Air Pollution on Plants and Animals, April 22–27, 1968, Wageningen. Centre for Agricultural Publishing and Documentation, Wageningen, the Netherlands.

[4] Nováková, E., and B. Tremmlová. 1973. Influence de la pollution de l'air sur la syngamosis du faison commun. Union Internationale des Biologistes du Gibier. Actes du X Congrès. Paris, 1971, pp. 389–394.

[5] Snyder, R. L., and H. L. Ratcliffe. 1966. Primary lung cancers in birds and mammals of the Philadelphia Zoo. Cancer Res. 26:514–518.

The Effect of Synthetic Smog on Voluntary Activity of CD-1 Mice

SHERMAN F. STINSON and CLAYTON G. LOOSLI

Because the serious pathological and physiological effects caused by environmental air pollutants have been studied extensively, a comprehensive literature review on the subject is beyond the scope of this report. However, a more subtle type of morbidity—depression of spontaneous activity by gaseous pollutants—has received less thorough attention, although it is of considerable importance, hygienically and economically.

Boche and Quilligan (1960) were among the first to measure a decrease in voluntary activity of laboratory rodents that had been exposed to a mixture of ozonized gasoline vapor and to propose a relationship between the altered activity and air pollution. Subsequent studies have corroborated these findings in mice exposed to various concentrations of ozone and nitrogen dioxide (Murphy, 1964; Murphy et al., 1964), to ambient Los Angeles area smog (Emik and Plata, 1969; Emik et al., 1971), and to automobile exhaust (Hueter et al., 1966). A strong correlation between depressed activity and the concentration of oxidant gases has been established. Observations are by no means unique to laboratory animals; decreased physical performance associated with high concentrations of oxidant air pollutants has been reported in humans as well (Wayne et al., 1967).

The purpose of the study reported below was to define more clearly the individual and synergistic effects on spontaneous activity in mice following exposure to common gaseous pollutants in precisely controlled concentrations similar to those found in first-stage smog alerts.

233

MATERIALS AND METHODS

Young adult CD-1, specific pathogen free, white mice (Charles River Farms) were placed in individual exercise cages similar to those described by Murphy (1964), Murphy *et al.* (1964), and Emik and Plata (1969). The cages consisted of an enclosed wheel 15 cm in diameter and a small auxiliary rest area where the mice obtained fresh water and Purina Lab Chow *ad libitum*. A cam on the axle of each wheel depressed a bidirectional counter switch one time for each revolution. This activated a recorder in an adjacent room. Wheels were completely disassembled, cleaned, and oiled weekly to maintain maximum operational efficiency. In addition, they were checked twice daily. Other maintenance was performed as needed.

Two identical sealed chambers were used for the filtered air and synthetic smog environments (see Loosli *et al.*, 1972, for a detailed description). Each chamber was fitted with activated charcoal and particle filters on the air supply. Synthetic smog consisting of 0.3 ppm ozone, 1 ppm nitrogen dioxide, and 2 ppm sulfur dioxide, or each gas individually, was continuously metered into the exposure chamber, except for 1 to 2 h/day when animals and cages were serviced. Nitrogen dioxide and sulfur dioxide were supplied from pure compressed gas in cylinders. Ozone was generated electrically. Carbon monoxide, not being removed by the filters, was present in both chambers at the ambient air concentration. Concentrations of all of the pollutants were continuously monitored. The chambers were artificially illuminated between 0800 and 1700 daily.

Forty exercise cages and mice were placed in the filtered-air chamber for a period of 2 weeks to allow the animals to become acclimatized to the cages. Twenty of the cages were then transferred to the exposure chamber containing one of the atmospheres previously described. After 2 weeks of exposure, the animals in the filtered-air and exposure chambers were exchanged and were kept in the alternate chambers for another 2-week period. The group in the filtered-air chamber was then removed and replaced by the group from the exposure chamber, which remained there for a final 2 weeks. Each group, therefore, received a 2-week exposure followed by 2 weeks of recovery in filtered air.

The number of revolutions run by each mouse was recorded between the hours of 1700 and 0800 h the following morning. Daytime activity was disregarded due to extreme variability. A daily activity index, defined as the ratio of the number of revolutions run in the exposure chamber to the average number run each day during the last 7 days of the acclimation period (control), was calculated for each animal. The indexes of all the mice in each group were averaged every day for a mean daily activity index.

RESULTS

Activity of mice decreased by over 75% after 1 day in the synthetic smog mixture (Figure 1). The activity remained at this level through the third day, and then progressively increased to approximately 85% of the original activity by the fourteenth day of exposure. Five days following their return to filtered air, activity of the mice was not significantly different from the basal levels.

A similar response was elicited when mice were exposed to an atmosphere containing 0.3 ppm of ozone (Figure 2). An immediate

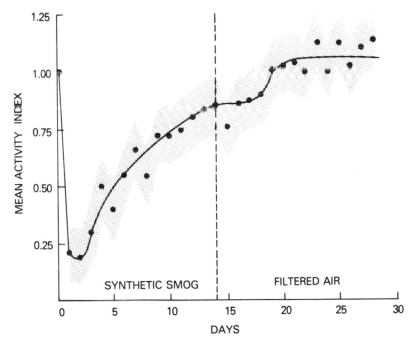

FIGURE 1 Daily mean activity index of mice in a synthetic smog mixture for 14 consecutive days and for another 14 days following their return to filtered air. The shaded area represents the standard error of the mean.

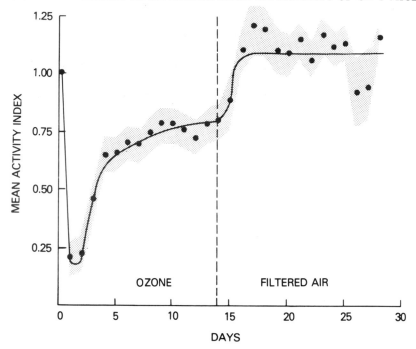

FIGURE 2 Daily mean activity index of mice in 0.3 ppm ozone for 14 consecutive days and for another 14 days following their return to filtered air. The shaded area represents the standard error of the mean.

decrease in activity for the first 3 days was followed by a progressive increase, plateauing at approximately 80% of the original activity after 14 days of exposure. A rapid increase in activity was observed on the first 2 days following the return of the animals to filtered air, but the indexes were not significantly different from the basal levels by the fifth day.

Changes in spontaneous activity were much less marked in mice exposed to 1 ppm of nitrogen dioxide or 2 ppm of sulfur dioxide (Figures 3 and 4). In both atmospheres, activity decreased regularly, reaching a minimum of between 50% and 75% of the original activity after 5 days. The activity was not reduced by the end of the 14-day exposure. Return of the animals to filtered air resulted in a transient, but significant (P < 0.05) increase in the activity of the mice from the nitrogen dioxide atmosphere. No such response was noted in the mice that had been exposed to sulfur dioxide.

DISCUSSION

Continuous exposure of mice to synthetic smog atmospheres containing a mixture of ozone, nitrogen dioxide, and sulfur dioxide resulted in an immediate decrease in spontaneous activity. Ozone alone produced a response that was very similar to that of the synthetic smog mixture, indicating that, with the concentrations of the gases used in this study, ozone had the major influence on depression of activity. This conclusion was reinforced by the demonstration of only moderate relative effects on activity of the nitrogen or sulfur dioxides.

Emik and Plata (1969) and Emik *et al.* (1971) have reached similar conclusions concerning activity of mice breathing ambient polluted air. Running activity was more highly correlated with oxidant gas concentration than any of the other pollutants or conditions that were studied.

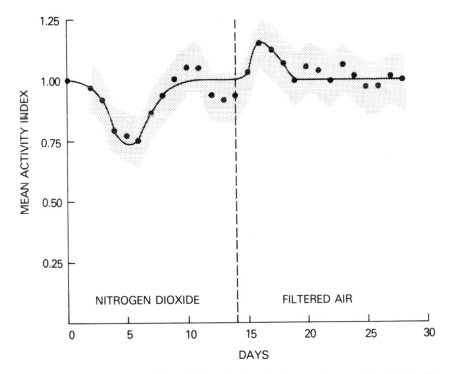

FIGURE 3 Daily mean activity index of mice in 1 ppm nitrogen dioxide for 14 consecutive days and for another 14 days following their return to filtered air. The shaded area represents the standard error of the mean.

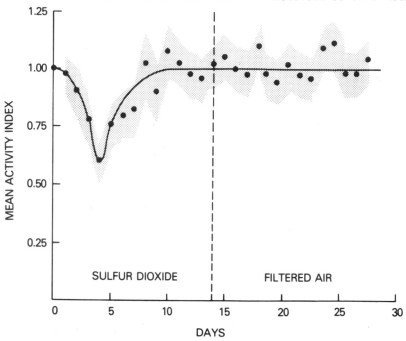

FIGURE 4 Daily mean activity index of mice in 2 ppm sulfur dioxide for 14 days and for another 14 days following their return to filtered air. The shaded area represents the standard error of the mean.

In humans as well, athletic performance was found to be highly correlated with the oxidant level in the atmosphere (Wayne *et al.*, 1967).

Although marked reductions in spontaneous activity were caused by the synthetic smog mixture, as well as by each of the component gases, the mice appeared to develop a tolerance to the pollutants. By the end of the 14-day exposure, activity was near normal in the synthetic smog or ozone, and recovery was complete in the nitrogen and sulfur dioxide atmospheres. Stokinger *et al.* (1956) observed a development of tolerance in rats forced to exercise in atmospheres containing potentially lethal levels of ozone. Tolerance developed within 24 h, but persisted for only 4 to 6 weeks. It is not known whether the observed tolerance to sublethal concentrations of pollutants would persist indefinitely or be a transient effect.

Measurement of changes in spontaneous activity was a useful and sensitive technique for the study of effects of smog constituents.

Although the mice showed no traditional clinical or histopathological indications of toxicity, significant depression of activity was present. This technique should have many applications in studies of environmental pollutants.

SUMMARY

The purpose of this study was to investigate the individual and synergistic effects of common environmental air pollutants on voluntary activity in mice. Two hundred CD-1 mice were placed in recording exercise cages and exposed continuously for 2 weeks to synthetic smog consisting of ozone (0.3 ppm), nitrogen dioxide (1 ppm), and sulfur dioxide (2 ppm), or to each component separately. Voluntary running activity was decreased by 80% after 1 day in synthetic smog, but progressively increased and was 20% depressed by 14 days. Ozone produced a very similar effect, but depression by nitrogen dioxide and sulfur dioxide was minimal. Voluntary activity provided a very sensitive monitor of the pollutants' effects.

ACKNOWLEDGMENTS

This study was supported in part by the Environmental Protection Agency (grants PH 86-68-43 and AP 1075), the Council for Tobacco Research, Inc., the Hughes' Employees' Give Once Club, and the Hastings Foundation Fund of the University of Southern California.

REFERENCES

Boche, R. D., and J. J. Quilligan. 1960. Effect of synthetic smog on spontaneous activity of mice. Science 131:1733–1734.

Emik, L. O., and R. L. Plata. 1969. Depression of running activity in mice by exposure to polluted air. Arch. Environ. Health 18:574–579.

Emik, L. O., R. L. Plata, K. I. Campbell, and G. L. Clarke. 1971. Biological effects of urban air pollution. Arch. Environ. Health 23:335–342.

Hueter, F. G., G. L. Contner, K. A. Busch, and R. G. Hinners. 1966. Biological effects of atmospheres contaminated by auto exhaust. Arch. Environ. Health 12:553–560.

Loosli, C. G., R. D. Buckley, M. S. Hertweck, J. D. Hardy, D. P. Ryan, S. F. Stinson, and R. Serebrin. 1972. Pulmonary response of mice exposed to synthetic smog. Ann. Occup. Hyg. 15:251–260.

Murphy, S. D. 1964. A review of effects on animals of exposure to auto exhaust and some of its components. J. Air Pollut. Control Assoc. 14:303–308.

Murphy, S. D., C. E. Ulrich, S. H. Frankowitz, and C. Xintaras. 1964. Altered function in animals inhaling low concentrations of ozone and nitrogen dioxide. Am. Ind. Hyg. Assoc. J. 25:246–253.

Stokinger, H. E., W. D. Wagner, and P. G. Wright. 1956. Studies of ozone toxicity I. Potentiating effects of exercise and tolerance development. AMA Arch. Ind. Health 14:158–162.

Wayne, W. S., P. F. Wehrle, and R. E. Carroll. 1967. Oxidant air pollution and athletic performance. J. Am. Med. Assoc. 199:151–154.

QUESTIONS AND ANSWERS

L. KOLLER: Were any of these mice examined for periods longer than 14 days?

S. STINSON: The animals do not do well in exercise wheels for longer than 1 to 1.5 months because they become caught in the wheels and injure their tails and other appendages. The prolonged hyperactivity also seems to have deleterious effects on their general health. For these reasons, this procedure does not really lend itself to protracted studies.

G. KOLAJA: Did you study any of the chemical parameters, such as blood glucose and other substances in the blood, of these animals when they were not exercising?

S. STINSON: No, we handled the animals as little as possible so as not to disturb them.

G. KOLAJA: Is it correct to assume that you maintained your animals in a controlled environment?

S. STINSON: Yes, control mice were kept in an environment identical to the exposure chamber, of course without the pollutants, in an effort to ascertain whether any changes would occur in them. There was very little change in the activity of the control group except for a slight but statistically insignificant increase in their activity; an increase of approximately 10% over the 14-day period.

H. CASEY: Did you carry out these experiments with alert levels of smog and did you vary the concentration of smog by increasing and decreasing it from these levels?

S. STINSON: No, these studies were done with pollutants at concentrations that ranged from alert to subalert levels. In studies where higher concentrations of gases were used, illness and death of the mice have given us inconclusive results. Lower levels were not used because it was thought that significant changes in activity would not be observed.

H. CASEY: Has anyone done life-time studies on such animals in this environment?

S. STINSON: Yes, and the animals maintained in such atmospheres did not show decreased survival times. Some pathological changes were observed in the life-time studies but only after very long periods, extending up to 2 years. The changes that were detected were primarily at the ultrastructural level with only minimal pathology observable at the light microscopic level. The ultrastructural changes included cellular proliferation in the bronchiolar and bronchial regions and also some focal fibrosis.

Canine Pulmonary Disease:
A Spontaneous Model
for Environmental Epidemiology

JOHN S. REIF and DANIEL COHEN

In the urban ecosystem humans are viewed as the dominant force, exerting profound influences on the structure and quality of their environment. Included in the biological sphere of this ecosystem are a number of animal populations, both domestic and feral. Human and animal populations coexist in intimate proximity, sharing the support systems of the environment as well as contributing to its contamination. This common utilization of the environment produces a series of mutually dependent relationships for members of the community.

Of prime concern is the concept that animals living in polluted environments may serve as sentinels of hazards to human health. Just as the miner carried a canary with him to warn of insufficient oxygen or impurities in the air, animals may be used to predict and assess the effects of environmental contamination. Animals may be more sensitive indicators of environmental hazards to health than humans themselves, or, at least, these relationships may be more conveniently studied in nonhuman populations.

The role of environmental factors in the production of chronic pulmonary disease in humans remains a field of intensive investigation despite many attempts to define their importance. Air pollution, cigarette smoking, and occupational exposure have been implicated in the production of chronic bronchitis, emphysema, and lung cancer (Ferris and Anderson, 1962). Attempts to quantitate the role of environmental pollutants are hampered by the geographical mobility of humans, by the effects of confounding variables such as cigarette smoke, and by the length of time elapsed between exposure and manifestation of disease.

We used chronic canine pulmonary disease as a comparative epidemiologic model, and radiographic examination of the chest as a screening device to determine the prevalence of pulmonary abnormalities (Zeidberg et al., 1967). The validity of radiographic examination for this purpose was measured by radiographic pathologic correlation. Sensitivity and specificity were comparable to that found with other commonly used screening procedures (Reif et al., 1970).

A radiographic pattern characteristic of chronic respiratory disease in the dog has been described by Reif and Rhodes (1966). This pattern, frequently seen in the lungs of older dogs, consists of a generalized increase in the number of fine linear and reticular markings, thickening and increased density of bronchial walls, pleural thickening, and nodular densities. Histologically, these changes were found to be associated with varying degrees of chronic inflammation of the tracheobronchial tree and pulmonary parenchyma. Focal interstitial fibrosis and scarring, chronic interstitial inflammation, focal destruction of alveolar walls, heterotopic bone formation, and pleural fibrosis were noted. Chronic bronchial and bronchiolar inflammation with accumulations of alveolar macrophages, inflammatory cells, and proliferating alveolar septal cells in peribronchial alveoli were consistent findings. Moderate accumulations of black pigment (anthracosis) were occasionally found in pulmonary lymphatics.

An earlier paper (Reif and Cohen, 1970) reported attempts to determine whether a relationship between chronic nonspecific canine pulmonary disease and the urban environment could be demonstrated. Chest radiographs made on 1,007 canine subjects at the University of Pennsylvania Veterinary Hospital in Philadelphia between 1965 and 1968 were reviewed. A relationship between the prevalence of chronic respiratory disease and urban residence was detected for dogs aged 7 to 9 and 10 to 12, but not for younger or older age strata.

In the study described below, the methodology used in Philadelphia has been applied to two other animal hospital populations; one located in Boston, Mass., the other in Ithaca, N.Y. The purpose of this work was to determine whether the associations detected in Philadelphia could be demonstrated in other communities.

METHODS AND MATERIALS

The canine populations studied were patients at the University of Pennsylvania Veterinary Hospital in Philadelphia and at the Cornell University Veterinary Hospital in Ithaca between 1965 and 1968, and at the Angell Memorial Animal Hospital in Boston between January 1968 and May 1970. The methods used to read and grade radiographs were similar at all three institutions and have been described by Reif and Cohen (1970).

Briefly, the radiologists' reports on all canine patients examined during the time frame were reviewed. The investigators eliminated animals suffering from obvious chest disease (pleural effusion, thoracic neoplasia, congestive heart failure, bronchopneumonia, dirofilariasis, etc.) that would obscure or complicate classification. The remaining radiographs were then graded, with respect to changes consistent with those of chronic pulmonary disease, without knowledge of the animals' age or residence. Three grades were assigned: "absent," "moderate," and "severe." Previous studies demonstrated that the absent and severe categories were clearly distinguishable and that errors in classification between "moderate" and "severe" occurred on less than 20% of interpretations (Reif *et al.*, 1970). The results of grading and the corresponding identification numbers were entered on a code sheet.

After the grading, the patient's demographic characteristics, including age, sex, breed, institution, year, residence, and reason for radiography, were recorded.

Environment

Ithaca   The patients at the Cornell University Veterinary Hospital came from the agricultural Finger Lakes region around Ithaca and from the college community itself. The area is essentially free of the manufacturing and industrial operations usually associated with urban air pollution. In 1970, the population of Ithaca was approximately 20,000. This community was considered the baseline for comparisons in the analysis.

Boston   Boston lies in a flat coastal basin surrounded by a semicircle of hills to the south, west, and north, and by Massachusetts Bay on the east. Because of the proximity of the ocean, there is a sea breeze with a diurnal variation in direction that promotes dispersion of pollutants.

The communities comprising metropolitan Boston, and from which the population at the Angell Memorial Animal Hospital is derived, were divided into high and low pollution segments according to Heimann (1970). The division was based on the measurement of total suspended particulates (TSP) and sulfur dioxide. The mean TSP in the high pollution zone was 89.5 $\mu g/m^3$; in the low pollution zone it was 53.5 $\mu g/m^3$. The data on community air pollution were furnished by 20 sampling stations of the Massachusetts Air Pollution Control District that were measuring TSP, sulfur dioxide, and soiling index on a daily basis.

Philadelphia   Philadelphia lies in the Delaware River Valley, a shallow river basin that runs in a southwesterly direction. The city experiences relatively frequent periods of temperature inversion and low wind velocity that create systems of trapped air allowing the concentrations of pollutants to rise.

The city was divided into zones of heavy and light air pollution according to data furnished from continuous air monitoring stations operated by the Philadelphia Department of Health. Zip codes were used to code residence within the city. The distribution of TSP and the sulfate index approximated each other. The zones of highest concentration were found nearest the Delaware River paralleling the concentration of heavy industry. Although data for air pollutants outside Philadelphia were largely unavailable, the majority of the industries of the types found in the high pollution zone within the city (chemical, petroleum, ore-processing, and electrical generating stations) are similarly concentrated along the Delaware River above and below the city, permitting a classification of communities in metropolitan Philadelphia.

RESULTS

A total of 1,994 chest radiographs of dogs living in Ithaca, Philadelphia, and Boston were examined. In all three communities, there was a general trend of increasing prevalence of chronic pulmonary disease (CPD) with increasing age.

In Philadelphia, there were differences in the prevalence of CPD between dogs living in heavy pollution zones and those living in a less polluted environment. Significant differences were measured only in dogs aged 7 to 9 and 10 to 12. None were detected in younger or older age strata (Table 1). Significant breed and sex differences were not detected. The majority of dogs (73%) were radiographed as part of a routine clinical workup and had no clinical signs of chest disease.

In Boston, significant differences in the prevalence of CPD were not detected between dogs living in the heavy and light pollution zones. A relative risk of 2 or greater was calculated for the 10- to 12- and the ≥13-year age-groups (Table 1). The failure to detect intracity differences in CPD in Boston, as in Philadelphia, may reflect a lower level of atmospheric pollution in Boston and a smaller disparity in air quality between the heavy and light zones.

TABLE 1 Proportion of Dogs with Radiographic Evidence of CPD, by Environment, in Boston and Philadelphia from 1965 to 1968

| | Pollution Level | | | | |
| | Lightest Pollution | | Heaviest Pollution | | |
Age, yr	No. at Risk	% Dogs Showing CPD	No. at Risk	% Dogs Showing CPD	Relative Risk
Boston					
0–3	127	16.5	147	19.7	1.2
4–6	95	25.3	51	31.4	1.4
7–9	110	36.4	65	29.2	0.7
10–12	59	35.6	50	52.0	2.0
≥13	28	28.6	40	50.0	2.5
TOTAL	419		353		
Philadelphia					
0–3	105	9.5	136	11.8	1.3
4–6	94	20.2	87	26.4	1.4
7–9	106	19.8	150	43.3	3.1[a]
10–12	94	31.9	156	51.9	2.3[b]
≥13	26	46.2	53	47.2	1.0
TOTAL	425		582		

[a] P ≤ 0.001.
[b] P ≤ 0.01.

In both cities, there was a tendency for purebred dogs to reside in low pollution "rural" areas and for mixed breed dogs to live in the more industrialized and heavier polluted areas. In Boston and Philadelphia, 84% of the dogs living in low pollution areas and 65% of those living in heavily polluted areas were purebred (P < 0.001). In Ithaca, 79% of the dogs were purebred. This observation raises an important issue: namely, that observed differences in the prevalence of CPD may be associated with socioeconomic differences. The patterns of ownership of purebred dogs and place of residence reflect socioeconomic factors. More affluent families tend to live in less industrialized areas and own purebred dogs, while poorer families are more likely to live in heavily polluted sections of the city and own a mongrel. This fact has plagued other investigators who have attempted to associate air pollution and respiratory disease in humans (Ferris and Whitten-

TABLE 2 Proportion of Dogs with Radiographic Evidence of CPD and Relative Risk, by Environment

Age, yr	Proportion of Dogs, %, and Relative Risk, R	Ithaca (N = 321)	Boston, Light Pollution (N = 419)	Philadelphia, Light Pollution (N = 425)	Boston, Heavy Pollution (N = 353)	Philadelphia, Heavy Pollution (N = 582)
0–3	%	11.5	16.5	9.5	19.7	11.8
	R	1.0	1.5	0.8	1.9	1.0
4–6	%	21.2	25.3	20.2	31.4	26.4
	R	1.0	1.3	0.9	1.7	1.3
7–9	%	33.7	36.4	19.8	29.2	43.3
	R	1.0	1.1	0.5	0.8	1.5
10–12	%	29.4	35.6	31.9	52.0	51.9
	R	1.0	1.3	1.1	2.6[a]	2.6[b]
≥13	%	37.9	28.6	46.2	50.0	47.2
	R	1.0	0.7	1.4	1.6	1.5

[a] P ≤ 0.01.
[b] P ≤ 0.001.

246

berger, 1966). While at first glance such distinctions might seem irrelevant, careful analysis shows that animals living in the industrialized city may receive a suboptimal level of medical care.

Intercommunity comparisons for the prevalence of CPD were analyzed by the chi-square test. Relative risks were calculated on age-stratified crude rates (Table 2). Ithaca was used as a baseline for comparison because of its relatively uncontaminated atmospheric environment.

In dogs from birth to 3 years of age, the prevalence of CPD was between 10% and 20% (mean = 14.3%); from 4 to 6 years of age, it was between 20% and 31% (mean = 24.5%). No significant differences were detected between the urban communities and Ithaca. In dogs aged 7 to 9 years the prevalence of CPD was 20% to 43% (mean = 33.7%). The low pollution area of Philadelphia was unusually free of CPD. In dogs aged 10 to 12 years, the range in prevalence of CPD was 29% to 52% (mean = 41.2%). The heavy pollution zones of both Philadelphia and Boston had higher CPD prevalence than Ithaca. In addition, the heavy pollution area of Philadelphia was higher than the light pollution areas of Philadelphia ($P < 0.01$) and Boston ($P < 0.05$), and the heavy pollution area of Boston was higher than the light pollution area of Philadelphia ($P < 0.05$).

In the oldest age-group (≥ 13 years), the prevalence of CPD ranged between 29% and 50% (mean = 43.2%), and significant differences were not found between communities.

DISCUSSION

The results of this study support and expand findings in earlier studies (Reif and Cohen, 1970). First, differences in the prevalence of CPD between areas of heavy and light air pollution are not detected in dogs less than 7 years of age. This may be the latent period required for the cumulative effect of environmental hazards to manifest itself in intercommunity comparisons.

Second, a significant intracity gradient in CPD exists in Philadelphia dogs aged 7 to 9 years and 10 to 12 years. Third, 10- to 12-year-old dogs in Philadelphia and in Boston had significantly higher CPD prevalence than Ithaca dogs of the same age. This finding shows that the urban–rural gradient is not limited to a unique situation that exists in Philadelphia, but is demonstrable in other communities that exist under divergent air quality conditions.

Finally, as noted previously for Philadelphia, intercommunity differences in CPD were not detected for the oldest age-group. This may be a manifestation of differential mortality, thus minimizing differences that would have been observed had the most severely affected dogs survived.

The data support the contention that an "urban factor" is partially responsible for the production of chronic pulmonary disease. Despite socioeconomic differences in the population samples and the origin of the dogs examined, the likelihood that air pollutants may influence the production of certain pulmonary disease appears to be strengthened by epidemiological evidence. The dog, intimately sharing his master's environment, may be a good indicator for some of the undesirable side effects of contemporary urbanization. More sophisticated studies of the mechanisms involved in the production of pulmonary changes by atmospheric pollutants should further establish the dog as a model in comparative environmental health research.

SUMMARY

Chest radiographs of 1,892 dogs living in Ithaca, in Boston, and in Philadelphia were reviewed to determine whether a relationship between chronic canine pulmonary disease and the urban environment could be demonstrated. Radiographs were graded for evidence of pulmonary changes without knowledge of the animal's age or residence. Philadelphia and Boston were divided into urban and rural segments based on available atmospheric pollution data and the concentration of industrialization. A distinct relationship between increasing age and pulmonary changes was evident. A rural to urban gradient in the prevalence of pulmonary disease was detected in older age-groups.

REFERENCES

Ferris, B. G., and D. O. Anderson. 1962. The prevalence of chronic respiratory disease in a New Hampshire town. Am. Rev. Resp. Dis. 86:165–185.

Ferris, B. G., and J. L. Whittenberger. 1966. Environmental hazards. Effects of community air pollution on prevalence of respiratory disease. N. Engl. J. Med. 275:1413–1419.

Heimann, H. 1970. Episodic air pollution in metropolitan Boston. A trial epidemiological study. Arch. Environ. Health 20:230–251.

Reif, J. S., and D. Cohen. 1970. II. Retrospective radiographic analysis of pulmonary disease in rural and urban dogs. Arch. Environ. Health 20:684–689.

Reif, J. S., and W. H. Rhodes. 1966. The lungs of aged dogs: A radiographic-morphologic correlation. J. Am. Vet. Radiol. Soc. 7:5–11.

Reif, J. S., W. H. Rhodes, and D. Cohen. 1970. Canine pulmonary disease and the urban environment. I. The validity of radiographic examination for estimating the prevalence of pulmonary disease. Arch. Environ. Health 20:676–683.

Zeidberg, L. D., R. J. Horton, and E. Landau. 1967. The Nashville air pollution study. Arch. Environ. Health 15:214–224.

QUESTIONS AND ANSWERS

D. KELLY: Is there any information on the morphological background of the radiographic changes that you have assessed in these animals? For example, are some of these changes associated with excessive mucus gland secretion and peribronchial fibrosis?

J. REIF: I do not have that information for individual animals. However, we have attempted to create a composite of the pathologic changes associated with the three grades of radiographic disease I have described. In the most severely affected dogs, those having chronic pulmonary disease, morphological changes in the lung consisted of interstitial pulmonary scarring, bronchial calcification, and some chronic peribronchial and peribronchiolar inflammation. The degree of mucus gland hyperplasia has not been quantitated by the Reid index. The degree of alveolar wall destruction, which would parallel human emphysema, has also not been quantitated by any morphometric studies. However, there is some loss of alveolar wall structure and some replacement with connective tissue in these areas in the lung.

Those of you who have studied canine lungs have seen these types of lesions I have just described. It is difficult to ascribe all of these changes to environmental pollutants since some of these animals are undoubtedly suffering from an infectious disease such as bronchitis due to a bacterial infection.

B. ZOOK: Some dogs live almost all of their lives inside an apartment. Was there any attempt made to separate indoor dogs from outdoor ones?

J. REIF: There was not. Your question is certainly a cogent one since the effects of air conditioning and of cigarette smoke in the household are factors that have to be taken into account. In retrospect, the data presented in a 1971 Surgeon General's report[1] on cigarette smoking suggests that such effects should be considered in future studies on canine pulmonary disease. The assumption we made originally was that, since the dog was not the primary inhaler of cigarette smoke, the presence of cigarette smoke in the environment could be discounted. I believe now that the secondary inhalation of cigarette smoke by the dog is a factor that we should keep in mind. We do not have a residence history on these dogs and this certainly would be an interesting bit of data to have.

J. LAST: If the current view that atmospheric pollutants do indeed cause chronic pulmonary disease is correct, one would anticipate that cities, such as Boston or Philadelphia, with largely reducing smogs should predispose dogs to develop chronic bronchitis rather than alveolar emphysema and fibrosis. It might be interesting if you selected the cases of bronchitis out of your total study to look at this particular point.

[1] U.S. Department of Health, Education, and Welfare. 1971. The Health Consequences of Smoking. Washington, D.C.

J. REIF: Yes, this would be an interesting question to ask, but I do not know if we can do this, since we have only looked at the total distribution of chronic pulmonary disease as characterized by a particular set of radiographic patterns. In this way it is somewhat similar to a pneumoconiosis index that would be done on a population of coal miners.

D. SCARPELLI: Some years ago, Shabad of the Soviet Union[2] reported at a meeting on the high incidence of carcinoma of the lung in city dogs as compared to rural ones. Has that been substantiated by subsequent studies in the United States or other countries?

J. REIF: No, other than the Russian report to which you referred, Dr. Scarpelli, there have not been any other studies supporting an urban–rural gradient for pulmonary neoplasia in the dog. In some independent work related to this project, we investigated the question of an urban–rural gradient in canine lung cancer. In 1967, Ragland and Gorham[3] reported that tonsillar carcinoma in the dog was a relatively rare event in the animal hospital at Washington State University. When it was seen, it was almost exclusively in dogs that had been referred from larger industrialized areas like Los Angeles or Seattle. We attempted to look at the residence history of the dogs at the hospital over a 15-year period to determine whether there was any association between urban residence and lung cancer (Reif and Cohen, 1971).[4] For a control group, we selected animals with gastrointestinal neoplasia that were subject to the same referral biases as the lung cancer group. There were no differences detected in the origins of the dogs with lung cancer as compared to those with gastrointestinal cancer. However, there was a difference in the origins of the dogs with tonsillar carcinoma. These dogs came predominantly from urban environments. It may well be that canine tonsillar carcinoma (squamous cell carcinoma of the tonsil) is a sensitive indicator of carcinogenic pollutants and their effects on the respiratory system in this species.

D. HASTINGS: Dogs with a gene defect reflected in low alpha$_1$-antitrypsin inhibitor levels could be ideal animals to serve as monitors of environmental air pollutants. Have you ever looked at alpha$_1$-antitrypsin inhibitor levels in these dogs?

J. REIF: Such mutants would make excellent models, but I have not studied these levels.

[2] Shabad, L. M. 1965. Pp. 165–180 in Lung Tumors in Animals. Proceedings of the 3rd Quadrennial Conference on Cancer. University of Perugia, Italy.

[3] Ragland, W. L., and J. R. Gorham. 1967. Tonsillar carcinoma in rural dogs. Nature 214:925–926.

[4] Reif, J. S., and D. Cohen. 1971. The environmental distribution of canine respiratory tract neoplasms. Arch. Environ. Health 22:136–140.

FLUORIDE

Fluoride Toxicosis in Wild
Ungulates of the Western United States[1]

J. L. SHUPE, H. B. PETERSON, and A. E. OLSON

Fluorine rarely occurs in its free state in nature. It combines chemically to form fluorides. Fluorides are widely distributed in nature and various amounts are universally present in soil, water, the atmosphere, vegetation, and animal tissue (National Academy of Sciences, 1971). The terms "fluorine" and "fluoride" are often used interchangeably in the literature in referring to fluoride compounds or to the element fluorine (F).

Animals routinely ingest small amounts of fluoride with no adverse effects (National Academy of Sciences, 1974; Shupe, 1967). Small amounts of fluoride may be beneficial, but greater amounts induce fluoride toxicosis (McClure, 1970; Roholm, 1937; Schmidt et al., 1954; Shupe, 1972; Shupe and Alther, 1966; Shupe and Olson, 1971; WHO, 1970). Wild as well as domestic animals are susceptible to potential undesirable effects during prolonged ingestion of elevated levels of fluorides (Karstad, 1967; Newman and Yu, 1976; Shupe et al., 1972). The most common sources of excessive fluoride intake by animals are forages contaminated by airborne fluorides emitted in excess quantities by nearby industrial operations; water with a naturally high fluoride content; forages contaminated by soils with high concentrations of fluoride; feed supplements and mineral mixtures containing too much fluoride; or any combination of these. Figure 1 illustrates possible routes in which animals encounter fluorides.

[1] Published with the approval of the Utah State University Agriculture Experiment Station, Logan, Utah, as Journal Paper No. 2203.

253

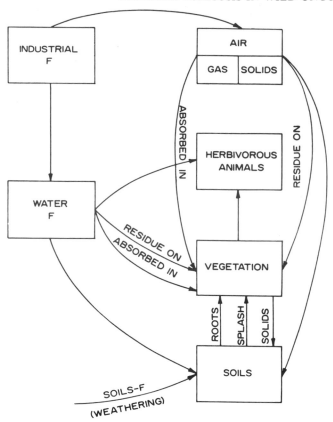

FIGURE 1 Sources and routes of fluorides that affect wildlife.

In some industrial operations, ores of moderate to high fluoride content are ground and/or heated during processing. Unless control mechanisms are installed and functioning properly, some of the fluorides may escape into the environment as hydrogen fluoride and particulate matter, which can contaminate vegetation eaten by animals.

Some waters, most often those from geothermal sources, contain higher than tolerable concentrations of fluorides (Neely and Harbaugh, 1954; Rand and Schmidt, 1952). The current impetus to locate, develop, and utilize geothermal energy sources will very likely increase water-associated environmental fluoride hazards that previously existed only with certain springs and geysers. Data in Table 1 indicate that geothermal waters of the western United States tend to have a high

TABLE 1   Fluoride in Geothermal Spring and Well
Waters of Western United States

Location	Ranges of Fluoride Concentration, mg/l	Frequency in Range of Fluoride Content, mg/l		
		0–2	2.1–10	10.1–30
California	0.1–12	31	34	2
Idaho	0.1–30	35	44	45
Nevada	0.1–19	6	30	9
Oregon	0.8–21	7	19	6
Utah	0.3–14	22	23	2
Wyoming	2.0–10	2	4	1
TOTAL SAMPLES		103	154	65

fluoride content. A total of 68% of the samples contained more than 2.1 mg/l of fluoride.

Some soils throughout the world have a high natural fluoride content (Hobbs and Merriman, 1962; Merriman and Hobbs, 1962). Vegetation growing on these soils may be contaminated by windblown or rain-splashed soil. In such cases, most of the fluorides remain on the surface of the vegetation and are not incorporated within it. However, some plant species are able to translocate appreciable amounts of fluoride from the soil. This ability varies with the plant and soil conditions. Wild and domestic animals have access to such sources of fluoride.

Excessive levels of fluoride in some mineral mixes used as livestock feed supplements have been incriminated in some cases of fluoride toxicosis in domestic animals. These mixes would not normally be available to wildlife.

INCIDENCE OF TOXICOSIS IN WILDLIFE

In the geothermal areas of Yellowstone Park, buffalo, elk, and deer have evidenced symptoms and various degrees of fluoride toxicosis. In these instances, it is assumed that the fluoride came primarily from the water. Additionally, however, when forage is scarce during the winter, the animals cluster around the warm springs where the snow is melted. Not only do the animals drink the water but they may also ingest fluorides by grazing on the vegetation. Some of the waters in the park contain high concentrations of fluoride (Table 2).

TABLE 2   Fluoride Content of Waters in the
Yellowstone Park Area

Water Source	Concentration of Fluoride, mg/l[a]
Upper Basin–Yellowstone Park (near Old Faithful)	28, 27, 25
Yellowstone River	1.2, 0.8
Hayden Valley–Hot Springs	25, 22, 18
Mammoth Hot Springs	26, 25, 22
Gardner River	0.3
Flagg Ranch Hot Springs	10
Norris Basin	4.5
Madison River	0.5, 0.4

[a] Two or three concentrations given for one source reflect results obtained from different samples.

Several cases of fluoride toxicosis in deer, elk, and moose have been identified in areas of industrial pollution. Some vegetation collected in such an area, where deer exhibited symptoms of fluoride damage, had high fluoride contents (Table 3). Because the animals had access to many possible sources of good water, it is assumed that the high concentrations of ingested fluoride came from the vegetation.

Most current fluoride toxicosis problems are attributable to some form of environmental pollution. Historically, fluoride sources have included volcano eruptions (an early environmental pollutant), hot water springs or geysers, and effluents from phosphate deposits (Velu, 1932). Rarely were these substantial enough to endanger wildlife. However, with the expansion of industrialization into some wildlife habitats, fluoride toxicosis has been diagnosed and will likely become more prevalent.

DISTRIBUTION OF INGESTED FLUORIDES

Ingested fluorides are absorbed primarily in the stomach and intestines. Normally only trace or insignificant amounts of fluorides are absorbed through the respiratory system. Total diet composition greatly affects fluoride absorption in the gastrointestinal tract. Large amounts of calcium, aluminum, and magnesium ions decrease the amount of fluoride absorbed by forming relatively insoluble fluoride complexes (Shupe *et al.*, 1962).

Most fluoride accumulated by the body is incorporated in the bones and developing teeth. Only very small amounts of fluoride are found in soft tissues even when dietary intake is high (Shupe *et al.*, 1963; Suttie *et al.*, 1958). Approximately 96% of the fluoride in an animal body is in the bones. The accumulation process can continue throughout life if the animal ingests a constant or increasing amount of fluoride. The major mode of fluoride excretion is in the urine.

FLUORIDE TOXICOSIS

Excessive fluoride ingestion can induce either an acute or chronic response. The acute disease is relatively rare and most often results from accidental ingestion of large amounts of toxic compounds, such as sodium fluorosilicate or sodium fluoroacetate (used as rodenticides); sodium fluoride (an ascaricide in swine); or water, vegetation, and feeds containing extremely high concentrations of fluoride.

Chronic toxicosis is the condition most commonly seen in wildlife. The condition usually develops gradually and insidiously. In chronic fluoride toxicosis, symptoms and lesions are often not evident until several months after excessive ingestion has occurred. This delay further complicates an accurate diagnosis and proper evaluation. Confirmation and evaluation of clinical diagnosis of fluoride toxicosis require substantiation through necropsy findings and chemical analyses of tissues.

Fluoride toxicosis in both domestic and wild animals can be influenced by the amount of fluoride ingested; the duration of fluoride

TABLE 3 Fluoride in Vegetation from a Deer Study Area

Plant	Fluoride Content, ppm
Western wheatgrass (*Agropyron spicatum*)	60.2
Mullein (*Verbascum thapsus*)	48.0
Sunflower (*Helianthus annuus*)	127, 50, 76.7
Salt grass (*Distichlis strica*)	82.1
Cheatgrass (*Bromus tectorium*)	69.2
Mean of all plants over 2-year period	47.4
Cache Valley vegetation (control—low fluoride area)	8.0

ingestion; fluctuations in fluoride intake with time (often seasonal); solubility of fluoride (toxicity usually increases with solubility); species of animal involved; age of animal at time of fluoride ingestion; general level of nutrition (malnutrition intensifies toxicity); stress factors; and individual biological response.

Although the tolerance levels shown in Table 4 are for domestic animals, they are indicative of tolerance levels for related wildlife species.

FLUORIDE TOXICOSIS LESIONS

Teeth

Fluoride dental lesions are induced if animals ingest excessive soluble fluoride while their teeth are forming and mineralizing. The characteristic lesions are visible in the teeth after they erupt (Brown *et al.*, 1960; Garlick, 1955). The nature and severity of fluoride-induced dental lesions can be correlated with the level of fluoride ingestion (Greenwood *et al.*, 1964). Continuously erupting teeth (as those of rodents and lagomorphs) similarly reflect the level of the animal's fluoride intake at the time of formation and mineralization. Specific lesions involve chalkiness (dull white chalklike appearance); mottling (white, chalklike horizontal striations or patches); hypoplasia (defective development); and hypocalcification (defective calcification). The severity of these lesions may vary from tooth to tooth depending on the

TABLE 4   Fluoride Tolerance Levels in Feed and
Water of Domestic Animals[a]

	Fluoride Tolerance	
Animal	Feed, ppm	Water, mg/l
Heifers, dairy and beef	30	2.5–4
Beef cattle, mature	50	4–8
Breeding ewes	60	5–8
Horses	60	4–8

[a] The values must be reduced proportionally when both water and feed contain appreciable amounts of fluorides. The average ambient air temperature and the physical and biological activity of the animals influence the amount of water consumed and, hence, the wide range of tolerance levels suggested. For active animals in a warm climate the lower values should be used as critical-level indicators.

fluoride ingested during the formation and mineralization of the particular teeth. Teeth evidencing moderate to severe degrees of effects are subject to increased attrition and, in some cases, to an erosion of the enamel from the dentine. Examples of incisor lesions in several species are shown in Figure 2.

Incisor Teeth

Fluoride-induced dental lesions of incisor teeth can be classified into the following numerical categories according to degree or severity: 0,

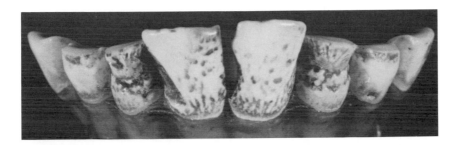
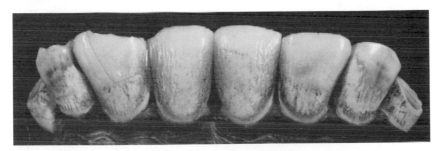
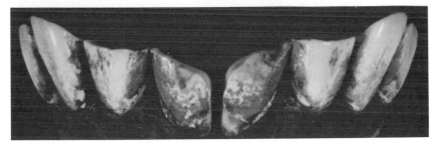

FIGURE 2 Dental fluorosis in permanent incisor teeth: a, mule deer; b, bison (buffalo); c, wapiti (elk).

normal; 1, questionable; 2, slight; 3, moderate; 4, marked; and 5, severe effects. Examples of these categories in permanent incisor teeth of buffalo are shown in Figure 3.

Cheek Teeth

Molar and premolar fluorosis are classified in terms of degree of abrasion alone because of various examination and interpretation problems. These characteristic lesions can be correlated with the incisor lesions. Typical abnormal and selective wear patterns are shown in Figure 4.

Bone

In animals on a normal, low-level fluoride intake, bone fluoride content will increase slightly throughout life. This gradual increase occurs without any demonstrable changes in bone structure or function. If, however, fluoride intake exceeds normal levels for any appreciable period, bone changes will occur. The exact nature and expression of these changes will vary according to the nine factors listed previously as influencing fluoride toxicosis. Osteofluorotic changes in bones have been classified as osteosclerosis, osteoporosis, hyperostosis, os-

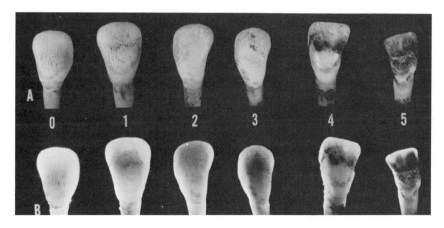

FIGURE 3 Clinical standards used in diagnosing dental fluorosis. Teeth shown are permanent bison incisors: A, photographed with front lighting; B, photographed with back (transmitted) lighting. 0, normal; 1, questionable; 2, slight effect; 3, moderate effect; 4, marked effect; and 5, severe effect.

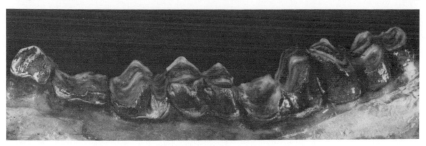

FIGURE 4 Dental fluorosis in elk mandibular premolar and molar teeth. Note uneven wear due to excessive selective abrasion. These are from same animal as incisors in Figure 2-C. Note correlations of lesions between incisors and cheek teeth.

teophytosis, and osteomalacia (Johnson, 1965). Their incidence depends upon the interacting factors influencing the degree of fluoride toxicosis.

Slightly elevated, long-term intake of soluble fluorides will result in hardened or sclerotic bone. Extremely high, long-term ingestion of soluble fluorides results in excessive periosteal proliferation and porous and malacic bone.

The severity of fluoride-induced bone lesions will vary with the bone and with parts of the same bone according to its structure and function and the stress placed on the bone.

Bones that are heavily involved in locomotion, chewing, or breathing are more affected than those that are primarily protective in function. Areas of tendon or muscle attachment are more involved than are adjacent nonattachment areas. Metabolically active bone, especially near growth areas, will show greater effects and much higher accumulations of fluoride than will less active areas. This phenomenon must certainly be borne in mind when bones are being sampled to identify pathologic changes and analyzed for fluoride content. Samples must be taken consistently from the same bone areas to facilitate accurate interpretations and evaluations.

Grossly, bones that are severely affected by fluoride appear chalky white with a roughened irregular periosteal surface. They are larger in diameter and heavier than normal (Figure 5). In advanced cases, osteofluorotic lesions are bilateral and clinically palpable. Fluorotic bone lesions are not primarily associated with articular surfaces.

Characteristic histological changes are associated with the various degrees of osteofluorosis (Figure 6). These have been described and illustrated by Johnson (1965) and Shupe and Alther (1966). Some of the

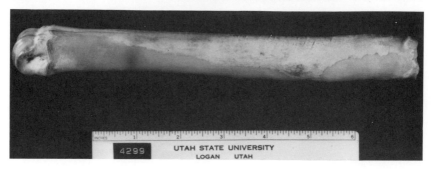

FIGURE 5 Osteofluorosis in a mule deer metatarsal bone. Note lighter colored area of abnormal, irregular periosteal hyperostosis.

gross and histological changes induced in bones by fluoride toxicosis may resemble lesions and alterations associated with other bone diseases. Therefore, bone abnormalities must be correlated carefully with other lesions and symptoms before making a definite diagnosis of fluoride toxicosis.

Intermittent lameness or stiffness is often seen in advanced cases of fluoride toxicosis. The exact causes of these problems are not fully understood, but they appear to be associated with mineralization of periarticular structures and tendon insertions.

Soft Tissues

Most assimilated fluoride is either stored in an animal's bones and teeth, or excreted in the urine. Only small amounts of fluoride are found in the soft tissues. This small quantity does not induce any characteristic structural changes in any of the soft tissues or organs.

Reproduction

Fluorides appear to have no direct or primary effects on reproduction in animals. Some secondary effects such as general unthriftiness or an alteration of systemic functions may, however, indirectly affect reproduction.

General Condition

A generalized unthriftiness characterized by weight loss, rough dry hair coat, thick nonpliable skin, and lethargy is sometimes seen in animals suffering from severe fluoride toxicosis. These symptoms are not unique to fluoride toxicosis and are only apparent after more specific symptoms and lesions are observed.

DISCUSSION

The above findings relate to all herbivores whether domestic or wild. The basic data on fluoride toxicosis have been substantiated regarding toxic levels, timing of insult, and other general concepts using domestic laboratory animals. However, enough data have been collected from several wild species so that the accumulated information can be extrapolated to those species.

Diagnosticians evaluating fluoride toxicosis in wild animals should thoroughly examine all available specimens. When considering teeth, they should remember that, while permanent dentition presents the most readily detectable signs of fluoride toxicosis, the dental effects will be apparent *only* if excessive fluorides were ingested during the period of tooth formation and mineralization. Severe cases of fluoride toxicosis have been correctly diagnosed in animals with normal dentition. These animals had ingested excessive fluorides later in life.

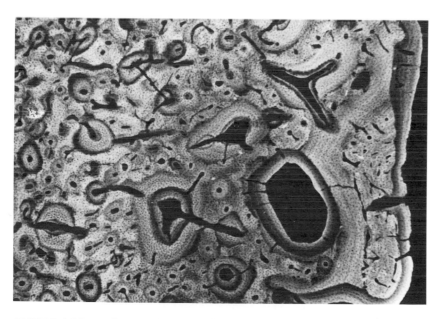

FIGURE 6 Microradiograph of fluoride-affected mule deer metatarsal bone. (×33.) Note excessive porosity and irregular, poorly organized bone structure.

Clinical findings should be confirmed by chemical analyses, radiographs, and histopathological evaluation—particularly of bone.

All possible sources of fluorides should be analyzed. These include vegetation, water, soil, and any mineral sources that might be ingested by the animal as a dietary supplement. Because fluorides are cumulative in chronic cases of fluoride toxicosis, the possible contribution of all sources to total long-term fluoride intake should be evaluated.

The final evidence for the diagnosis and evaluation for fluoride toxicosis depends on the clinical symptoms and lesions in the animals. Based on our knowledge of life habits and anatomical features of wildlife herbivores, one could assume that their tolerance levels for fluoride probably are somewhat similar to domestic livestock.

SUMMARY

Fluoride toxicosis has been diagnosed in many species of animals. There is an optimal level of fluoride ingestion, and safety margins have been established for animals. Sources of fluoride include industrial emissions that contaminate vegetation and water, some natural water supplies, and soil and mineral supplements with high fluoride content. The multiple sources and combinations thereof complicate the identification of sources in some cases. Primary sites of fluoride effects are developing teeth and bones. No characteristic or distinctive effects on soft tissues have been seen. Characteristic dental lesions will develop if the optimal ingestion level is greatly exceeded during the formation of permanent teeth. Excessive fluoride ingestion will induce abnormal bone changes at any age.

REFERENCES

Brown, W. A. B., P. V. Christofferson, M. Massler, and M. B. Weiss. 1960. Postnatal tooth development in cattle. Am. J. Vet. Res. 21:7–34.

Garlick, N. L. 1955. The teeth of the ox in clinical diagnosis. IV. Dental fluorosis. Am. J. Vet. Res. 16:38–44.

Greenwood, D. A., J. L. Shupe, G. E. Stoddard, L. E. Harris, H. M. Nielsen, and L. E. Olson. 1964. Fluorosis in cattle. Utah State University Agricultural Experiment Station Special Report 17. Utah State University, Logan. 36 pp.

Hobbs, C. S., and G. M. Merriman. 1962. Fluorosis in beef cattle. Tennessee Agricultural Experiment Station Bulletin No. 351. University of Tennessee, Knoxville. 183 pp.

Johnson, L. C. 1965. Histogenesis and mechanisms in the development of osteofluorosis. Pp. 424–441 in J. H. Simons, ed. Fluorine Chemistry, Vol. 4. Academic Press, New York.

Karstad, L. 1967. Fluorosis in deer (*Odocoileus virginianus*). Bull. Wildl. Dis. Assoc. 3:42–46.

McClure, F. J. 1970. Water fluoridation. Pp. 29–35 in The Search and the Victory. National Institutes of Health, Public Health Service, Department of Health, Education, and Welfare, Washington, D.C.

Merriman, G. M., and C. S. Hobbs. 1962. Bovine fluorosis from soil and water sources. Tennessee Agricultural Experiment Station Bulletin No. 347. University of Tennessee, Knoxville. 46 pp.

National Academy of Sciences, Committee on Biologic Effects of Atmospheric Pollutants. 1971. Fluorides. National Academy of Sciences, Washington, D.C. 295 pp.

National Academy of Sciences, Committee on Animal Nutrition, Subcommittee on Fluorosis. 1974. Effects of Fluorides in Animals. National Academy of Sciences, Washington, D.C. 70 pp.

Neeley, K. L., and F. G. Harbaugh. 1954. Effects of fluoride ingestion on a herd of dairy cattle in the Lubbock, Texas, area. J. Am. Vet. Med. Assoc. 124:344–350.

Newman, J. R., and M. H. Yu. 1976. Fluorosis in black tailed deer. J. Wildl. Dis. 12:39–41.

Rand, W. E., and H. J. Schmidt. 1952. The effect upon cattle of Arizona water of high fluorine content. Am. J. Vet. Res. 13:50–61.

Roholm, K. 1937. Fluorine intoxication: A clinical–hygienic study with a review of the literature and some experimental investigations. H. K. Lewis & Co., Ltd., London. 364 pp.

Schmidt, H. J., G. W. Newell, and W. E. Rand. 1954. The controlled feeding of fluorine, as sodium fluoride, to dairy cattle. Am. J. Vet. Res. 15:232–239.

Shupe, J. L. 1967. Diagnosis of fluorosis in cattle. Pp. 15–30 in IVth International Meeting of the World Association for Buiatrics, 4–9 August 1966. Publ. No. 4. World Association for Buiatrics, Zurich.

Shupe, J. L. 1972. Clinical and pathological effects of fluoride toxicity in animals. Pp. 357–388 in Carbon–Fluorine Compounds; Chemistry, Biochemistry and Biological Activities. CIBA Foundation. Associated Scientific Publishers, Amsterdam.

Shupe, J. L., and E. W. Alther. 1966. The effects of fluorides on livestock, with particular reference to cattle. Pp. 307–354 in O. Eichler, A. Faran, H. Herken, A. D. Welch, and F. A. Smith, eds. Handbook of Experimental Pharmacology, Vol. 20, Pt. 1. Springer-Verlag, New York.

Shupe, J. L., and A. E. Olson. 1971. Clinical aspects of fluorosis in horses. J. Am. Vet. Med. Assoc. 158:167–174.

Shupe, J. L., M. L. Miner, L. E. Harris, and D. A. Greenwood. 1962. Relative effects of feeding hay atmospherically contaminated by fluoride residue, normal hay plus calcium fluoride, and normal hay plus sodium fluoride to dairy heifers. Am. J. Vet. Res. 23:777–787.

Shupe, J. L., M. L. Miner, D. A. Greenwood, L. E. Harris, and G. E. Stoddard. 1963. The effect of fluorine on dairy cattle. II. Clinical and pathological effects. Am. J. Vet. Res. 24:964–984.

Shupe, J. L., A. E. Olson, and R. P. Sharma. 1972. Fluoride toxicity in domestic and wild animals. Clin. Toxicol. 5:195–213.

Suttie, J. W., P. H. Phillips, and R. F. Miller. 1958. Studies of the effects of dietary sodium fluoride on dairy cows. III. Skeletal and soft-tissue fluorine deposition and fluorine toxicosis. J. Nutr. 65:293–304.

Velu, J. 1932. Le darmous (ou dermes) fluorose spontanée des zones phosphatées. Arch. Inst. Pasteur d'Algerie 10:41–118.

World Health Organization (WHO). 1970. Fluorides and Human Health. World Health Organization, Geneva.

QUESTIONS AND ANSWERS

J. NEWMAN: Some game managers believe fluorosis is only a cosmetic problem in wildlife. Fluorosis in cattle is associated with lameness. Is there any information that fluorosis affects the survivability of wildlife?

J. SHUPE: I cannot speak directly to your question about the possible number of animals that may appear on a given range at a given time. We have seen various degrees of fluoride toxicosis in deer. It was not as severe in southern Germany and northern Switzerland along the Rhine River at Mohlin and Rheinfelden as we have seen in various places in the United States. It is extremely difficult to see adequately and observe properly lameness and some of the symptoms of fluoride toxicosis in wildlife that we customarily encounter in domestic animals. I would anticipate lesions of varying degrees of severity in wild animals according to species and habitat differences. Suffice it to say that we have seen bone changes characteristic of chronic fluoride toxicosis in wildlife that are very similar to those observed in domestic animals.

H. CASEY: What happens to rodents suffering from chronic fluorosis since they have continual tooth eruption or teeth that continue to grow?

J. SHUPE: In those species you see dental lesions characteristic of fluorosis that reflect the amount of fluoride intake at the time that portion of the tooth was forming and mineralizing. In the large ungulates that we have studied, characteristic fluoride-induced lesions do not occur once the enamel has been formed by the ameloblasts and the permanent teeth have erupted. It is important to point out that such anatomical differences do occur among species and that these differences are reflected in the varying responses to chronic fluoride toxicosis that have been observed.

In bone, the type of lesions that develop depend not only on the amount of fluoride that is assimilated but also on the duration of exposure and several other factors that are known to influence the degree of fluoride toxicosis in animals. For example, more severe bone lesions develop in young animals than in older ones ingesting the same level of fluoride. It is extremely important to standardize bone sampling procedures for fluoride analysis, not only by taking samples from specific bones, but also by taking samples from specific areas of the specific bones, because they differ in structure and function and thus differ in fluoride content.

PESTICIDE
TOXICITY

The Effects of Organochlorines on Reproduction of British Sparrowhawks (*Accipiter nisus*)

J. A. BOGAN and I. NEWTON

Bird-eating raptors are more highly contaminated by organochlorines than other raptors. The British sparrowhawk (*Accipiter nisus*) is one of the best-suited avian species for detailed study of the effects of organochlorine compounds on breeding. Of the other most common British raptors, the peregrine (*Falco peregrinus*) and merlin (*Falco columbarius*) are not sufficiently numerous or accessible for a detailed study. Like other birds of prey, the sparrowhawk showed a marked decline in both breeding success and population about the time that organochlorine compounds came into widespread use (Prestt, 1965).

This study was designed to investigate various parameters of the sparrowhawk's breeding performance and to examine these in relation to the organochlorine content of their eggs.

MATERIALS AND METHODS

Detailed information was collected from 315 clutches obtained during 1971 to 1974 from several areas in Scotland and Northern England. Only fresh eggs were obtained from some clutches; addled (unsound, unproductive) and unhatched eggs were also taken from others.

The principal organochlorines found, in order of concentration in eggs, were dichlorodiphenylethane (DDE), polychlorinated biphenyls (PCB's), and dieldrin (HEOD). Their concentrations were determined by standard analytical methods (Newton and Bogan, 1974) on the basis of μg/g of lipid in the eggs. Shell index was determined by the method of Ratcliffe (1970). The shell index is calculated by dividing eggshell weight (mg) by the product of the length (mm) and breadth (mm).

The field methods used were those described by Newton (1976). They entailed frequent visits to each nest during the breeding season. Full data were obtained from the majority of nests.

These data were analyzed with Fortran and Genstat computer programs.

RESULTS

Sparrowhawks normally lay three to six eggs. Twenty full clutches were obtained: 15 fresh clutches from deserted nests and 5 addled but incubated clutches. When the organochlorine content and shell indices of these eggs were expressed as percentages of clutch means, the extent of variation within clutches was derived from standard deviations: DDE, ± 12 (P < 0.05 = 76–124); PCB, ± 20 (P < 0.05 = 60–140); HEOD, ± 19 (P < 0.05 = 62–138); shell index, ± 6 (P < 0.05 = 88–112). There was greater variation in PCB and HEOD within clutches, but all values were greatly influenced by a few eggs that differed markedly from the rest of their clutch. This was not due to faulty chemical analysis. The lipid in any egg is derived not only from triglyceride mobilized from adipose tissue, but also from triglyceride from recent food intake. Because of the varied contamination of different prey, a single egg in the wild may be less representative of a clutch mean or the parents' body burden than an egg in many experimental situations where there is a more uniformly contaminated diet.

In five clutches, where the sequence of laying was known, both organochlorine levels and shell indices fluctuated irregularly (Table 1). This is in contrast to the findings of Lincer (1972). In his experiments, successive eggs appeared to have increasing amounts of or-

TABLE 1   Total Organochlorine Concentrations in Lipid (and Shell Indices) in Eggs of Five Clutches

| Clutch Number | Organochlorine Concentrations, ppm, in Eggs, in Known Laying Sequence | | | | |
	1	2	3	4	5
1	167(1.31)	140(1.29)	141(1.46)	—	—
2	184(1.06)	172(1.12)	224(0.95)	—	—
3	198(1.33)	188(1.24)	168(1.20)	217(1.22)	—
4	215(1.12)	210(1.11)	233(1.11)	264(1.05)	—
5	169(1.34)	186(1.37)	215(1.33)	156(1.33)	186(1.32)

ganochlorine. Nevertheless, an analysis of variance on more than one egg from each of 61 clutches showed the "within-clutch" variance to be small relative to that of "between clutches" (Table 2). This allowed us to interpret the organochlorine contents and shell indices as representative of their clutches.

These analyses were restricted to undeveloped eggs. Comparison of eggs from the same clutch at different stages of incubation showed that the concentration of organochlorine in lipid did not increase markedly until late in development (Table 3). Therefore, we used data from eggs with embryos up to half-grown in our further analyses. Where more than one egg was obtained from any clutch, mean values were used.

Eggs were obtained from seven known (ringed) females in more than 1 year. The data were insufficient for analysis of variance but differed greatly from year to year (Table 4), with no consistent trends in either organochlorine content or shell index.

Sparrowhawks nest in the same, well-defined territories each year. Their nests are usually new but built adjacent to an old nest. It was surprising to us that the influence of territory (Table 5) was unimportant in relation to other factors. The "within territory" variance was almost as large as the "between territory" variance.

TABLE 2 Analyses of Variance in 61 Clutches from Which at Least Two Undeveloped Eggs, a Total of 185, Were Obtained

	Source of Variation	Degrees of Freedom	Sum of Squares	Mean Squares	Variance Ratio
DDE	Between clutches	60	884,451.07	14,740.85	18.83[a]
	Within clutches	124	97,068.93	782.81	
	Totals	184	981,519.99		
PCB	Between clutches	60	902,725.21	15,045.42	13.73[a]
	Within clutches	124	135,883.02	1,095.83	
	Totals	184	1,038,608.22		
HEOD	Between clutches	60	59,041.45	984.02	35.68[a]
	Within clutches	124	3,419.62	27.58	
	Totals	184	62,461.07		
Shell index	Between clutches	53	1.34	0.03	2.253[b]
	Within clutches	111	1.25	0.01	
	Totals	164	2.59		

[a] $P < 0.001$.

[b] $P < 0.01$.

TABLE 3 Organochlorine Levels and Shell Indices in Eggs from 33 Clutches in Which More Than One Stage of Development Was Represented[a]

Total Embryonic Development, %	Number of Eggs	DDE		PCB		HEOD		Shell Index	
		Mean	Variance	Mean	Variance	Mean	Variance	Mean	Variance
Zygote–25	16	1.03	0.039	1.06	0.099	1.11	0.080	0.99	0.017
25–50	6	1.17	0.014	1.31	0.139	1.28	0.062	0.90	0.008
50–75	11	2.27	0.922	2.17	0.879	1.73	0.246	1.02	0.036
75–100	5	2.33	0.285	2.18	0.156	1.91	0.425	0.98	0.019

[a] For ease of comparison organochlorine levels and shell indices are expressed as a ratio to those in unfertilized and addled eggs from the same clutch in which there was no development.

TABLE 4 Total Organochlorine Contents (and Shell Indices) of Eggs from Seven Known Females in Four Different Years

Female Number	Organochlorine Concentration (and Shell Indices), ppm			
	1971	1972	1973	1974
In same territory				
1	185(1.09)	313(0.87)	—	—
2	174([a])	—	198(1.24)	—
3	—	—	400(1.16)	209(1.21)
4	—	—	158(1.39)	169([a])
In different territories				
5	—	186(1.16)	157(1.10)	—
6	—	—	634(1.08)	221(0.92)
7	317(1.19)	—	158(1.39)	—

[a] Not measured.

TABLE 5 Analysis of Variance in Results from 48 Territories in Which Undeveloped Eggs (112) Were Obtained in More Than One Year

Compound	Location	Degrees of Freedom	Sum of Squares	Mean Squares	Variance Ratios
DDE	Between territories	47	301,474.36	6,414.35	1.40 (NS)[a]
	Within territories	64	293,420.63	4,584.70	
	Totals	111	594,894.99		
PCB	Between territories	47	84,229.25	1,792.11	1.47 (NS)
	Within territories	64	77,996.56	1,218.70	
	Totals	111	162,225.81		
HEOD	Between territories	47	16,290.60	346.61	1.54 (NS)
	Within territories	64	14,376.46	224.63	
	Totals	111	30,667.06		
Shell index	Between territories	55	0.98	0.02	1.51 (NS)
	Within territories	72	0.85	0.01	
	Totals	127	1.83		

[a] NS = not significant.

In all eggs, the mean ratio of DDE:PCB:HEOD was 1:0.53:0.19, and the concentration of each compound in each egg was correlated with the others (Table 6).

The distribution of organochlorines within our sample was not statistically normal. In all further tests, $\log_{10}$ values were used to produce a log-normal distribution.

The shell index was inversely correlated with DDE only, as has been demonstrated for many species in both wild and experimental situations, but not with PCB and HEOD.

In this study, shell index $= K - 0.15$ log DDE, where K varied between 1.41 and 1.55 for nine smaller subareas in the sample. This equation is consistent with that found earlier (Newton and Bogan, 1974) in the same study area for fewer eggs. After allowing for DDE, no relationship between shell index and PCB or HEOD was found.

When other aspects of breeding performance were considered, egg breakage was related to shell index. Clutches were classified on a none/some/all-broken basis and also, therefore, to DDE content (Table 7).

Addled eggs had watery contents and no visible embryonic development. Clutches were again classified as none/some/all-addled. Egg addling was correlated with DDE and even more so with PCB.

Egg hatching in a clutch on the basis of none/some/all-hatched was also correlated with DDE, PCB, and shell index. This was expected, since both egg breakage and egg addling affected hatchability. The only other relationship found was that yearling females laid eggs with less organochlorine than older birds.

All of these correlations were observed without any allowances being made for the other organochlorines in the analysis. When this was done, the only relationships that remained significant were those

TABLE 6   Extent of Correlation (r Values) Among Concentrations of Different Organochlorine Compounds in Eggs

	DDE	PCB	HEOD
DDE	1.000		
PCB	0.563[a]	1.000	
HEOD	0.320[a]	0.255[b]	1.000

Degrees of freedom = 271.
[a] P < 0.001.
[b] P < 0.01.

TABLE 7   Significance of Relationships of DDE, PCB, HEOD, and Shell Index with Aspects of Breeding Performance Without Allowing for Other Organochlorines. Based on Analyses of Variance in Which for Each Organochlorine Any Effects of Years or Areas Were Allowed for

	DDE	PCB	HEOD	Shell Index
Eggs broken	P < 0.01	NS[a]	NS	P < 0.01
Eggs addled	P < 0.05	P < 0.01	NS	P < 0.01
Eggs with dead embryos	NS	NS	NS	NS
Eggs deserted	NS	NS	NS	NS
Eggs hatched	P < 0.01	P < 0.05	NS	P < 0.01
Nestling survival	NS	NS	NS	NS
Laying date	NS	NS	NS	NS
Clutch size	NS	NS	NS	NS
Age of female	P < 0.01	P < 0.01	P < 0.05	NS

[a] NS = not significant.

of DDE and shell index with egg breakage. The significance of the other relationships disappeared, probably because the different organochlorines were so well correlated with each other. It is not necessarily more valid to allow for other organochlorines since synergistic or additive effects could thereby be missed. There are other major factors that may also affect breeding. Failure to lay eggs after having built a nest was the greatest cause of breeding failure, accounting for 43% of all failures, whereas egg breakage accounted for only 31% (Newton and Bogan, 1974). Organochlorines might have been responsible for this aberrant behavior (failure to lay) by delaying the birds' development into a breeding condition. It was interesting, therefore, that the date on which the first egg was laid was not related to organochlorine content. This failure to lay might still, of course, in some way be related to organochlorine content.

Organochlorines could also reduce the sparrowhawk population by killing birds directly. Like other species, sparrowhawks are most at risk from organochlorines at times of food shortage. We investigated the distribution of organochlorines in the organs of 18 sparrowhawks found dead and which, from their body weight, represented the extremes of body condition found in this species. As adipose tissue is utilized, the concentration of DDE and other organochlorines in other body compartments increases. However, we found that brain retains its lipid content much better than other tissues. The brain lipid content in the five birds with the lowest body lipid content (<1.5%) was not significantly different from that in those with the highest (>4.0%) (8.32% versus 8.36%). Thus, a starving bird is doubly at risk in that the brain, which is the target organ for organochlorines, not only received increased organochlorines released from adipose tissue, but also accumulated more than its "share" of this released organochlorine because of its relatively high lipid content (Figure 1).

SUMMARY

From 1971 to 1974 the significance of the relationship of DDE, PCB, and HEOD with different aspects of breeding performance was followed in 315 clutches from British sparrowhawks in the wild. When variance and regressions were analyzed to assess the influence of any one organochlorine, allowances were made for possible effects of other organochlorines, of areas, or of year. Investigators examined shell index; number of eggs broken; eggs with dead-in-shell embryos; deserted, hatched, and addled eggs; nestling survival; date first egg was laid; the size of clutch; and the age of the female. Results indicate a

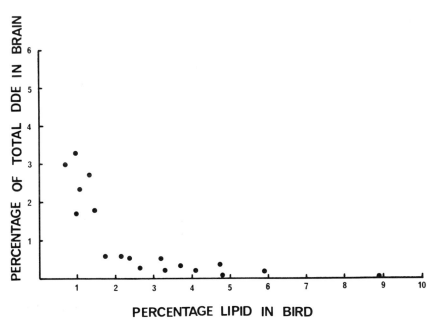

FIGURE 1 Percentage of total DDE in sparrowhawks located in brain with different amounts of total body lipid.

correlation, without allowances for other organochlorines in the analysis, for the three different compounds. The only other relationship was that yearling females laid eggs with less organochlorine than older birds.

REFERENCES

Lincer, J. L. 1972. The effects of organochlorines in the American kestrel. Ph.D. thesis. Cornell University, Ithaca, N.Y.

Newton, I. 1976. Breeding of sparrowhawks (*Accipiter nisus*) in different environments. J. Anim. Ecol. 45:831–849.

Newton, I., and J. A. Bogan. 1974. Relationships between organochlorine residues, eggshell thinning and hatching success in British sparrowhawks. Nature (London) 249:582–583.

Prestt, I. 1965. An enquiry into the recent breeding status of some of the smaller birds of prey and crows in Britain. Bird Study 12:196–221.

Ratcliffe, D. A. 1970. Changes attributable to pesticides in egg breakage frequency and eggshell thickness in some British birds. J. Appl. Ecol. 7:67–115.

QUESTIONS AND ANSWERS

P. WEIS: Dr. Bogan, when new nestlings died did you analyze them for residues of organochlorine compounds? The reason I ask this is that you showed data indicating that the concentration of the pesticides in the yolk approximately doubled in the latter half of development, and that apparently the embryo or the nestlings were not assimilating the pesticide selectively.

J. BOGAN: At our current state of knowledge it is difficult to detemine whether a developing embryo would be subjected to a greater concentration of organochlorine compounds. The amount of organochlorine in each egg is a constant and from our study there is no evidence of its further metabolism.

P. WEIS: How would you account for the doubling of the concentration of organochlorine in the lipid?

J. BOGAN: As the embryo develops, it uses the lipid to form body tissues, and the concentration of organochlorine in the lipid increases simply because the lipid is being used up. There is no increase in the total organochlorine content in the yolk.

P. WEIS: Wouldn't the last bit of yolk that is being taken up by the chick contain very high levels of organochlorine compounds, since their concentration in the yolk doubles towards hatching?

J. BOGAN: Yes, but there is more chick to absorb and distribute this burden of organochlorine compounds.

P. WEIS: I'm having a problem understanding the dynamics of the metabolism of pesticide residue during the last days of incubation, since it is highly concentrated by that time.

J. BOGAN: The last small bit of yolk will have a very high concentration of organochlorine, but the fact that the embryo is large and can distribute this through much greater amounts of tissue is probably the reason that there isn't more risk at this stage of development.

J. ZINKL: What about the migration pattern of British sparrowhawks? Do they migrate to countries that are still using organochlorine pesticides, and could this account for the fact that you did not observe a decline in the organochlorine levels in these birds?

J. BOGAN: No, British sparrowhawks stay close to their nesting sites. The great majority of ringing recoveries are within 50 km of their original nests. There are a few recoveries of birds that have migrated to other European countries, but there is certainly no migration to the contaminated Third World. I do not believe that migration patterns can explain our failure to detect decreases in organochlorine residues in these birds. Our findings are in contrast to a number of studies in other countries. Indeed, studies in the United States have shown that pesticide residues have declined. Our results would suggest that either there have been gross underestimates of the environmental half-life of organochlorine compounds or else that there is still a substantial input into the British environment from some source.

J. ZINKL: C. Henny at the University of Oregon at Corvallis is studying two groups of peregrines (*Falco peregrinus*)—those that migrate to South America and those that remain in North America—to determine if there is a difference in DDT residues in these birds. The results of his studies, which should be available in a year or so, should determine whether DDT contamination originating from South America is present in some groups of peregrines and Swainson's hawks (*Buteo swainsoni*), which migrate to South America.

J. BOGAN: The amount of DDT currently used in the world is difficult to estimate but it appears that the total amount has actually increased due to its increased usage in Africa and South America.

G. FOX: We have approximately 8 years of continuous data on a population of merlin (*Falco columbarius*) in Western Canada. The merlin represent a resident population that migrates as far south as Colorado and then returns to the prairies of Saskatchewan for breeding. The residues we encounter in this species appear to have increased slightly in recent years. Merlins feed on horned larks (*Eremophila alpestris*) and other insectivorous birds with which they migrate in winter. Since we have not applied organochlorines of any type on the prairies of North America over the past 5 years, it would appear that the prolonged half-life of these compounds is responsible for the increased level. In our most recent peregrine survey the birds, which summer in the Arctic and winter in South America, showed a marked increase in the levels of organochlorine pesticides over the levels of peregrines nesting in Colorado. This difference between populations indicates that the South American migrants are importing organochlorines.

S. SLEIGHT: Are British sparrowhawks feeding on migratory birds coming from other areas?

J. BOGAN: No, they eat small thrushes and finches that are local residents. They eat few migratory birds.

A. deFREITAS: The concentration of stable residues such as organochlorine compounds will undoubtedly increase as a bird becomes ill, stops feeding, and loses body weight. How confident are you that the organochlorine residues represented poisonous levels? Did you examine these birds for any pathological changes?

J. BOGAN: We based our judgment simply on the levels of organochlorine compounds present in the birds and considered dieldrin as the causative agent. We considered levels greater than 6 ppm in the brain as possibly fatal in the birds. We did not see these birds *in extremis,* and we have no evidence that they definitely died as a direct result of the organochlorine levels.

D. JONES: Could you expand your observation that addled eggs underwent early embryonic death due to PCB's? A number of studies suggest that avian embryos die late in development; in fact, many die while pipping out of the egg.

J. BOGAN: We've tried to relate the concentrations of PCB's in the egg with each reproductive parameter, but we have done this only in a statistical sense. There was no evidence that developed embryos were dying in the shell once they developed beyond an early stage.

D. JONES: Can these embryos metabolize PCB residues to the hydroxylated metabolites?

J. BOGAN: This would be very difficult to show. With the exception of dieldrin, there is no evidence that any metabolism of organochlorine compounds occurs in early embryos. Even in the case of dieldrin, only small amounts are metabolized. We haven't looked for hydroxylated PCB's.

D. JONES: I ask this because hydroxylated PCB's are approximately 6 times more toxic than the parent compounds.

J. BOGAN: Yes, that is correct. However, the degree of metabolism would have to be fairly extensive, say greater than 10%, to increase significantly the toxicity of the PCB's. I think we would have noticed this extent of metabolism had it occurred.

Herring Gulls (*Larus argentatus*) as Monitors of Contamination in the Great Lakes

ANDREW P. GILMAN, DAVID B. PEAKALL,
DOUGLAS J. HALLETT, GLEN A. FOX,
and ROSS J. NORSTROM

Environmentalists have had to seek out more effective methods of evaluating the extent and effects of the increasingly complex global contamination. One approach to environmental monitoring is the use of indicator species. Although the earth presently sustains some 3 million species in a diversity of ecosystems, the basic biochemical mechanisms supporting earth's fauna have fundamental similarities. Man can only hope to study in detail a few members of most ecosystems; hence, an indicator species must be selected with care and the limitations of its usefulness must be well documented.

The herring gull (*Larus argentatus*) has many points in its favor as an indicator of contamination of the Great Lakes of North America. First, it feeds at the highest trophic level of both aquatic and terrestrial food chains. Although fish are the single most important food item of its diet, this gull consumes insects, earthworms, crustaceans, small birds, mammals, and a large variety of decaying organic debris. Its ability to bioaccumulate organochlorines was shown in one of the earliest studies on dichlorodiphenyltrichloroethane (DDT) and its metabolites, which was carried out in Lake Michigan during the mid 1960's (Hickey *et al.*, 1966). While the levels of a given organochlorine compound are virtually undetectable in water, i.e., in the parts per trillion range, they can be detected readily in the same ecosystem in herring gull eggs, where they are in the parts per million range.

A second factor that makes the herring gull a useful monitor species is its year-round residence in the Great Lakes (Moore, 1976). In general, there is little movement of these gulls from lake to lake, although Lake Superior gulls have been shown by banding recoveries to overwinter in Lake Michigan (Gilman *et al.*, 1977). Within each lake the herring gulls are wide-ranging, and the organochlorine concentrations in their eggs vary little from site to site within a given lake (Gilman *et al.*, 1977; Norstrom, personal communication, 1977). On a whole-lake basis, the herring gull appears to be an integrator of pollution, largely from aquatic food chains.

Third, the herring gull nests colonially. Thus, its entire breeding population in the Great Lakes can be counted at one time. Colonial birds are probably the only type of organism for which this information can be readily obtained. Studies of their reproductive success, behavior, and levels of contamination can all be assessed more easily. Our monitoring has been based on analyses of herring gull eggs rather than on the tissues of the gulls themselves. Eggs are more convenient to collect, reflect adult contaminant levels, and can be replaced by the bird during the breeding season, thus reducing the effect of sampling on reproduction.

A fourth advantage of the herring gull as a monitor species is its wide Holarctic distribution. This distribution enables investigators to compare contaminant levels, reproductive rates, and behavioral characteristics between Great Lakes gulls and those from coastal and European populations. Additionally, the large amount of scientific literature available on the herring gull and other larids provides a base for both historical and interspecific comparison.

Thus, the herring gull meets all the requirements of an indicator species laid down a decade ago by Moore (1966) except that gulls cannot be aged readily after their fourth year.

During 1974 and 1975 we determined the levels of some more common organochlorine compounds and mercury in herring gull eggs from several colonies in the Great Lakes and one colony in New Brunswick (Table 1). Levels in the Great Lakes eggs were at least an order of magnitude higher than those found in eggs from New Brunswick and those reported by Bjerk and Holt (1971) off the coast of Norway. Herring gull eggs collected from the Baltic (Jørgensen and Kraul, 1974) were as contaminated as those from the Great Lakes. Within the Great Lakes a rather complex pattern of egg residue levels was found. In all eggs the quantitatively dominant residues were the polychlorinated biphenyls (PCB's). The highest levels of these industrial pollutants were found in Lake Ontario followed by Lake Michigan. The levels of PCB's in eggs from Lake Superior were high considering the low PCB content of fish in this lake. Presumably, much of the contamination occurred while the gulls were wintering in Lake Michigan. The levels of

TABLE 1 Concentrations of Contaminants in Eggs of Herring Gulls Collected in 1974 and 1975[a]

Contaminant	Median and Ranges of Contaminant Concentrations, ppm Wet Weight, by Location					
	Lake Ontario N=39	Lake Erie N=42	Lake Huron N=40	Lake Superior N=39	Lake Michigan[b] N=10	New Brunswick[c] N=5
DDE	22.60 (8.8–35.1)	7.04 (3.8–14.3)	13.80 (5.4–41.9)	18.60 (8.6–47.1)	31.80 (15.8–145)	1.59 (0.85–2.17)
DDD	0.08 (trace–0.24)	0.09 (trace–0.83)	0.10 (trace–0.38)	0.15 (trace–0.04)	trace (trace–0.07)	trace
DDT	0.09 (0.02–1.04)	0.04 (0.01–0.15)	0.08 (0.01–0.32)	0.12 (0.02–0.58)	0.13 (0.07–0.39)	0.01 (trace–0.03)
Dieldrin	0.37 (0.08–1.08)	0.30 (0.10–0.69)	0.41 (0.13–0.87)	0.39 (0.13–1.35)	0.48 (0.3–0.92)	0.05 (0.03–0.06)
Heptachlor epoxide	0.12 (0.01–0.36)	0.14 (0.04–0.28)	0.12 (0.04–0.26)	0.14 (0.07–0.38)	0.16 (0.11–0.60)	0.02 (0.02–0.04)
Mirex	5.06 (1.95–18.6)	0.31 (0.14–2.19)	0.56 (0.06–6.92)	0.66 (0.2–5.17)	trace (trace–2.47)	not detected
Hexachloro-benzene	0.19 (0.01–0.72)	0.11 (0.06–0.31)	0.14 (0.05–0.42)	0.11 (0.02–0.33)	0.04 (0.02–0.14)	0.01 (0.01–0.05)
PCB[d]	142.0 (73.8–261)	65.8 (41.2–110)	51.5 (15.4–118)	60.0 (33.4–148)	91.3 (55.1–395)	7.8 (4.11–11.1)
Mercury	0.51 (0.29–1.47)	0.22 (0.11–0.35)	0.23 (0.11–0.50)	0.39 (0.16–0.63)	not determined	0.08 (0.04–0.10)

[a] Taken in part from Gilman et al., 1977.
[b] Samples collected in 1975 only.
[c] Samples collected in 1976 only.
[d] Based on 1:1, Aroclor 1260:1254.

dichlorodiphenyldichloroethylene (DDE) were highest in eggs collected from Lake Michigan and lowest in eggs from Lake Erie. Levels of dieldrin and heptachlor epoxide in the eggs were remarkably similar throughout the lakes. Mirex residues were higher by an order of magnitude in Lake Ontario eggs than in eggs from the other lakes. Mercury levels, which were measured in all eggs but those from Lake Michigan, were highest in eggs from Lake Ontario. Levels of all the above compounds in Lake Ontario herring gull eggs have remained high and relatively constant since 1972 (Hallett *et al.*, 1976).

We have compared the ability of the Lake Ontario herring gull to accumulate pollutants with another Great Lakes predator, the coho salmon (*Oncorhyncus kisutch*). This salmon consumes alewife (*Alosa pseudoharengus*) and American smelt (*Osmerus mordax*) in Lake Ontario as does the herring gull. Analyses of gull eggs, salmon, smelt, and alewives indicated that four residues—PCB's, DDE, mirex, and photomirex—were predominant in all species. Levels of these four organochlorines were 2 to 3 times higher in the coho muscle and 37 to 60 times higher in gull eggs than in the alewives or smelt (Table 2). Recently reported studies by Anderson and Hickey (1976) showed that captive herring gulls fed Lake Michigan alewives accumulated organochlorine residues as high as those found in wild Lake Michigan gulls. The similarities between the residue ratios in our study of coho salmon and gull eggs and the accumulation of residues from fish that

TABLE 2   Major Organochlorine Residues in Lake Ontario Biota[a]

	Organochlorine Concentrations, Arithmetic Means in ppm, Wet Weight			
	Alewife and Smelt[b] (Pooled Sample of 50)	Coho Salmon (Individual Analysis of 28)		Herring Gull Eggs (Six Pools of 9 to 10 Eggs Each)
		Muscle	Liver	
Organochlorine				
PCB[c]	2.21	5.77	2.31	138
DDE	0.47	0.97	0.41	17.4
Mirex	0.09	0.23	0.10	4.4
Photomirex[d]	0.03	0.11	0.04	1.6
% lipid	2.34	8.17	6.16	6.29

[a] Summarized from Norstrom *et al.*, 1978.
[b] Stomach contents of coho salmon.
[c] Calculated as 1:1 Aroclor 1254:1260.
[d] 8-Monohydromirex.

was reported by Anderson and Hickey (1976) indicate the importance of fish in the gulls' feeding ecology and the usefulness of gulls as monitors of low-level lake contamination.

The use of herring gulls and their eggs as indicators of trace contaminants has facilitated the identification of a number of compounds not listed in Tables 1 or 2 because the compounds were at higher levels than in other ecosystem components. Thirty polynuclear aromatic hydrocarbons, including some known carcinogens, have been identified by capillary gas chromatographic and mass spectrographic analysis (Hallett *et al.*, 1977a). Surveillance of these compounds in other lakes is under way. In addition, chlordanes and several chlordane breakdown compounds, methoxychlor, hydroxylated PCB metabolites (Hallett *et al.*, 1977b), and chlorobenzenes (Hallett, personal communication, 1977) have been found.

The effects of organochlorines on avian reproductive physiology have been reported by several authors and reviewed by Peakall (1975). Since the reproductive behavior of herring gulls is well documented, we studied several aspects of gull reproduction and attempted to relate them to contaminant levels. In 1975, we studied reproductive success of herring gulls on four of the Great Lakes (Table 3). A significant reduction in the number of eggs per nest was observed in the Lake Ontario colony. Only 20% of the eggs hatched and few of these chicks survived. In the Lake Erie colony, fewer eggs hatched than in either of the Lake Huron or Lake Superior colonies but a larger proportion of the chicks survived. Thus, overall productivity in Lakes Erie, Huron, and Superior was similar. The reproductive success rates in these three lakes, the least contaminated of the five Great Lakes, were within the range reported for coastal colonies (Kadlec and Drury, 1968; Haycock and Threlfall, 1975).

During our study of reproductive success on Lake Ontario, we

TABLE 3   Reproductive Success of Great Lakes Herring Gulls in 1975[a]

Lake	Mean Number of Eggs Observed per Clutch	Mean Number of Eggs Hatched per Nest	Mean Number of Chicks Surviving 21 Days per Nest
Superior	2.98	2.38	1.38
Huron	2.84	2.38	1.48
Erie	2.90	1.93	1.41
Ontario	2.43	0.59	0.15

[a] Summarized from Gilman *et al.* (1977).

observed two embryos (of 24 hatched) with structural abnormalities. One embryo was hydrocephalic and unilaterally macrophthalmic. Its upper mandible was shorter than the lower one. The other embryo had a twisted mandible. Gilbertson *et al.* (1976) have recently reported that the incidence of abnormalities in young of Lake Ontario colonial birds is higher than elsewhere in the Great Lakes. Hays and Risebrough (1972) suggested that an increase in the numbers of abnormal tern chicks (*Stena hirundo*) on the eastern U.S. seaboard was linked to environmental contaminants. However, there is still no firm evidence on the cause of these abnormalities in Lake Ontario gulls.

We attempted to identify the factors contributing to the reproductive failure of the Lake Ontario herring gulls. The high rate of egg loss from their nests together with the lack of nest defense observed in the Lake Ontario colonies indicated the potential importance of adult behavioral changes (Gilman *et al.*, 1977). Telemetered eggs were used to monitor nest attentiveness, incubation temperature, and the rate of egg heating and cooling in Lake Ontario colonies and in one control colony in New Brunswick (Fox *et al.*, 1978). We found that the time off the nest averaged 45 min/day in the Lake Ontario colonies compared to 14 min/day in the control colony. The average temperature measured by the telemetered eggs was significantly lower in Lake Ontario nests than that measured in New Brunswick. More important, however, was the fact that egg temperatures ranging from 13°C to 27°C were recorded in Lake Ontario nests, whereas no temperatures below 27°C were recorded in the New Brunswick colony. The high rate of egg loss in the Lake Ontario colonies can probably be explained by the poor nest attentiveness of the adults. High early embryonic mortality, characteristic of the egg failure, can be explained in part by the variation in nest incubation temperatures.

Behavioral abnormalities in incubating adults do not explain all of the embryonic mortality observed in Lake Ontario herring gull eggs. Artificially incubated eggs from Lake Ontario, Lake Huron, and the Atlantic Coast also die during early embryonic development or fail to hatch (Gilbertson and Fox, 1977; Gilman *et al.*, 1977). We attempted to establish a cause-and-effect relationship between embryonic failure and levels of organochlorine contamination in eggs (Gilman *et al.*, 1978). Uncontaminated herring gull eggs from New Brunswick were injected with extracts of Lake Ontario eggs and then incubated naturally by adult gulls in New Brunswick. Some synthetic mixtures of PCB's, DDE, mirex, photomirex, and hexachlorobenzene, mimicking environmental egg concentrations, were also injected into some New Brunswick gull eggs.

The embryonic uptake of the injected contaminants proceeded almost identically with the uptake of contaminants by embryos exposed to environmentally deposited organochlorines. However, no increase in embryonic or chick mortality was observed in any group injected with contaminants compared to the control group, which was injected with solvent alone. The data suggested that the presence of the organochlorine contaminants listed above in the yolk of the egg during incubation was not the direct cause of embryonic mortality observed in Lake Ontario herring gull colonies. Alternatively, the high embryonic mortality observed in Lake Ontario may result from the effects of toxic chemicals on the eggs before they are laid. Changes in sex cell viability, chromosomes, yolk quality, and shell porosity may all contribute to embryonic failure and be caused by environmental contaminants.

We are attempting to focus our monitoring program onto aspects of reproduction, behavior, eggshell structure, and egg quality to determine more accurately how environmental contamination affects the ecosystem. However, when using indicator species one must always keep in mind the problems of species variation. For example, the amount of DDE in eggs associated with 20% thinning is six times greater in the herring gull than in the double-crested cormorant (*Phalacrocorax auritus*) (Gilman *et al.*, 1977; Keith and Gruchy, 1972). This lower sensitivity of the herring gull may well be advantageous, as Moore (1966) forecast, since the cormorant has ceased to breed on Lake Ontario.

SUMMARY

The herring gull (*Larus argentatus*) is at the top of several food chains. Adults of this species are essentially year-round residents of the Great Lakes. The levels of contaminants found in them are higher than those in other Laridae in the Great Lakes. Their ability to bioaccumulate high loads of persistent contaminants allows investigators to identify compounds that would be difficult to determine in lower trophic levels. Since there are high levels of many contaminants in this species, it is likely that detrimental biological effects will also be found, although the possibility that other species will be more sensitive should be borne in mind.

Chlorinated hydrocarbons and other persistent pollutants accumulate to high levels in gull tissues and are deposited into the eggs. Egg contaminant levels reflect the levels of lake contamination. High levels of PCB's, DDE, and mirex in Lake Ontario herring gulls and their association with early embryonic mortality, chick deformity, and aberrant adult behavior were examined. In an attempt to establish cause-

and-effect relationships, relatively uncontaminated gull eggs were injected with organochlorine compounds and incubated naturally. It appears that the major effects of toxic chemicals or embryonic mortality may be manifested before the eggs are laid. The behavioral and reproductive anomalies described in this paper indicate that the environmental quality of Lake Ontario is unacceptable. We believe that these findings have direct implications for other species, including man.

REFERENCES

Anderson, D. W., and J. J. Hickey. 1976. Dynamics of storage of organochlorine pollutants in herring gulls. Environ. Pollut. 10:183–200.

Bjerk, J. E., and G. Holt. 1971. Residues of DDE and PCB in eggs from the herring gull (*L. argentatus*) and the common gull (*L. canus*) in Norway. Acta Vet. Scand. 12:429–441.

Fox, G. A., A. P. Gilman, D. B. Peakall, and F. W. Anderka. 1978. Behavioral abnormalities of nesting Lake Ontario herring gulls. J. Wildl. Manage. 42(3):477–483.

Gilbertson, M. R., and G. A. Fox. 1977. Pollutant-associated embryonic mortality of Great Lake herring gulls. Environ. Pollut. 12:211–216.

Gilbertson, M. R., D. Morris, and R. A. Hunter. 1976. Abnormal chicks and PCB residue levels in eggs of colonial birds on the lower Great Lakes. Auk 93:434–442.

Gilman, A. P., G. A. Fox, D. B. Peakall, S. M. Teeple, T. R. Carroll, and G. T. Haymes. 1977. Reproductive parameters and contaminant levels of Great Lakes herring gulls. J. Wildl. Manage. 41(3):458–468.

Gilman, A. P., D. J. Hallett, G. A. Fox, L. J. Allan, W. J. Learning, and D. B. Peakall. 1978. Effects of injected organochlorines on naturally incubated herring gull eggs. J. Wildl. Manage. 42(3):484–493.

Hallett, D. J., R. J. Norstrom, F. I. Onuska, M. E. Comba, and R. Sampson. 1976. Mass spectral confirmation and analysis by the Hall detector of mirex and photomirex in herring gulls from Lake Ontario. J. Agric. Food Chem. 24:1189–1193.

Hallett, D. J., R. J. Norstrom, F. I. Onuska, and M. E. Comba. 1977a. Analysis of polynuclear aromatic hydrocarbons and organochlorine pollutants in Great Lakes herring gulls by high resolution gas chromatography. Pp. 115–125 in Proceedings of the Second International Symposium on Glass Capillary Chromatography. Institute of Chromatography. Bad Dürkheim, W. Germany.

Hallett, D. J., R. J. Norstrom, F. I. Onuska, and M. E. Comba. 1977b. Mirex, chlordane, dieldrin, DDT, and PCBs: Metabolites and photoisomers in L. Ontario herring gulls. Pp. 183–191 in G. W. Ivie and H. W. Dorough, eds. Pesticide Metabolism in Large Animals. Academic Press, New York.

Haycock, K. A., and W. Threlfall. 1975. The breeding biology of the herring gull in Newfoundland. Auk 92:678–697.

Hays, H., and R. W. Risebrough. 1972. Pollutant concentrations in abnormal young terns from Long Island Sound. Auk 89:19–35.

Hickey, J., J. A. Keith, and F. B. Coon. 1966. An exploration of pesticides in a Lake Michigan ecosystem. J. Appl. Ecol. 3(Suppl.):141–154.

Jørgensen, O. H., and I. Kraul. 1974. Eggshell parameters and residues of PCB and DDE in eggs from Danish herring gulls, *Larus a. argentatus*. Ornis Scand. 5:173–179.

Kadlec, J. A., and W. H. Drury. 1968. Structure of the New England herring gull population. Ecology 49:644–676.

Keith, J. A., and I. M. Gruchy. 1972. Residue levels of chemical pollutants in North American birdlife. Pp. 437–454 in K. H. Voous, ed. Proceedings of the 15th International Ornithological Congress, The Hague, 30 August–5 September 1970. E. J. Brill, Publishers, Leiden.

Moore, F. R. 1976. The dynamics of seasonal distribution of Great Lakes herring gulls. Bird Banding 47:141–159.

Moore, N. W. 1966. A pesticide monitoring system with special reference to the selection of indicator species. J. Appl. Ecol. 3(Suppl.):261–269.

Norstrom, R. J., D. J. Hallett, and R. A. Sonstegardo. 1978. Coho salmon (*Oncorhynchus kisutch*) and herring gulls (*Lavus argentatus*) as indicators of organochlorine contamination in Lake Ontario. J. Fish. Res. Board Can. 35:1401–1409.

Peakall, D. B. 1975. Physiological effects of chlorinated hydrocarbons on avian species. Pp. 343–360 in R. Hague and V. H. Freed, eds. Environmental Dynamics of Pesticides. Plenum Press, New York.

QUESTIONS AND ANSWERS

J. WEIS: Is it possible that eggs removed from the Lake Ontario nests that failed to hatch after incubation may have been exposed to cooler temperatures as a result of adults spending more time off the nest before you collected the eggs?

A. GILMAN: That is an interesting idea. We tried to collect fresh eggs. In a species such as the herring gull, no significant incubation begins until the third egg has been laid. We monitored the nests daily and collected the first egg that was laid. That egg is essentially unincubated and remains at ambient temperature with neither the male nor female incubating it.

M. FRIEND: Has anyone looked at the distribution of influenza virus in these birds? I ask this because a great upsurge of virus isolations from migratory birds has been documented[1] in recent years. One of the manifestations of influenza infection in birds is reproductive failure. We have been routinely isolating influenza viruses from migratory waterfowl at the rate of approximately 5%, and in some populations as high as 25%, of the birds sampled.

A. GILMAN: We have not looked at the association of virus with the reproductive failure of Lake Ontario herring gulls. One aspect that might be considered is the influence of pollutant residue levels on both reproduction and the appearance of influenza viruses. For example, herring gull and ring-billed gull (*Larus delawarensis*) populations exist side by side. The ring-billed gull is expanding and reproducing whereas the herring gull is not. In examining the pollutant residues of these two species, we found that the herring gull exhibited approximately five times the pollutant residue level of the ring-billed gull. The low pollutant residue level of the ring-billed gulls may be due to the fact that they winter in the Gulf area of Central and South America. If

---

[1] Easterday, B. C., and B. Tumova. 1978. Pp. 549–573 in M. S. Hofstad, B. W. Calnek, C. F. Helmboldt, W. M. Reid, and H. W. Yoder, Jr. Diseases of Poultry. Iowa State University Press, Ames.

influenza viruses are a factor, we would be surprised that they were not present in both species if they were present in one.

J. ZINKL: Your report of the fivefold increase in organochlorine compound residues in the muscle of coho salmon is of interest since it brings to mind the effects of low levels of PCB's in mink as described recently by Ringer *et al.* (1972).[2] His studies on ranch mink (*Mustela vison*) suggest that they are one of the most susceptible species to the effects of PCB. The paper emphasizes the need to take a closer look at wild Mustelids for effects of environmental pollutants.

[2] Ringer, R. K., R. J. Aulerich, and M. Zabik. 1972. Effects of dietary polychlorinated biphenyls on growth and reproduction of mink. Proc. Am. Chem. Soc. 12:149–154.

Acute Foodborne Pesticide Toxicity in Cormorants (*Phalacrocorax* sp.) and Seagulls (*Larus californicus*)

EDWIN B. HOWARD, GERALD N. ESRA, and
DAVID YOUNG

Dichlorodiphenyltrichloroethane (DDT) has not been in general use as a pesticide since a ban was imposed in 1970. Prior to that time, production of this widely used chemical was extensive. Because DDT is not readily biodegradable, much of the chemical used prior to 1970 remains in the environment. There is still a great deal to be learned about its biological effects. DDT has been incriminated as the etiological agent in reproductive failure in several avian species (Heath *et al.*, 1969) and has been suggested as a cause of the population decline in brown pelicans (*Pelicanus occidentalis*) along the southern California coast (Keith *et al.*, 1970). However, results of a recent survey by the California Fish and Game Department suggest that the brown pelican population is increasing in this area.

This report describes the poisoning of cormorants (*Phalacocorax* sp.) and seagulls (*Larus californicus*) at the Los Angeles City Zoo following a diet of bottom-feeding fish that were harvested in the coastal waters off the San Pedro harbor of Los Angeles County. It also records the analysis for DDT, its degradation product dichlorodiphenyldichloroethylene (DDE), and other compounds such as polychlorinated biphenyls (PCB's), not only in the tissues of birds that died, but also in fish fed to them. A limited analysis of tissues from California sea lions (*Zalophus californianus*) residing in this area is presented as well.

MATERIALS AND METHODS

Dead birds were necropsied and various tissues were collected for routine histopathological examination and toxicological studies. Various lot samples of fish fed to different animals and birds at the Los Angeles Zoo were also collected for toxicological tissue examination. The fish that were studied included surf smelt (*Hypomesus pretiosus*) and mackerel (*Scomber scombrus*), obtained from a northern California distributor, and queenfish (*Seriphus politus*), obtained from a San Pedro, California, distributor. Additionally, liver tissues were analyzed from some California sea lions that were necropsied as a part of the continuing Marine Mammal Disease Surveillance Program (Schroeder *et al.*, 1973) aimed at determining the cause of death of stranded marine mammals along the beaches of the Los Angeles County coast.

Tissues were processed by fixation in 10% buffered formalin, followed by paraffin embedding, sectioning, and staining with hematoxylin and eosin.

Bird, fish, and mammal tissues were analyzed by homogenizing them into fine paste and extracting the chlorinated hydrocarbons with acetonitrile. This effectively extracted 99% of the chlorinated hydrocarbons from the tissues. The acetonitrile solution was filtered and extracted with hexane, which recovers about 90% of the chlorinated hydrocarbons. The hexane solvent was concentrated by evaporation and run through a Tracor 220 gas chromatograph equipped with an electron capture detector. The carrier gas was nitrogen (20 ml/min through the column; 60 ml/min through the purge). The column was 180 cm long, 6 mm outside diameter, and 2 mm inside diameter, packed with 1.5% OV-17 + 1.95% QF-1 on 90/100 mesh Gas-chrom-Q. Injector, column, and detector temperatures were 235°C, 200°C, and 285°C, respectively. Tissue extracts were analyzed for total chlorinated hydrocarbons—DDT, DDE, dichlorodiphenyldichloroethane (DDD), and PCB (Aroclor 1242, 1254, 1260).

RESULTS

A collection of 16 California gulls and five cormorants (two Brandt's [*Phalacrocorax penicillatus*] and three Guanay [*Phalacrocorax bougainvillii*]) had been maintained in captivity and exhibited in the zoo. The gulls had been recovered from the Santa Barbara oil spill of 1969, and the cormorants had been purchased from the San Francisco Zoo in 1971. Since the beginning of their captivity, they had been housed in the aquatic section of the zoo, and had been fed a queenfish diet exclusively.

Beginning on May 22, 1976, the first gull showed signs of illness. Between this time and June 10, all of the gulls exhibited similar signs, followed the same clinical course, and died. The cormorants died

TABLE 1 Mean Concentrations of Chlorinated Hydrocarbons in Tissues Analyzed for Pesticide Residue

Tissue	Concentrations of Chlorinated Hydrocarbons, ppm Wet Weight				Total Pesticide (DDE, DDT, DDD) Concentration, ppm Wet Weight	Total Chlorinated Hydrocarbon Concentration, Including PCB, ppm Wet Weight
	DDE	DDT	DDD	PCB		
Bird liver composite	770.0	62.9	9.3	135.0	842.2	977.2
Gulls						
liver	3,057.0	127.0	—	150.0	3,184.0	3,334.0
muscle	292.0	7.0	—	29.5	299.0	328.5
brain	428.0	16.0	—	66.1	444.0	510.1
Cormorant						
liver	746.0	61.0	—	134.6	807.0	941.6
muscle	487.0	17.0	—	91.8	504.0	595.8
brain	217.0	11.0	—	48.6	228.0	276.6
Queenfish	3.06	0.12	0.15	1.13	3.33	4.46

between June 12 and 24, 1976. Only one cormorant from the collection has survived. Clinically, the birds became anorectic and, within 1 or 2 days, would begin trembling and die within minutes with severe muscle fasciculations. Attempts to treat the birds with atropine and supportive therapy did not alter the course of the disease.

Gross examination of the birds revealed that they were all in poor nutritional condition and that their digestive tracts were nearly empty. No other gross lesions were found. Histopathological examination of the tissues from these birds revealed marked hyperemia of the lung, liver, kidney, and other visceral tissues. There was considerable congestion and scattered petechial hemorrhages of the brain. Perivascular edema was observed in the cerebral tissues and focal coagulation necrosis was observed in the liver, which had no particular lobular distribution.

Analysis of a composite of liver tissue from several of the cormorants and gulls revealed very high concentrations of total chlorinated hydrocarbons (977 ppm wet weight). DDE predominated (770 ppm) and there was 135 ppm of PCB (Table 1).

Subsequent examination of various tissues from the gulls that died revealed total pesticide residues of 299 ppm in the skeletal muscle tissue, 3,184 ppm in the liver, and 444 ppm in the brain (Table 1). Analysis of tissues from the cormorants revealed 504 ppm total pesticide residues in the muscle, 807 ppm in the liver, and 228 ppm in the brain. PCB concentrations were also high in the tissues of both the gulls and cormorants.

In an attempt to determine the source of the toxic substances, several types of fish used for feed at the zoo were analyzed in the same manner as the bird tissues. Mackerel samples contained only 0.2 to 0.38 ppm total chlorinated hydrocarbons, and surf smelt from 0.05 to 0.06 ppm, based on the whole fish tissues. On the other hand, various queenfish contained from 1.8 to 7.5 ppm total chlorinated hydrocarbons, most of which was DDE. The average concentration in queenfish tissues was 4.46 ppm wet weight, without the heads (Table 1).

Total pesticide residues in gull livers averaged 0.08 g. Based on the average pesticide residue of 3.33×10^{-6} g/g of queenfish, it was calculated that gulls would have to consume 24 kg of queenfish to accumulate those liver residues, i.e.,

$$\frac{0.08 \text{ g pesticide}}{3.33 \times 10^{-6} \text{ g/g fish}} = 0.024 \times 10^{6} \text{g, or 24 kg of fish.}$$

Liver tissues from 10 sea lions were examined. Pesticide residue levels ranged from 7.6 ppm to 31 ppm.

A trace on the queenfish suppliers revealed that the fish had been packed in San Pedro, California. A comprehensive tissue analysis of these fish, including trace element analysis, indicated that the fish were taken from the Palos Verdes Sector area, off the Los Angeles County coast by San Pedro (Figure 1).

DISCUSSION

Examination of the tissues from the birds and the fish that they were fed demonstrated that the gulls and cormorants died from an accumulation of DDT (especially the degradation product DDE) in the tissues, which, over a period of time, reached a lethal level and caused acute

FIGURE 1 Map of Los Angeles County, showing coastline with sewage plants (1, 2, 3) along Palos Verdes Peninsula area. Shaded area off the coast shows the relative concentration of DDT and PCB in the sediment.

neurological death. It is not known how long the birds were fed DDT-contaminated fish while at the zoo, but continuing analysis of sediment and bottom-feeding fish from this area (by Young, an author of this paper) has shown the continuing presence of DDT and PCB in the effluent of sewage treatment facilities located along the Los Angeles coast (Figure 1).

The retention time of DDT appears to be shorter than its degradation product, DDE, which has a half-life of 250 days in the pigeon (*Columba livia*) (Bailey *et al.*, 1969). Experimental feeding of a diet containing 2.8 ppm of DDE was lethal to American kestrels (*Falco sparverius*) (Porter and Weimeyer, 1972) after 14 to 16 months. Their brains had DDE residues of 213 to 301 ppm. The lethal range of DDE toxicity in the brain tissue of those birds was estimated to be between 200 and 300 ppm. The gulls and cormorants in this study contained a comparable range of DDE residue. Since they ingested a diet containing an average of 3.33 ppm of DDE and DDT, at an average daily intake of 149 to 225 g, the lethal level could easily have been reached in less time than reported in the American kestrels.

The pathological lesions associated with DDT poisoning have not been examined comprehensively (Innes and Saunders, 1962). Investigators have not reported any pesticide-specific lesions. Congestion of lungs and petechial hemorrhages in various tissues have been observed (Smith *et al.*, 1972). The presence of lesions in the brain is inconsistent in previous reports. Both neuronal degeneration and demyelination have been described, while other investigators have found no lesions in the central nervous system except congestion. The toxic necrotic lesions that were found in the livers of the gulls and cormorants have also been reported in other DDT-poisoned animals (Smith *et al.*, 1972), and focal muscle necrosis and renal tubular necrosis are described in some cases. Congestion and petechia, as well as edema of the brain tissue, were also seen in these birds, as were neuronal degeneration and necrosis.

The pathological findings, correlated with the clinical history and tissue analysis of affected birds and fish used for feed, indicated that the birds died as a result of acute neurological toxicity resulting from a progressive accumulation of DDE in the tissues, especially in the brain.

This episode clearly points to a need to monitor potentially toxic levels of pesticide residues in fish that are destined to be used as the principal diet of captive birds. Contaminated fish might also be destined for use as food for captive marine mammals in seaquariums. The potential long-term effects of such chlorinated hydrocarbons in these species have not been investigated.

SUMMARY

This report describes the acute neurological death of captive seagulls and cormorants due to the ingestion of bottom-feeding fish that were contaminated with pesticides. The report also describes the presence of organochlorine compounds in marine mammals stranded on the Los Angeles coast. Data on pesticide concentration in bird and fish tissues are presented.

REFERENCES

Bailey, S., P. J. Bunyan, B. D. Rennison, and A. Taylor. 1969. The metabolism of 1,1-di(p-chlorophenyl)-2,2-dichloroethylene and 1,1-di(p-chlorophenyl)-2-chloroethylene in the pigeon. Toxicol. Appl. Pharmacol. 14:13–22.
Heath, R. G., J. W. Spann, and J. F. Kreitzer. 1969. Marked DDE impairment of mallard reproduction in controlled studies. Nature 224:47–48.
Innes, J. R. M., and L. Z. Saunders. 1962. Pp. 700–704 in Comparative Neuropathology. Academic Press, Inc., New York.
Keith, J. O., L. A. Woods, Jr., and E. G. Hunt. 1970. Reproductive failure in brown pelicans on the Pacific Coast. Pp. 56–63 in Transactions of the Thirty-Fifth North American Wildlife and Natural Resources Conference: Man's Stake in a Good Environment. Palmer House, Chicago, Ill., March 22–25, 1970. Wildlife Management Institute, Washington, D.C.
Porter, R. D., and S. N. Weimeyer. 1972. DDE at low dietary levels kills captive American kestrels. Bull. Environ. Contam. Toxicol. 8(4):193–199.
Schroeder, R. J., C. A. Delli Quadri, R. W. McIntyre, and W. A. Walker. 1973. Marine mammal disease surveillance program in Los Angeles County. J. Am. Vet. Med. Assoc. 163(6):580–581.
Smith, H. A., T. C. Jones, and R. D. Hunt, eds. 1972. Insecticides of the chlorinated hydrocarbon group. Pp. 963-968 in Veterinary Pathology, 4th ed. Lea and Febiger, Philadelphia.

QUESTIONS AND ANSWERS

G. MIGAKI: Would you comment about the finding of systemic mycoses in the animals you studied? This condition is the result of opportunistic fungi that can cause extensive disease under certain conditions. We have found such cases in marine mammals. Is there any correlation between the presence of these diseases and high levels of pesticides?

E. HOWARD: I would have to speculate. Dr. Koller has shown that these compounds are capable of suppressing the immune system. We have not yet done similar studies on sea lions. In addition to bacterial infections we have encountered asperigillosis, mucormycosis, and, more recently, coccidioidomycosis in a wild sea otter (*Enhydris lutris*).

A. GILMAN: Were your animals in breeding condition at the time of death?

E. HOWARD: No.

Acute Respiratory Distress in Dogs with Paraquat Poisoning

D. F. KELLY, D. G. MORGAN, and
V. M. LUCKE

Paraquat (1,1'-dimethyl-4,4'-dipyridylium dichloride):

is a broad-spectrum contact weedkiller and herbage desiccant that is widely used in agriculture and horticulture. The compound has the unique property of killing quickly by contact with leaf surfaces, but immediate inactivation by adsorption onto clay minerals in soil makes paraquat of considerable practical value in weed control.

The possible toxic action of the bipyridyls has been discussed by Conning et al. (1969), who suggested that reduction of the bipyridilium cation by chlorophyll produces the free radical ion. This is then reoxidized by molecular oxygen to give the original bipyridilium cation and hydrogen peroxide; the bipyridilium cation can reenter the reduction-oxidation cycle and the hydrogen peroxide may reach a concentration sufficient to damage a susceptible part of the plant. This mode of toxic action in plants is supported by the observation that the reaction is light- and oxygen-dependent. A similar mode of toxic action has been suggested for animal tissues (Clements and Fisher, 1970; Fisher et al., 1973; Rhodes et al., 1976).

Commercial formulations of paraquat contain between 5% and 20% of the ion. In 1972, there were several such formulations available in the United Kingdom: Dextrone X, Dexuron, Esgram, Gramonol,

297

Gramoxone, and Orvar (the sales of which were all regulated by the Pharmacy and Poisons Act of 1933); and Pathclear and Weedol, which were formulated for use by the home gardener. In the past 2 years, accidental (and possibly malicious) paraquat poisoning in dogs has been observed (Johnson and Huxtable, 1976; Darke et al., 1977). The paragraphs below describe the clinical and pathological features of acute respiratory distress in dogs when paraquat poisoning is either known or believed to be the cause.

CLINICAL PHARMACOLOGY

Paraquat is poorly absorbed from the gut and is rapidly excreted from the body (Conning et al., 1969). Absorption rates vary for different species between 1% and 20%. The peak concentration of paraquat in the blood is reached within a few hours; rapid excretion follows. For example, in the rat 90% to 100% of a systemic dose is excreted in urine after 48 h (Daniel and Gage, 1966). The oral LD_{50} varies widely among species (guinea pig 30 mg/kg; rat 125 mg/kg) (Conning et al., 1969). The lung tissue of dogs, monkeys, and rabbits can accumulate paraquat in vitro to concentrations nearly 10 times that of the medium. The rat lung can accumulate paraquat in vivo to concentrations 6 times that of plasma (Rose et al., 1976). This probably accounts in part for selective toxic effects of paraquat that produce pulmonary edema and hemorrhage followed by pulmonary fibrosis (Clark et al., 1966; D. F. Kelly et al., unpublished observations, 1977).

CLINICAL FEATURES OF PARAQUAT POISONING IN DOGS

A history of access to paraquat is not always available. Scavenging of carcasses baited with paraquat has been recorded as a source of poisoning (Darke et al., 1977). Johnson and Huxtable (1976) have reported that access to recently treated grass has been followed by signs of poisoning. In most instances, however, the source of poisoning is not known, although questioning of owners has revealed that paraquat has been used recently in the animal's vicinity.

Contact with large amounts of paraquat preparation may produce hyperemia, vesication, and ulceration of the mouth and pharynx. These inflammatory changes are attributed to irritation by spreading agents incorporated in the 20% solutions of paraquat (Matthew, 1971). In most instances, changes of this kind are not apparent and the earliest sign is vomiting, accompanied by anorexia and depression. Where the time of ingestion is known or can be inferred, there is usually a delay of

1 to 3 days before the onset of clinical signs. Dyspnea develops from 12 h to several days later. Arterial PO_2 values remain low (e.g., 5.60 kPa in a normal range of 10.6 to 13.3 kPa) in spite of controlled ventilation and administration of high concentrations of inspired oxygen (Darke *et al.*, 1977).

Dyspnea is the most dominant clinical feature and progresses to cyanotic respiratory failure. The extreme severity of the dyspnea is disproportionate to the unspectacular radiographic appearance of the lungs in most dogs with acute respiratory distress. Some dogs, however, have striking bullous pulmonary emphysema, pneumomediastinum, and subcutaneous emphysema. This is presumed to be a sequel to severe pulmonary parenchymal damage. Similar findings have been described in dogs whose lungs were damaged by *N*-nitroso-*N*-methylurethane (Ryan *et al.*, 1976). There are few contributory laboratory findings. Hematological examinations reveal only changes consistent with a degree of dehydration resulting from the combination of vomiting and decreased water intake. The only striking biochemical abnormality is azotemia with blood urea values ranging between 16 and 90 mmol/l (normal <7 mmol/l); this results from dehydration and toxic tubular damage to the kidney. Death may occur after 24 h—presumably when a large dose of paraquat has been ingested—but survival for 8 days has also been recorded. Recognition of the clinical entity and its hopeless prognosis once dyspnea occurs usually justifies earlier euthanasia on humane grounds. Paraquat may be detected in the urine and tissues of dogs with acute respiratory distress, using a modification of the technique of Calderbank and Yuen (1965). The rate of paraquat detection seems to be related to the duration of the illness; paraquat is more likely to be identified in the early stages of the clinical disease (Table 1). Although published data concerning the rate at which dogs excrete paraquat are not available, the detection rates may reflect rapid elimination of the compound from the body.

TABLE 1   Detection of Paraquat in Postmortem Samples from Dogs with Acute Respiratory Distress

Duration of Illness, days	Frequency of Detection in Samples Analyzed
1–3	2/2
4–5	2/5
>5	2/6

PATHOLOGICAL FEATURES OF ACUTE RESPIRATORY DISTRESS
IN DOGS

In all cases the lungs are heavy, dark, and congested with subpleural petechial and ecchymotic hemorrhages. Gross pulmonary edema may occur in dogs that die early in the disease following ingestion of large amounts of paraquat. Bullous emphysema and pneumomediastinum may be present.

The microscopic pulmonary changes vary with the duration of dyspnea. Early changes include marked alveolar septal capillary congestion, pulmonary edema, and extravasation of red blood corpuscles (Figure 1). Inflammatory changes in the lung are usually minimal. There is extensive alveolar collapse, contrasting with overdistension of alveolar ducts and terminal bronchioles (Figure 2). Focal necrosis of bronchiolar epithelium and desquamation of debris into bronchiolar lumina also occur. In the early stages, electron microscopy of the lung reveals, in addition to the changes described above, complete loss of

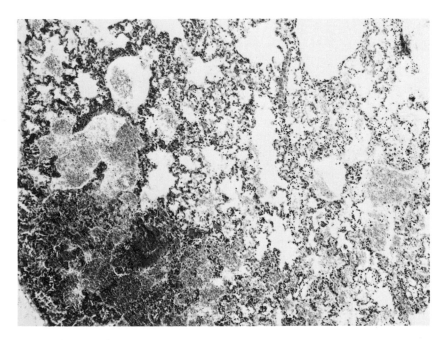

FIGURE 1 Section of lung from a dog with acute respiratory distress showing congestion, hemorrhage, and partial collapse. (Hematoxylin and eosin. ×88.)

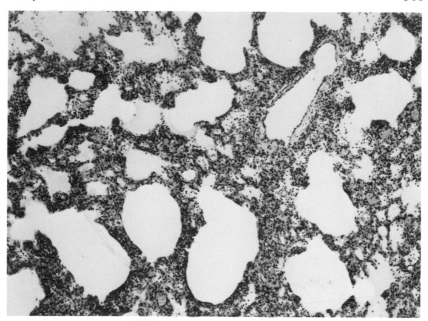

FIGURE 2 Lung of dog with acute respiratory distress. There is overdistension of alveolar ducts and partial alveolar collapse. (Hematoxylin and eosin. ×66.)

both types of pneumonocytes and exposure of the underlying alveolar septal basement membrane (Figure 3). No distinct ultrastructural lesion of endothelial cells has been recognized.

In the later stages of acute respiratory distress, there is both histological and ultrastructural evidence of pulmonary regeneration and repair. The most impressive feature is alveolar fibrosis (Figure 4): numerous large immature fibroblasts are present close to the alveolar septa and within alveolar lumina. The presence of these cells is associated with increasing amounts of collagen fibers (Figure 5). Bronchiolar epithelial regeneration is striking, often producing irregular thickening and papillary elevation of the lining of the airway (Figure 6). Alveolar epithelialization results from proliferation of immature type II pneumonocytes (Figure 7). Alveolar epithelial hyperplasia and intra-alveolar fibrosis combine to produce a progressively solidified lung, thus producing the morphological background to the marked impairment of gas exchange, which leads to cyanotic respiratory failure.

Microscopic extrapulmonary lesions may be present in dogs with

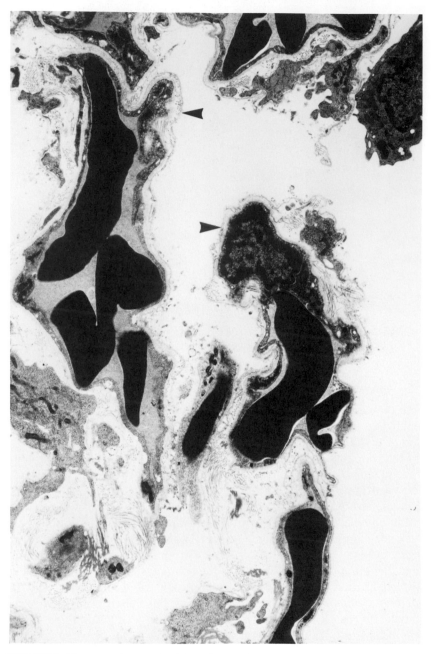

FIGURE 3 Electron micrograph of lung from a dog with acute respiratory distress. There is total loss of types I and II pneumonocytes, with exposure of the underlying basement membrane (arrows). (×6,100.)

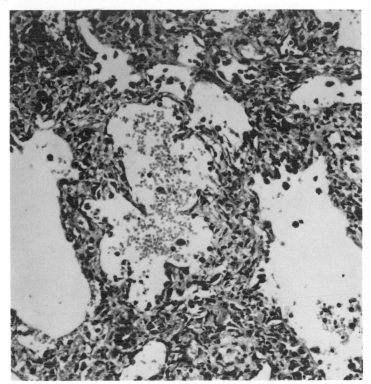

FIGURE 4 Section of lung from a dog with acute respiratory distress, showing extensive alveolar fibrosis. (Hematoxylin and eosin. ×200.)

acute respiratory distress. These are summarized in Table 2. Some of the changes, such as necrosis of the renal tubular epithelium and of the zona glomerulosa in other species poisoned by paraquat, have been described by Clark *et al.* (1966), but the focal cardiomyopathy and coronary arterial necrosis appear not to have been recognized previously.

DISCUSSION

Paraquat has extensive practical applications for weed control, in both farm and garden. Fortunately, rapid inactivation occurs when paraquat contacts the soil so that toxic effects from accidental contact with treated plants are uncommon. Consumption of large amounts of paraquat, particularly in its concentrated form, may follow accidental

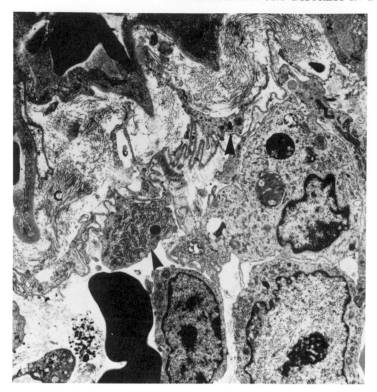

FIGURE 5 Electron micrograph of lung from dog with acute respiratory distress, showing alveolar collapse, fibroblasts (arrows), and intra-alveolar collagen (C). (×6,100.)

access to carelessly stored containers and access to large unguarded amounts of the compound, such as in carcasses "treated" with paraquat and used as bait for vermin such as foxes.

The observations recorded in this paper concern the acute disease following ingestion of large amounts of paraquat. Although toxicological confirmation of paraquat involvement has not always been obtained, we assume that the acute respiratory distress is a sequel to ingestion of this chemical (Darke *et al.*, 1977).

Early clinical and pathological changes resulting from paraquat poisoning are nonspecific. They may be related to mild renal damage. The later changes involve progressive and severe damage in the lung. There is extensive degeneration of pneumonocytes and alveolar collapse. This is followed by progressive intraalveolar fibrosis and

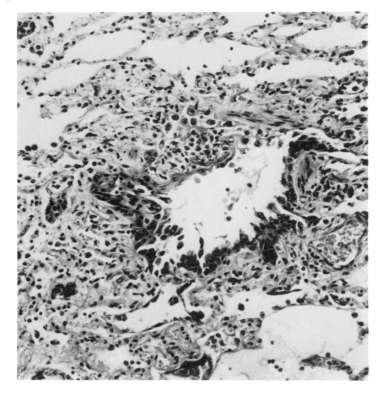

FIGURE 6 Bronchiole showing irregular epithelial hyperplasia following toxic damage. (Hematoxylin and eosin. ×165.)

epithelialization of the lung, which leads to severe respiratory distress. Recognition of the range of clinical and pathological features in acute respiratory distress is important for both lay persons and veterinarians. The life styles of pet animals are such that circumstantial evidence of paraquat ingestion may not be available. In addition, definitive toxicological diagnosis is often difficult if the animal survives for more than a few days.

SUMMARY

During 1975–1976, 14 dogs with a severe illness characterized by anorexia, emesis, and depression were examined at the University of Bristol Veterinary School. Subsequently, dyspnea, leading to cyanotic respiratory failure, developed. Pathological changes in the lungs of

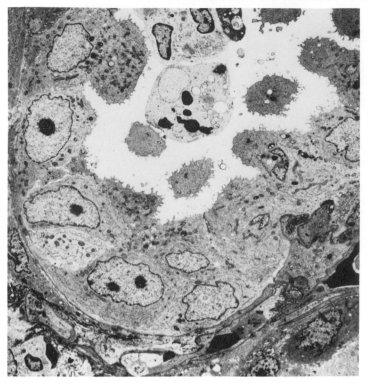

FIGURE 7   Electron micrograph of lung from a dog with acute respiratory distress. An alveolus is lined by cuboidal epithelial cells. (×2,200.)

these dogs were congestion, hemorrhage, edema, and collapse. This was associated with degeneration of alveolar and bronchiolar epithelial cells. Later changes included alveolar epithelialization, intraalveolar fibrosis, and bizarre hyperplasia of the bronchiolar epithelium. Some dogs had necrotizing lesions in renal tubules, adrenal cortex, coronary arteries, and myocardium.

The herbicide paraquat was detected in six of these dogs. Paraquat-induced respiratory failure in dogs is a relatively new hazard of the agricultural and horticultural use of this potent herbicide.

ACKNOWLEDGMENTS

Paraquat analyses were conducted by Dr. T. J. Iswaran, Imperial Chemical Industries, Ltd. Christine Lambert assisted with electron microscopy and photography.

TABLE 2　Extrapulmonary Lesions in Dogs with
Acute Respiratory Distress Associated with Paraquat
Poisoning

Site	Microscopic Features	Number of Lesions Found in Animals Examined
Pharynx	Epithelial necrosis, ulceration, pharyngitis	1/12
Stomach	Submucosal edema	1/12
Kidney	Focal tubular necrosis, epithelial regeneration	10/14
Adrenal glands	Focal necrosis of zona glomerulosa	6/12
Heart	Focal myocardial necrosis, fibrinoid necrosis of tunica media in coronary arteries	4/6

REFERENCES

Calderbank, A., and S. H. Yuen. 1965. An ion-exchange method for determining paraquat residues in food crops. Analyst (London) 90:99–106.

Clark, D. G., T. F. McElligott, and F. W. Hurst. 1966. The toxicity of paraquat. Br. J. Ind. Med. 23:126–132.

Clements, J. A., and K. Fisher. 1970. The oxygen dilemma. N. Engl. J. Med. 282:976–977.

Conning, D. M., K. Fletcher, and A. A. B. Swan. 1969. Paraquat and related bipyridyls. Br. Med. Bull. 25:245–249,

Daniel, J. W., and T. C. Gage. 1966. Absorption and excretion of diquat and paraquat in rats. Br. J. Ind. Med. 23:133–136.

Darke, P. G. G., C. Gibbs, D. F. Kelly, D. G. Morgan, H. Pearson, and B. M. Q. Weaver. 1977. Acute respiratory distress in the dog associated with paraquat poisoning. Vet. Rec. 100:275–277.

Fisher, H. K., J. A. Clements, and R. R. Wright. 1973. Enhancement of oxygen toxicity by the herbicide paraquat. Am. Rev. Resp. Dis. 107:246–252.

Johnson, R. P., and C. R. Huxtable. 1976. Paraquat poisoning in a dog and cat. Vet. Rec. 98:189–191.

Matthew, H. 1971. Paraquat poisoning. (Clinicopathological Conference.) Scott. Med. J. 16:407–421.

Rhodes, M. L., D. C. Zavala, and D. Brown. 1976. Hypoxic protection in paraquat poisoning. Lab. Invest. 35:496–500.

Rose, M. S., E. A. Lock, L. L. Smith, and I. Wyatt. 1976. Paraquat accumulation: Tissue and species specificity. Biochem. Pharmacol. 25:419–423.

Ryan, S. F., A. L. L. Bell, and C. R. Barrett. 1976. Experimental acute alveolar injury in the dog. Morphologic-mechanical correlations. Am. J. Pathol. 82:353–364.

QUESTIONS AND ANSWERS

s. NIELSEN: How does paraquat get into the body? Does it penetrate through skin, or is it inhaled after spraying?

D. KELLY: One of the paradoxes of paraquat poisoning is that it is not highly toxic when administered by inhalation. Although inhalation produces changes related to its direct irritant effect, these are strikingly dose-related. It is, however, toxic when it is given by the oral or parenteral routes and then secondarily concentrated in the lungs.

DDT-Induced Reduction in Eggshell Thickness, Weight, and Calcium Is Accompanied by Calcium ATPase Inhibition

GERALD J. KOLAJA and DAVID E. HINTON

Environmental exposure to the pesticide dichlorodiphenyl-trichloro-ethane (DDT) causes a variety of avian species to produce thin-shelled eggs (Blus et al., 1971, 1974; Cooke, 1973; Peakall, 1970; Prestt et al., 1970; Risebrough et al., 1968). Such eggs are deficient in calcium in spite of normal serum calcium levels (Bitman et al., 1969). Thus, it appears that the defect in calcium transport occurs within the egg-shell gland (Miller et al., 1975), that portion of the reproductive tract responsible for deposition of calcium into the forming eggshell (Hohman and Schraer, 1966). High concentrations of calcium-dependent ATPase (Ca–ATPase) have been found in the microsomal fraction of the eggshell gland mucosa and are thought to be responsible for calcium transport (Pike and Alvarado, 1975). In vitro inhibition of this enzyme has been demonstrated using DDT (Kolaja and Hinton, 1977) and its metabolite dichlorodiphenyldichloroethylene (DDE) (Miller et al., 1976). The study was undertaken to determine the effect of DDT on Ca–ATPase activity in vivo and to demonstrate with cytochemical techniques the sites of Ca–ATPase activity in the eggshell gland epithelium.

MATERIALS AND METHODS

Mature domestic mallard ducks (Anas platyrhynchos) were randomly assigned to two cages with five hens and one drake per cage. The ducks were acclimated to the housing area for 1 month.

During this time they were fed poultry reproductive mash (Purina

309

Chow) *ad libitum*. Beginning in the second month, the treated ducks were fed poultry mash containing 50 ppm of DDT. Controls continued on the poultry mash alone. After 5 months of feeding, egg production was induced by increasing the photoperiod from 8 to 16 h/day. The eggs were collected daily, and the shell weight, length, width, and thickness were determined. To correct for changes in shell thickness due to changes in size, a ratio of the weight/length × width (r-value) was calculated (Ratcliffe, 1967). After 6 months, the hens were killed by cervical dislocation.

Sections of eggshell glands were rapidly excised and placed in ice-cold 0.25-M sucrose. The epithelium was removed by scraping with a clean razor blade and 1 g was diluted to 10 ml in ice-cold 0.25-M sucrose. The mixture was homogenized in a Potter Elvehjem tissue homogenizer at 400 rpm. The homogenate was centrifuged at 18,000 × G in a Beckman L-2 ultracentrifuge at 4°C, resulting in supernatant, which was recentrifuged at 98,000 × G, and pellets, which were suspended in 0.25-M sucrose to a final volume of 3 ml. The purity of the microsomal pellets was assessed with an electron microscope. Aliquots of microsomal suspension were fixed in an equal volume of 8% glutaraldehyde in 0.4-M phosphate buffer at pH 7.2. Following sedimentation in a clinical centrifuge at 1,500 rpm for 1 min, the pellets were processed for electron microscopic examination by routine methods (Hayat, 1970). Microsomal preparations were stored at −60°C until the time of assay (1 to 2 weeks). No detectable change in activity of Ca–ATPase was detected during this storage.

Ca–ATPase activities were determined by the method of Rorive and Kleinzeller (1974); the amount of phosphate liberated, by the method of Stanton (1968). Ca–ATPase was expressed as μmol Pi (inorganic phosphate) liberated per mg of protein per 30 min incubation. Protein from the microsomal pellets was measured by the method developed by Lowry *et al.* (1951).

Calcium was extracted by ashing a 0.5 g sample of eggshell at 450°C for 12 h. The ash was dissolved in concentrated hydrochloric acid, and calcium was measured by atomic absorption spectrophotometry (Kopito, 1970).

Additional sections of eggshell gland were trimmed into 0.5 cm sections and fixed by immersion in 3% glutaraldehyde in 0.5-M sodium cacodylate buffer at pH 7.2 for 20, 30, and 60 min. Fixed tissues were rinsed in three changes of 0.1-M cacodylate buffer in 7.5% sucrose, pH 7.2, over a 24-h period. Fresh tissue pieces were frozen and sectioned (10 μm) in a cryostat at −20°C. Tissue sections were mounted on glass slides and incubated at room temperature in Coplin jars. Ca–ATPase was localized by a slight modification of the method reported by Koenig and Vial (1973). The incubation solution contained 20 ml of 125 mg/100 ml ATP (tris salt), 20 ml of 0.02-M tris maleate buffer (pH 7.2), 5 ml of 0.1-M calcium chloride, 3 ml of 2% lead nitrate, and 2 ml of distilled water. Control incubations contained no substrate. Tissues fixed for 20 min showed adequate fixation and reaction product. These tissues were used for subsequent enzyme localization. Following incubation, tissues were examined by light microscopy to determine the optimal incubation period. To visualize the reaction product (lead), the slides were immersed for 1 min in a 2% solution of ammonium sulfide. Light microscopic incubation times were reduced by one-third for cytochemical localization within 40 μm sections. Following incubation, tissue sections were rinsed in three changes of cold 0.1-M cacodylate buffer for an additional 48 h. Tissues were postfixed in 1% phosphate-buffered osmium tetroxide, dehydrated in graded alcohol solutions, cleared in propylene oxide, and embedded in Epon (Luft, 1961). For correlation between light and electron microscopy 1 μm semithin sections were cut and stained with toluidine blue (Trump *et al.*, 1961). Thin sections were stained with uranyl magnesium acetate and lead citrate (Frasca and Parks, 1965) and viewed with an AEI 6B electron microscope.

RESULTS

Ducks fed 50 ppm of DDT and then brought into egg production had eggshell weights reduced by 12%, thicknesses by 18%, and r-values by 16% (Table 1). There were no changes in length and width. The total eggshell calcium, but not the calcium concentration, was altered by the exposure (Table 2).

The ultrastructural composition of a representative microsomal pellet is shown in Figure 1 The principal organelles were rough endoplasmic reticulum, smooth endoplasmic reticulum, and elements of the plasma membrane (e.g., microvilli and cilia).

An analysis of specific activities of microsomal Ca–ATPase from eggshell glands of both control and DDT-fed ducks revealed a significant reduction of Ca–ATPase ($P < 0.01$) after exposure. As shown in Table 3, enzyme activity in exposed ducks was 65% of control values.

The ultrastructural localization of Ca–ATP is shown in Figure 2. The reaction product can be seen on the cilia and microvilli. Cytoplasmic localization of the reaction product occurred near the apical surface of Type II cells. Type I cell cytoplasm was devoid of the reaction product. Mitochondria and rough endoplasmic reticulum were also free of the

TABLE 1   Eggshell Thickness, Weight, and r-Value
from Control and DDT-Fed Ducks[a]

Measured Variables	Control	50 ppm DDT[b]
Eggshell thickness, mm	0.412 ± 0.004	0.337 ± 0.005
Eggshell weight, g	5.30 ± 0.11	4.68 ± 0.14
r-value[c]	0.236 ± 0.004	0.198 ± 0.006

[a] Each value represents a minimum of 19 eggshells $\pm$ 1 SE.
[b] All three values are significant at $P < 0.01$.
[c] $r = \dfrac{\text{Weight}}{\text{Length x Width}}$

reaction product. Tissue from a control incubation is shown in Figure
3. No reaction product occurred in the controls.

DISCUSSION

The results of this study clearly demonstrate that the chronic feeding of
50 ppm of DDT to ducks causes alteration in eggshell quality as
evidenced by decreases in shell thickness, weight, and total calcium.
Bitman *et al.* (1969) demonstrated a reduction in the calcium content
within eggshells during the time that normal serum calcium levels were
maintained. This further indicates that the site of inhibition of calcium
transport by DDT and its metabolites is within the eggshell gland.

Morphological examination of the eggshell glands during DDT exposure revealed marked edema of glandular epithelium (Kolaja and
Hinton, 1976), indicating that changes occur within the gland during
production of the thin eggshells. Hohman and Schraer (1966) studied

TABLE 2   Eggshell Weight, Calcium Concentration,
and Total Calcium from Control and DDT-Fed Ducks[a]

Measured Variables	Control	50 ppm DDT
$\overline{X}$ eggshell wt, g	5.50	4.26[b]
$\overline{X}$ Ca^{++}/g eggshell, g	0.384 ± 0.02	0.371 ± -0.02
$\overline{X}$ total eggshell Ca^{++}, g	2.12 ± 0.29	1.57 ± 0.08[b]

[a] Each group represents a minimum of six eggshells $\pm$ 1 SD.
[b] Significant at $P < 0.01$.

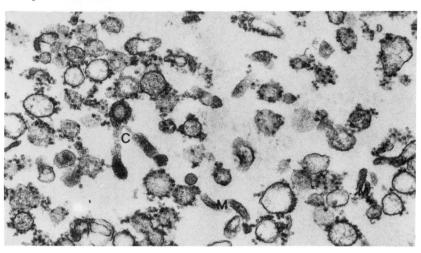

FIGURE 1 Ultrastructural appearance of microsomal fraction obtained from eggshell gland. Fragments of rough endoplasmic reticulum, smooth endoplasmic reticulum, cilia, and microvilli can be seen. (×60,000.)

the sites of calcium within the eggshell gland. They reported that calcium was released from the mitochondria to the microsomal fraction upon entry of the egg into the eggshell gland. Pike and Alvarado (1975) showed that the microsomal fraction has a high activity of Ca–ATPase.

Although the effect of DDE on mitochondrial calcium ATPase has been studied with *in vitro* preparations by Miller *et al.* (1976), we focused our attention upon the microsomal Ca–ATPase and its relationship to changes in eggshell quality upon exposure to DDT. Miller *et al.* (1976) found the mitochondrial Ca–ATPase to be the most sensitive to DDE; however, our histochemical preparations showed no activity over

TABLE 3   Microsomal Calcium ATPase Values from Eggshell of Control and DDT-Fed Ducks[a]

	Control	50 ppm DDT
Ca–ATPase[b]	9.75 ± 1.01	6.36 ± 0.66[c]
Control activity, %	100	65

[a] ± 1 SD.
[b] μmol Pi liberated/mg/protein/30 min incubation.
[c] Significant at P < 0.01.

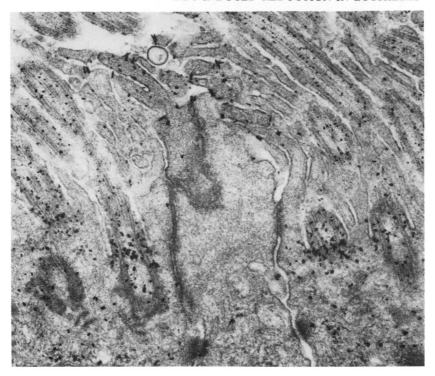

FIGURE 2 Cytochemical reaction for Ca–ATPase. The reaction product is seen on the cilia, microvilli, and in the cytoplasm near the luminal surface of Type II cells. (×25,000.)

mitochondria. The Ca–ATPase reaction product localized over apical cytoplasm of Type II cells and over plasma membrane structures such as microvilli and cilia.

Jacobsen and Jorgenson (1969) performed a quantitative biochemical and histochemical study by the lead method for localization of ATPases in another transporting epithelium, the renal tubular cell. They concluded that the plasma membrane staining, as visualized histochemically, is caused by sodium (Na)-ATPase and Ca-ATPase contained within the microsomal fraction. Further evidence that plasma membrane components could be responsible for the bio-chemical Ca–ATPase in this study was provided by the electron microscope analysis of microsomal pellets, which revealed plasma membrane constituents, e.g., microvilli and cilia (Figure 1).

Moreover, Beaufay et al. (1974) have conclusively established that up to 50% of the plasma membrane from the liver is contained within

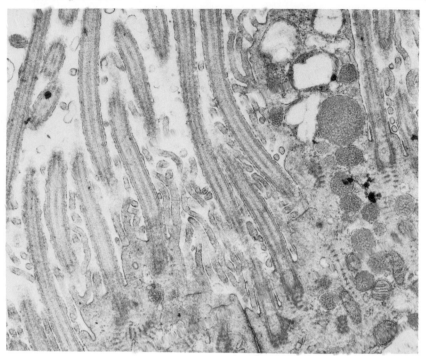

FIGURE 3 Cytochemical localization of Ca–ATPase, control. The incubation mixture contained no ATP. The shell gland is free of reaction product. (×25,000.)

the microsomal fraction. Thus, our inability to demonstrate cytochemically the reaction product over endoplasmic reticulum may be explained by the presence of the enzyme within the plasma membrane. The localization of Ca–ATPase to the apical surface appears to be an ideal location for the transport of calcium into the forming eggshell since the distance between this site and the forming eggshell is minimal. Thus, from the data resulting from this study, we propose that the plasma membrane is one of the sites involved in the DDT-induced inhibition of calcium transport in the eggshell gland.

SUMMARY

Effects of DDT ingestion on eggshells and eggshell glands were studied in the mallard duck. DDT caused a decrease in eggshell thickness, weight, and total calcium. Alterations in morphology of the eggshell gland accompanied thin-shelled eggs. Microsomal Ca–ATPase from eggshell gland epithelium was inhibited in ducks fed DDT or when DDT was added to an *in vitro* system. Cytochemical localization of Ca–ATPase to cilia and microvilli in lateral cell membranes suggests that these areas are sites of DDT-induced alterations in calcium transport.

REFERENCES

Beaufay, H., A. Armar-Costesec, D. Thines-Sempoux, M. Wibo, M. Robbi, and J. Berthet. 1974. Analytical study of microsomes and isolated subfractionation of the microsomal fraction by isopycnic and differential centrifugation in density gradients. J. Cell Biol. 61:213–231.

Bitman, J., H. C. Cecil, S. J. Harris, and G. G. Fries. 1969. DDT induces a decrease in eggshell calcium. Nature 224:44–46.

Blus, L. J., R. G. Heath, C. D. Gish, A. A. Belisle, and R. M. Prouty. 1971. Eggshell thinning in the brown pelican: Implications of DDE. BioScience 21:1213–1215.

Blus, L. J., S. N. Burkett, A. A. Belisle, and R. M. Prouty. 1974. Organochlorine residues in brown pelican eggs: A relation to reproductive success. Environ. Pollut. 7:81–91.

Cooke, A. S. 1973. Shell thinning in avian eggs by environmental pollutants. Environ. Pollut. 4:85–152.

Frasca, J. M., and V. R. Parks. 1965. A routine technique for double-staining ultrathin sections using uranyl and lead salts. J. Cell Biol. 25:157–165.

Hayat, M. A. 1970. Pp. 13–118 in Principles and Techniques of Electron Microscopy, Vol. 1. Van Nostrand Reinhold Company, New York.

Hohman, W., and H. Schraer. 1966. The intercellular distribution of calcium in the mucosa of the avian shell gland. J. Cell Biol. 30:317–331.

Jacobsen, N. O., and P. L. Jorgenson. 1969. A quantitative biochemical and histochemical study of the lead method for localization of adenosine triphosphate-hydrolyzing enzymes. J. Histochem. Cytochem. 17:443–453.

Koenig, C. S., and J. D. Vial. 1973. A critical study of the histochemical lead method for localization of Mg-ATPase at cell boundaries. Histochemistry 5:503–518.

Kolaja, G. J., and D. E. Hinton. 1976. Morphologic lesions in the eggshell gland accompanying DDT induced eggshell thinning. Environ. Pollut. 10:225–231.

Kolaja, G. J., and D. E. Hinton. 1977. *In vitro* inhibition of microsomal Ca–ATPase from eggshell gland of mallard duck. Bull. Environ. Contam. Toxicol. 17:591–594.

Kopito, L. 1970. Pp. 95–98 in Atomic Absorption Methods Manual, Vol. III. Fisher Scientific Co. Waltham, Mass.

Lowry, O. H., N. J. Rosebrough, L. A. Farr, and R. J. Randal. 1951. Protein measurement with the folin phenol reagent. J. Biol. Chem. 193:265–275.

Luft, J. H. 1961. Improvements in epoxy resin embedding methods. J. Biophys. Biochem. Cytol. 9:409–414.

Miller, D. S., D. B. Peakall, and W. B. Kinter. 1975. Biochemical basis for DDE induced eggshell thinning in ducks. Fed. Proc. 34:811.

Miller, D. S., W. B. Kinter, and D. B. Peakall. 1976. Enzymatic basis for DDT induced eggshell thinning in a sensitive bird. Nature 259:122–124.

Peakall, D. B. 1970. p,p' DDT: Effect on calcium metabolism and concentration of estradiol in the blood. Science 168:592–594.

Pike, W. J., and R. H. Alvarado. 1975. Ca²⁺–Mg²⁺ activated ATPase in the shell gland of Japanese quail (*Coturnix coturnix japonica*). Comp. Biochem. Physiol. 51B:119–125.

Prestt, I., D. J. Jefferies, and N. W. Moore. 1970. Polychlorinated biphenyls in wild birds in Britain and their avian toxicity. Environ. Pollut. 1:3–26.

Ratcliffe, D. A. 1967. Decrease in eggshell weight in certain birds of prey. Nature 215:208–210.

Risebrough, R. W., P. Rieche, D. B. Peakall, S. G. Herman, and M. N. Kerven. 1968. Polychlorinated biphenyls in the global ecosystem. Nature 220:1098–1102.

Rorive, G., and A. Kleinzeller. 1974. Ca^{++} activated ATPase from renal tubular cells. Pp. 303–307 in Methods of Enzymology, Vol. 32, Academic Press, New York.

Stanton, M. 1968. Colorimetric determination of inorganic phosphate in presence of biological material and adenosine triphosphate. Anal. Biochem. 22:27–34.

Trump, B. F., E. A. Smuckler, and E. P. Benditt. 1961. A method for staining epoxy sections for light microscopy. J. Ultrastruct. Res. 5:343–348.

QUESTIONS AND ANSWERS

W. COLLINS: Does DDT have any effect on vitamin D or its synthesis to the more active form of 1,25-dihydroxycholecalciferol and also would this have any effect on lowering the uptake of calcium by the eggshell gland?

G. KOLAJA: Vitamin D is known to induce Ca–ATPase both in the intestine and in other organs responsible for calcium transport. I know of no reports detailing the effects of DDT on vitamin D activity per se or its conversion to its more active form.

D. MILLER: As far as I know such studies have only been done in chickens, a species that is insensitive to DDE. Therefore these results probably have little bearing on the phenomenon of shell thinning. We have done some preliminary cell fractionation studies, and we consistently find high Ca–ATPase activity in the mitochondrial fraction.

G. KOLAJA: There is some controversy regarding the precise localization of Ca–ATPase. Despite the fact that the histochemical technique employing lead precipitant leads to some inhibition of Ca–ATPase, our results suggest that the enzyme activity is localized on the plasma membrane and the endoplasmic reticulum adjacent to it.

D. MILLER: Have you looked for any mitochondrial enzyme marker in your microsomal fraction?

G. KOLAJA: No, however, we assessed the purity of our fraction by electron microscopy, and after looking at several grids I could only find mitochondria in the fraction examined.

D. MILLER: There may still be fragments of mitochondrial membranes. We consistently find cross contamination in our fractions.

G. KOLAJA: That is a possibility.

D. SCARPELLI: Do you have any information on the concentration of DDT or its metabolites in the gland?

G. KOLAJA: Unfortunately we did not look at this particular aspect. We studied body fat and did indeed find concentrations as high as 1,800 ppm at the time of sacrifice, compared to a control concentration of 20 ppm in fat.

D. MILLER: We have looked at ducks in this regard after feeding a diet containing 40 ppm of DDE for 1.5 to 4 weeks. This diet resulted in approximately 1.5 ppm wet weight of DDE in the shell gland.

G. KOLAJA: It is possible that DDT reaches the eggshell gland from the egg when it passes down the reproductive tract.

D. SCARPELLI: Yes, that is an important point. The membrane components of cells, mitochondria, and microsomal fraction may accumulate DDT and its metabolites on the basis of their solubility in lipids.

Correlations Between Residues of Dichlorodiphenylethane, Polychlorinated Biphenyl, and Dieldrin in the Serum and Tissues of Mallard Ducks (*Anas platyrhynchos*)

MILTON FRIEND, MAX A. HAEGLE, DENNIS L. MEEKER, RICK HUDSON, and C. HAROLD BAER

Traditionally, levels of exposure of animals to chlorinated hydrocarbon pollutants have been measured in adipose tissue or subsamples of a composite of whole carcasses. As more sensitive analytical techniques became available, epidemiologists investigating exposure of humans to pesticides used blood to analyze residue (Davies and Edmundson, 1972; Radomski *et al.,* 1971a,b). This transition from adipose tissue to blood has been readily accepted for residue studies in humans because blood samples can be more easily obtained. However, no similar transition has taken place for pesticide–wildlife investigations. This paper supports the usefulness of blood for sampling residues of chlorinated hydrocarbons in birds.

MATERIALS AND METHODS

A series of studies was conducted with mallard ducks (*Anas platyrhynchos*) to determine the relationships among residues of selected environmental pollutants in the fat, serum, and whole bodies of the test birds. Two of these studies are reported here. In the first study, the relationships were studied during exposure of adult mallards to a single pesticide, and the second, during the degradation and elimination of three compounds given in mixture to juvenile mallards.

Blood samples were allowed to clot at room temperature for 1 to 2 h, refrigerated overnight, and then centrifuged for 20 min at 2,200 rpm. The serum was pipetted off and stored in acetone-washed glass shell vials. Fat samples were placed directly on preweighed pieces of aluminum foil, weighed, and then wrapped in this foil and stored at −68°C, along with the serum samples, until analyzed for dichlorodiphenyldichloroethylene (DDE) residues.

Juvenile Mallards Fed DDE, PCB, and Dieldrin

Thirty male and 30 female 12-week-old mallards were randomly divided into two groups of 30 each (15 of each sex). Both groups received rations containing p,p'-DDE, the polychlorinated biphenyl (PCB) Aroclor 1254, and dieldrin, in a ratio of 100:100:20 ppm (dry weight basis), respectively, for one group and 250:250:50 ppm for the other. After 2 weeks of continuous exposure, the contaminated rations were replaced by "clean rations" for the duration of the study.

Three males and three females randomly selected from each treatment group were decapitated 2 weeks after the contaminated rations were withdrawn. Six additional birds were similarly sampled 4, 6, 10, and 18 weeks after being placed on clean rations.

Adult Mallards Fed DDE

Twenty-four separately caged adult male mallards were randomly divided into two groups of 12 each. For 1 year, one group received 2.5 ppm (dry weight basis) of p,p'-DDE (the principal metabolite of dichlorodiphenyltrichloroethane [DDT] degradation in birds) in their feed; the other group received 25 ppm in the same manner. Within each group, six birds were given feed and water *ad libitum* and six were maintained on a rotating schedule of 14 days with feed and water followed by 7 days with water only (interrupted feeding). Each 28 days, blood samples of 5 to 8 ml were drawn from either the brachial or jugular veins of all birds. In addition, half of the birds on each treatment and feeding schedule (the same 12 each month) were biopsied for fat samples. Their feathers were removed along the midventral portion of the body from the base of the sternum to the vent. The area was disinfected with chlorhexidine solution and then infiltrated subcutaneously by syringe and needle with 1.5 to 2 ml of 1% lidocaine hydrochloride (a local anesthetic). An incision was made along the midline of this area, a 200- to 250-mg sample of abdominal fat was excised near the gizzard, the incision was sutured, and the bird was returned to its cage. (The study began in January; biopsies were taken each month from April to the following January, except August.)

At the time of death, 10- to 12-ml blood samples were collected by cardiac puncture from each bird, and a fat sample of about 200 mg was removed from the gizzard area. The remainder of each carcass was prepared for analysis of whole body residues by removing the feathers, the bill, the wings (at the elbow), and the legs (at the knees). Carcasses and fat samples were weighed, wrapped in aluminum foil, and stored at $-20°C$ until homogenized and were then analyzed. Blood samples were treated and stored as described above.

All residue analyses were conducted at the Denver Wildlife Research Center by two-column gas chromatography with electron-capture detection. Values given are on a wet-weight basis unless indicated otherwise.

RESULTS

Adult Mallards Fed DDE

Generally, DDE residues increased progressively throughout the study, but were slightly higher in birds fed intermittently (subjected to periodic starvation) than in birds fed continuously (mean in sera, 1.94 versus 1.82 ppm), and slightly higher in birds subjected to the stress of monthly surgery than in those not biopsied (mean in sera, 2.19 versus 1.67 ppm). However, in neither group were these differences significant ($P > 0.05$; four-factor analysis of variance with repeated measurements on the fourth factor, time). Monthly sampling of birds on interrupted feeding resulted in three different experimental sampling times—10 days after pesticide feeding was initiated, 3 days after cessation of feeding, and 3 days after resumption of feeding. There were no consistent trends in residues associated with these sampling times.

As expected, both the higher DDE exposure (25 ppm) and increasing length of exposure resulted in significant elevations in DDE residues. For sera, both differences were significant at $P < 0.001$ ($F = 55.08$, 1, 10 df, for exposure level; $F = 13.95$, 12, 120 df for time exposed). We did not anticipate that the 10-fold difference between the two DDE exposure levels would be reflected in an almost exact 10-fold difference between the mean DDE residues for the two groups at every sampling period. Overall means were 0.33 versus 3.42 in sera (Table 1) and 82.6 versus 871.5 ppm in fat (Table 2). For both exposure levels, residues peaked in September and appeared to have leveled off when the final samples were taken in January.

One hundred eleven paired serum and fat samples from 10 sampling times were available for comparison of DDE residues. The correlation coefficient (r) for these samples was 0.929 ($r^2 = 0.863$). The relatively

TABLE 1   DDE Residues in the Serum of Adult Male
Mallards Exposed to 2.5 or 25 ppm (Dry Weight) of
p,p'-DDE in Their Feed for 1 Year

Date of Sample	DDE Residues (Wet Weight) by Exposure Group					
	2.5 ppm (Dry Weight)			25 ppm (Dry Weight)		
	$\overline{X}$	SD	N	$\overline{X}$	SD	N
18 Jan. 1973	0.02	0.04	12	0.13	0.43	12
7 Feb.	0.10	0.06	12	1.09	0.95	12
7 Mar.	0.14	0.05	12	1.78	0.76	11
14 Apr.	0.21	0.08	12	1.81	0.63	11
12 May	0.29	0.15	12	2.87	1.44	11
30 May	0.31	0.14	12	3.48	1.16	11
27 June	0.28	0.09	12	2.49	0.65	11
25 July	0.32	0.17	12	2.66	1.07	10
31 Aug.	0.34	0.12	12	4.90	1.73	10
13 Sept.	0.49	0.16	12	5.95	2.48	10
10 Oct.	0.47	0.10	11	5.45	1.45	10
7 Nov.	0.45	0.15	11	4.99	1.72	10
5 Dec.	0.52	0.20	10	4.67	1.69	9
2 Jan. 1974	0.48	0.13	9	4.42	1.89	9

narrow dispersal of data points about the calculated regression line and
the distinct separation between data points for the 2.5- and 25-ppm
treatments indicate that DDE residue concentrations in the sera of
mallards are a good index to DDE concentrations in the fat of these
birds and, therefore, to the intensity of DDE exposure. Correlation
coefficients for the individual sampling times (N = 7 to 12) ranged from
0.885 (May 2) to 0.991 (October 10). There was no evidence of seasonal
changes in these relationships.

In addition to blood and fat, whole bodies and brain tissues were
available for analysis on January 2, when 15 of the 18 survivors were
killed. Again, residues contained in the different tissues were strongly
correlated. Correlation coefficients were 0.926 for sera and fat, 0.833
for sera and whole body, 0.936 for fat and whole body, and 0.830 for
brain and whole body.

Juvenile Mallards Fed DDE, PCB, and Dieldrin

The correlations for DDE, PCB, and dieldrin residues in sera and whole
body compared closely with those for fat and whole body (Table 3).

These data complement those of the previous study (adult mallards fed DDE) and suggest that, in addition to DDE, residues of PCB's and dieldrin in the sera of mallards are also a good index to whole body burdens of those compounds. In all cases, either sera or fat adequately reflected whole body burdens of these residues (Table 3).

Correlation coefficients were also determined for the two different exposure levels. The higher exposures resulted in slightly higher r values for the fat and whole body residues for each compound. Slightly higher r values were obtained at the lower exposures for the sera and whole body residue for DDE and PCB, and at the higher exposures for dieldrin. In all instances these differences were too slight (range of 0.01 to 0.10 r^2 units) to have importance.

DISCUSSION

Dale *et al.* (1966) reported that residues of p,p'-DDT, o,p'-DDT, p,p'-DDE, beta-BHC (beta hexachloride [hexachlorocyclohexane]), dieldrin, and heptachlor epoxide, environmental contaminants that are routinely found in fat tissue of the human population, could also be detected easily in blood plasma. Therefore, they concluded that plasma or serum values were satisfactory for toxicological studies. This conclu-

TABLE 2   DDE Residues in Abdominal Fat of Adult Male Mallards Exposed to p,p'-DDE in Their Feed from 18 January 1973 to 2 January 1974

| | DDE Residues (Wet Weight) by Exposure Group | | | | | |
| | 2.5 ppm (Dry Weight) | | | 25 ppm (Dry Weight) | | |
Date of Sample[a]	X̄	SD	N	X̄	SD	N
4 Apr.	40.17	15.99	6	304.00	98.62	6
2 May	78.33	56.78	6	725.00	111.85	6
30 May	65.00	19.66	5	648.33	80.35	6
27 June	58.33	8.12	6	586.00	123.00	5
25 July	77.83	12.70	6	622.00	24.90	5
13 Sept.	94.33	21.11	6	1,108.00	127.75	5
10 Oct.	96.00	15.23	5	992.00	254.79	5
7 Nov.	79.80	10.99	5	934.00	311.74	5
5 Dec.	97.50	26.25	4	1,000.00	203.14	4
1 Jan.	134.29	19.02	7	1,030.00	240.30	8

[a] No samples taken before January 4 or in August.

TABLE 3   Coefficients of Determination for Fat, Serum, Whole-Body Residues of Selected Environmental Pollutants Fed to 12-Week-Old Mallards

	Coefficients of Determination (r^2) of Residues of Compounds		
Tissues Compared[a]	DDE	Dieldrin	Aroclor 1254
Whole body and serum	0.885	0.929	0.869
Whole body and fat	0.805	0.886	0.943
Fat and serum	0.801	0.836	0.801

[a] N = 60 for each comparison.

sion has been supported by data showing that residues of dieldrin, lindane, DDT, and DDE are closely correlated in the blood and fat of humans (Radomski *et al.*, 1971b). Our studies indicate that residues of DDE, dieldrin, and PCB in the sera of mallards adequately reflect whole-body burdens of these compounds. The feasibility of using blood as a sample tissue for residue studies of chlorinated hydrocarbon pollutants in birds is demonstrated in these studies to a greater extent than ever before.

There are three decisive advantages to blood sampling over the lethal sampling methods generally used in investigations of environmental pollution:

1. Because animals do not have to be killed to obtain blood, large-scale sampling is possible, even with rare species, and discrete populations can be repeatedly sampled in sufficient numbers to establish exposure patterns and detect biologically significant changes in residue levels.

2. Multiple samples can be taken over time from the same individuals, thus making possible certain studies not feasible in the past—for example, laboratory investigations of the relation over time of pesticide residues in the bodies of individual animals to such functions as behavior, reproduction, or response to disease organisms. Also, samples obtained from captive wildlife in "natural environments" could be used to study the rate of residue accumulation from pollution.

3. The technique is economical. Blood samples are much easier to collect than fat samples, take less freezer space than whole-body samples, and can be analyzed faster and more easily than either.

Although our studies are not comprehensive, we believe that the strength of these data clearly demonstrate that contamination of wildlife by chlorinated hydrocarbons can be adequately monitored by serum analysis. This is especially true for field monitoring situations where it is necessary to know only whether an animal is carrying low, moderate, or high concentrations of residues.

SUMMARY

Adult male mallard ducks, *Anas platyrhynchos,* were subjected to 2.5 or 25 ppm (dry weight basis) DDE in their feed for 12 months. Paired blood samples were taken, and visceral fat was surgically removed from the same birds at nine monthly intervals. The correlation between DDE concentrations in these paired blood and fat samples was significant (r + 0.929, N = 111), leading us to conclude that DDE residues in blood serum were as good an index to DDE exposure as those in adipose tissue. Monthly correlations ranged from a low of 0.885 to a high of 0.991. Comparisons were also made of residue correlations in brain, whole body, fat, and blood at the end of the experiment.

Mallard ducklings maintained on diets contaminated with one of two concentrations of DDE, PCB, dieldrin, or a combination of all three contaminants were periodically sampled over a 1-year period. Exposure patterns consisted of continuous or interrupted presence of the compound in the rations provided, and periodic starvation periods during which rations were withdrawn. Correlations were determined between residues of these compounds in sera versus fat, sera versus whole body, and fat versus whole body during each sample period. In most instances, serum residues compared favorably with fat residues as a reflection of whole-body residues of these compounds.

REFERENCES

Dale, W. E., A. Curley, and C. Cueto, Jr. 1966. Hexane extractable chlorinated insecticides in human blood. Life Sci. 5:47–54.
Davies, J. H., and W. F. Edmundson, eds. 1972. Epidemiology of DDT. Futura Publishing Co., Inc., Mount Kisco, N.Y. 157 pp.
Radomski, J. L., E. Astolfi, W. B. Deichmann, and A. A. Rey. 1971a. Blood levels of organochlorine pesticides in Argentina: Occupationally and nonoccupationally exposed adults, children and newborn infants. Toxicol. Appl. Pharmacol. 20:186–193.
Radomski, J. L., W. B. Deichmann, A. A. Rey, and T. Merkin. 1971b. Human pesticide blood levels as a measure of body burden and pesticide exposure. Toxicol. Appl. Pharmacol. 20:175–185.

QUESTIONS AND ANSWERS

G. CHOULES: Is there a difference in the way fat and serum were extracted in this study?

M. FRIEND: Not really. It's a standard laboratory procedure using Peterson's method that was modified slightly for the serum.

J. ZINKL: You reported that there was a good correlation between the concentration of DDE in whole-body fat and in serum. Is the formula describing the slopes of the regression lines similar in both experiments?

M. FRIEND: I can only give you the formula for the regression line on the adult mallard duck study since I happen to have that data with me. I would not expect the regression line for the other study to be identical. For adult mallard ducks the relationship was $\hat{Y} = 0.0392 + 0.0050X$. Dr. David Capen[1] found essentially similar results in work with free-living ibis (*Plegadis chihi*) in Utah. There were strong correlations between DDE residues in fat and in serum.

J. BOGAN: Your correlation coefficients are impressive; however, they may be misleading since the kinetics of DDE movement between fat and serum will be slow. In continuous feeding experiments, the longer the experiment proceeds, the less effect 1 day's diet will have on the ratio. This may be the reason for the good correlation you have obtained. It would be interesting to know what the correlation coefficients are in a situation where an uncontaminated bird is given a single dose of a pesticide. Clarification of this point is important, because individuals may be inclined to take serum levels of DDE as an estimate of whether the bird died of DDT poisoning when in fact the equilibrium had not been established between fat and serum or between fat, serum, and brain.

M. FRIEND: Although I am still wary of these data, I was impressed by the fact that when we sampled birds 3 days after they had been on the pesticide-containing diet and compared them to ones sampled 10 days after being placed on this diet, there were no significant differences in the correlations. The best correlations we obtained were with birds on a continuous exposure diet and with birds fed a mixture of the three compounds. We fed these particular compounds simultaneously, because at the time of our study, all three commonly appeared together in the tissues of wildlife being sampled for the purpose of monitoring environmental contamination.

A. deFREITAS: It would improve your study if you were to pulse-feed your animals and then starve them for 24 h prior to taking a blood sample. This should improve the correlation since this maneuver would establish an equilibrium between the ingested material and its distribution in the blood. The foregoing is true for the distribution of any substance in blood that is lipid soluble.

[1] Capen, D. 1977. The impact of pesticides on the white-faced ibis. Ph.D. dissertation. Utah State University, Logan. 85 pp.

Ultrastructural and Biochemical Effects of Polychlorinated Biphenyls (PCB) on the Thyroid Gland of Osborne Mendel Rats

WILLIAM T. COLLINS, JR., LOUIS KASZA,
and CHARLES C. CAPEN

The widespread contamination of the environment and disease-producing capability of polychlorinated biphenyl (PCB) compounds have been well documented (Kimbrough, 1974). PCB residues have been detected in rivers and oceans and in the tissues of wildlife, cattle, and poultry (Hammond, 1972). Human beings and domestic animals have been intoxicated following the accidental contamination of foodstuffs with PCB (Kuratsunc et al., 1972; Platonow et al., 1971). Lesions resulting from PCB intoxication include the induction of hepatic microsomal enzymes and fatty degeneration and necrosis of hepatocytes (Kasza et al., 1976). Epidermal hyperplasia and hyperkeratosis, porphyria, and degeneration of lymphoid tissues and kidney have been reported following PCB intoxication (Vos and Beems, 1971; Vos and Koeman, 1970). PCB also has adverse effects on reproduction and growth in several different animal species (Hansen et al., 1975, 1976; Ax and Hansen, 1975).

PCB may cause alterations in thyroid structure and thyroxine metabolism. In birds administered PCB, there is enlargement of the thyroid gland and alteration of ^{131}I uptake by follicular cells (Hurst et al., 1974). The administration of PCB to rats resulted in an increased ^{131}I uptake, lowered serum thyroxine concentration, reduced protein-binding of thyroid hormone, and increased conjugation of thyroxine and excretion of thyroxine–glucuronide in the bile (Bastomsky, 1974; Bastomsky and Murthy, 1976; Bastomsky and Wyse, 1975; Bastomsky et al., 1976). These findings suggest that some of the metabolic alterations produced by PCB intoxication may be related to alterations in thyroid structure and function.

327

The objectives of this investigation were to evaluate the histopathological, histochemical, and ultrastructural changes in thyroid follicular cells produced by the acute and chronic administration of high and low doses of PCB to rats; to correlate the structural alterations in follicular cells with changes in serum thyroxine concentration; and to investigate the persistence of the effects of PCB on the ultrastructure of thyroid follicular cells and serum thyroxine levels.

MATERIALS AND METHODS

Eight-week-old, male, Osborne-Mendel rats (FDA colony) were fed Purina Rat Chow mixed with Aroclor 1254 (Monsanto Co., Inc., St. Louis, Mo.) at concentrations of 50 and 500 ppm for 4 and 12 weeks. The PCB was mixed into the pulverized feed using corn oil as a vehicle. Control rats received a similar diet with 3% corn oil but without PCB and were housed in a separate room. The delayed and long-term delayed effects of PCB on the thyroid were investigated by feeding rats 50 and 500 ppm of PCB mixed into rat chow for 12 weeks followed by 12- and 35-week periods without the compound prior to euthanasia. Rats in the high-dose groups received 500 ppm of PCB for the first 6 weeks but were switched to 250 ppm for the last 6 weeks due to anorexia and weight loss. All diets were stored at −25°C in hexane-cleaned metal containers until use. The animals were housed individually in stainless steel cages with wire mesh bottoms. They were exposed to 12 h of light followed by 12 h of darkness. Feed and water were available *ad libitum*.

Thyroid glands were collected from five rats from each exposure period (4 weeks, 12 weeks, 12 weeks with 12-week recovery, and 12 weeks with 35-week recovery) and dose level (0, 50, and 500 ppm of PCB) for ultrastructural evaluation. The rats were sacrificed immediately after each period by carbon dioxide asphyxiation. Thyroid glands were collected for electron microscopic, histopathological, and histochemical evaluation.

Tissue for electron microscopy was minced immediately under fixative into 0.5 to 1.0 mm³ blocks, fixed in cold 3% glutaraldehyde with 0.1 M sodium cacodylate, buffered at pH 7.4, postfixed in 1% osmium tetroxide in 2,4,6-trimethylpyridine, dehydrated in graded ethanols, transferred to propylene oxide, and embedded in Epon (Shell Chemical Company, New York, N.Y.). Thin sections were cut with a diamond knife on an LKB or Reichert OmU2 ultramicrotome and floated on a water bath buffered at pH 7.4. Sections were stained with uranyl acetate and lead citrate. They were examined with a Philips 200 or 300 electron microscope.

Five additional rats from each exposure period and dose level were used for histochemical evaluation of the thyroid gland and serum levels of thyroid hormone. Tissues for histochemistry were collected in dry ice at necropsy. Acid phosphatase naphthol-AS-B1 reaction was evaluated on selected thyroid sections from rats in all exposure groups (Burstone, 1958). Thyroid glands to be histopathologically evaluated were fixed in phosphate-buffered formalin and stained with hematoxylin, eosin, and periodic acid-Schiff stain. Serum thyroxine was determined in five rats from each exposure period and PCB dose by radioimmunoassay (Bio-Science Laboratories, Van Nuys, Calif.).

RESULTS

Histological and Ultrastructural Evaluation of Thyroid Glands

Low-Dose PCB (50 ppm) for 4 Weeks   Thyroid glands of rats receiving the low dose of PCB for 4 weeks were slightly enlarged. Follicular cells were larger and more columnar than in controls (Figure 1). Profiles of rough endoplasmic reticulum were numerous and their cisternae frequently were dilated with a finely granular material. The Golgi apparatuses were more prominent than in controls and associated with many small granules. The oval nucleus was placed basally in follicular

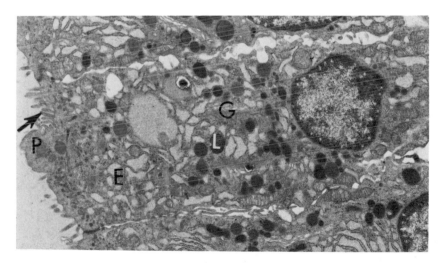

FIGURE 1 Columnar thyroid follicular cells with a prominent Golgi apparatus (G), profiles of dilated rough endoplasmic reticulum (E), and moderately increased numbers of lysosomal bodies (L). Microvilli are few in number (arrow) and portions of apical cytoplasm (P) project into the follicular lumen. Rat fed 50 ppm PCB for 4 weeks. (×7,000.)

cells. Microvilli were shortened, irregular in shape, and abnormally branched. Long projections of follicular cell cytoplasm often extended from the apical surface into the luminal colloid (Figure 1). Electron-dense apical vesicles appeared to accumulate immediately beneath the microvilli. Large, abnormally shaped lysosomal bodies with a hetero-geneous internal structure were present in increased numbers (Figure 2), and acid phosphatase activity was increased compared to control rats.

High-Dose PCB (500 ppm) for 4 Weeks Thyroid glands were enlarged and were composed of small, irregularly shaped follicles lined by a single layer of columnar cells. Papillary projections of hyperplastic follicular cells extended into the lumens of some follicles. The cyto-plasmic area contained long profiles of rough endoplasmic reticulum often irregularly dilated by a finely granular material. Mitochondria were frequently swollen and had disrupted cristae (Figure 3). The basally located nucleus was oval and contained coarsely granular chromatin. Striking changes were detected in the microvilli on the luminal border of follicular cells. Microvilli were shortened and irregu-larly branched (Figure 3). Areas of the luminal surface of follicular cells were completely devoid of microvilli. These areas often had large

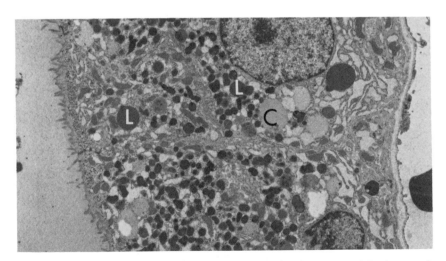

FIGURE 2 Thyroid follicular cells with expanded cytoplasmic area containing increased numbers of lysosomal bodies (L) and colloid droplets (C). Microvilli on the apical surface are present in normal numbers but are shorter than in control rats. Rat fed 50 ppm PCB for 4 weeks. (×5,500.)

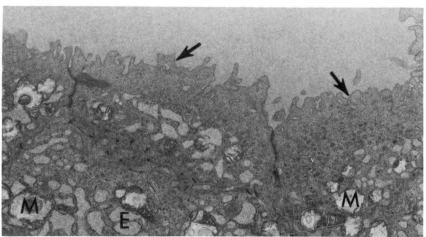

FIGURE 3 Apical surface of thyroid follicular cells with blunt, abnormally branched microvilli (arrows). The hypertrophied cytoplasmic area contains dilated profiles of rough endoplasmic reticulum (E) and numerous vacuolated mitochondria (M). Rat fed 500 ppm PCB for 4 weeks. (×8,000.)

cytoplasmic projections extending into the luminal colloid. Numerous apical vesicles were present immediately beneath the altered lumenal surface. There were increases in the number of membrane-limited colloid droplets and abnormally large electron-dense lysosomal bodies within the cytoplasm. Compared to controls, there was increased acid phosphatase activity in the cytoplasm of follicular cells.

Low-Dose PCB (50 ppm) for 12 Weeks Thyroid glands were enlarged. Follicles were lined by single or multiple layers of columnar cells. The rough endoplasmic reticulum was composed of small profiles that were irregularly dilated with finely granular material. The Golgi apparatus was compressed near the nucleus and was less extensive than in controls. Mitochondria were large but frequently were swollen and had disrupted cristae. There was a marked reduction in the number and length of microvilli on the luminal surface of follicular cells. The surface of the follicular cells was irregular and had large projections of apical cytoplasm that extended into the colloid. There was a marked increase in large, membrane-limited colloid droplets in the cytoplasm. In addition, lysosomal bodies were increased in follicular cells of rats of this group. They were extremely electron-dense, irregular in size and shape, and occasionally appeared to be fused with the colloid droplets. Acid phosphatase activity in follicular cells was strongly increased compared to control rats.

High-Dose PCB (500 ppm) for 12 Weeks   Thyroid glands had changes similar to but consistently more severe than those described in rats fed 50 ppm for 12 weeks. Thyroid follicles were irregular in size and lined by tall columnar cells that occasionally formed papillary projections into the lumen (Figure 4). The Golgi apparatus and rough endoplasmic reticulum were less well developed in follicular cells due to the abnormal accumulation of numerous lysosomal bodies and colloid droplets. The cytoplasmic area of some hypertrophied follicular cells appeared to be distended by the large numbers of lysosomal bodies with strong acid phosphatase activity, and colloid droplets. Mitochondria were more irregular and swollen with disrupted cristae than in rats from any of the other experimental groups. Microvilli on the luminal surfaces of most follicular cells appeared to be abnormally short, blunt, and branched. Extensive areas of the luminal surface of follicular cells were devoid of

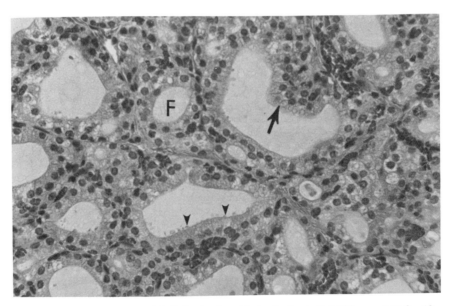

FIGURE 4 Irregularly sized follicles in thyroid gland of rats fed 500 ppm PCB for 12 weeks. Many follicles (F) were smaller than in control rats and lined by tall columnar epithelial cells. Papillary projections of hyperplastic follicular cells (arrow) and apical cytoplasmic processes (arrowheads) extended into the follicular lumens. (Hematoxylin and eosin. ×315.)

microvilli. Occasionally, unique cytoplasmic projections extended into the colloid.

Delayed Effects of Low-Dose (50 ppm) and High-Dose (500 ppm) PCB Thyroid follicles were enlarged and lined either by tall columnar or hypertrophied cuboidal cells, but there was less evidence of multiple layers of follicular cells than in rats receiving PCB for 12 weeks and then killed immediately. The rough endoplasmic reticulum was less well developed than in controls and the Golgi apparatus was small. Occasional mitochondria were swollen and had disrupted cristae. Microvilli on the luminal surface of some follicular cells were short and blunt; however, many cells had microvilli that appeared normal. Large cytoplasmic projections from the luminal surface were observed infrequently. Numerous irregular lysosomal bodies and colloid droplets also were present in follicular cells.

Long-Term Delayed Effects of Low- and High-Dose PCB Follicular cells of rats receiving either 50 or 500 ppm of PCB for 12 weeks followed by 35 weeks without PCB were similar to those of control rats both histologically and ultrastructurally. Thyroid follicles were large and lined by a single layer of cuboidal to low columnar cells. The endoplasmic reticulum was well developed and consisted of long profiles with narrow cisternae containing a finely granular material. Dilated profiles of endoplasmic reticulum were present, but they were considerably reduced compared to previous experimental groups. The Golgi apparatus was prominent and was associated with numerous dense granules. The number of mitochondria in follicular cells was similar to that in the control rats. Only an occasional mitochondrion was swollen with disruption of cristae. Microvilli on the luminal surface were numerous and had normal configurations. There were apical vesicles immediately beneath the microvilli. Lysosomal bodies and colloid droplets were markedly decreased in most follicular cells compared to previous experimental groups. The lysosomal bodies in the follicular cells resembled those in controls. They were round with a more homogeneous internal structure. Only a few follicular cells contained numerous, irregularly shaped lysosomes similar to those in the previous experimental groups. There was only a mild increase in acid phosphatase activity compared to controls. The residual ultrastructural alterations in follicular cells of rats from this group appeared to be minimal compared to all other PCB-treated rats.

Ultrastructural alterations related to PCB administration were not detected in thyroid C-cells.

TABLE 1 Serum Thyroxine of PCB-Treated and Control Rats Determined by Radioimmunoassay

Serum Thyroxine Concentrations, μg/dl

Rat Group	Acute Effects (4 Weeks PCB)	Chronic Effects (12 Weeks PCB)	Delayed Effects (12 Weeks PCB; 12 Weeks No PCB)	Long-Term Delayed Effect (12 Weeks PCB; 35 Weeks No PCB)
Controls	6.66 ± 0.3	7.18 ± 0.4	7.86 ± 0.8	6.18 ± 0.9
50 ppm of PCB	4.80 ± 0.3[a]	1.96 ± 0.2[b]	4.90 ± 0.1[c]	5.86 ± 1.2
500/250 ppm of PCB	2.10 ± 0.2[b]	1.78 ± 0.08[b]	3.02 ± 0.8[a]	6.02 ± 1.3

[a] $P < 0.005$.
[b] $P < 0.001$.
[c] $P < 0.025$.

334

Serum Thyroxine

Serum thyroxine concentration in control and PCB-fed rats was determined by radioimmunoassay. Groups were compared statistically using Students' t-test (Table 1). Thyroxine levels were significantly reduced in rats fed 50 and 500 ppm of PCB daily for 4 and 12 weeks. The greatest reduction in serum thyroxine levels occurred at 12 weeks in rats receiving 500 ppm PCB (Table 1). Serum thyroxine values returned toward normal in rats administered either high or low doses of PCB for 12 weeks followed by 12 weeks without PCB prior to euthanasia; however, the values remained significantly lower than in the control rats (Table 1). In rats receiving PCB for 12 weeks followed by 35 weeks without PCB, the serum thyroxine returned to the normal range (Table 1).

DISCUSSION

The striking ultrastructural alterations in thyroid follicular cells of rats following PCB administration contributed in part to the highly significant decrease in serum thyroxine levels in experimental rats. In spite of the alterations in microvillar structure, follicular cells appeared capable of responding to thyroid-stimulating hormones (TSH) and taking up colloid droplets by endocytosis. However, the increased lysosomal bodies with strong acid phosphatase activity in follicular cells of experimental rats appeared unable to interact with colloid droplets in a normal manner and to hydrolyze the cleavage of active thyroid hormone from the molecular structure of thyroglobulin. This resulted in a remarkable accumulation of colloid droplets and lysosomal bodies after chronic administration of PCB, resulting in a displacement of synthetic organelles such as the endoplasmic reticulum and Golgi apparatus.

In addition to the direct effect on follicular cells, PCB compounds significantly enhance the peripheral metabolism of thyroxine and reduce the binding of thyroid hormones to serum proteins. This results in a lowering of serum thyroxine and protein-bound iodine levels in rats (Bastomsky, 1974). The biliary excretion of thyroxine is enhanced (4- to 5-fold) by PCB and there is an increased proportion of biliary ^{125}I as thyroxine–glucuronide (Bastomsky *et al.*, 1976). The increased hepatic conjugation of thyroxine to glucuronic acid and excretion in the bile of rats receiving PCB probably are secondary to the induction of hepatic microsomal T_4–UDP (uridine diphosphate) glucuronyltransferase (Bastomsky and Murthy, 1976).

The results of this investigation and of studies reported in the literature demonstrate a highly significant reduction in serum thyroxine by PCB. The lowering of circulating thyroxine levels by PCB is dose- and time-dependent. It appears to be the combined result of a direct effect on thyroid follicular cells with an interference in hormone secretion plus an enhanced peripheral metabolism of thyroxine. Some of the metabolic alterations produced by PCB intoxication in laboratory animals and human beings, such as decreased weight gain (Kuratsune *et al.*, 1972; Hansen *et al.*, 1976), reduced feed efficiency (Hansen *et al.*, 1976), decreased reproductive performance (Ax and Hansen, 1975), and skin lesions with hyperpigmentation, may be related to an alteration in thyroid function.

SUMMARY

Polychlorinated biphenyls are widely used industrial chemicals that have gained wide access to the environment. The objectives of this investigation were to determine the acute and delayed effects of low- and high-dose exposure to PCB on the ultrastructure of thyroid follicular cells and to correlate the structural changes with alterations in serum thyroxine by radioimmunoassay. Results of the study suggest that PCB interferes with thyroxine biosynthesis in rats and that some metabolic alterations observed during PCB intoxication in animals and humans could be related to altered thyroid function.

ACKNOWLEDGMENTS

This subject was supported by grants GM-1052 and RR-05463 from the National Institutes of Health, U.S. Public Health Service, and contract 223-74-2094 from the Food and Drug Administration.

REFERENCES

Ax, R. L., and L. G. Hansen. 1975. Effects of purified polychlorinated biphenyl analogs on chicken reproduction. Poult. Sci. 54:895–900.

Bastomsky, C. H. 1974. Effects of a polychlorinated biphenyl mixture (Aroclor 1254) and DDT on biliary thyroxine excretion in rats. Endocrinology 95:1150–1155.

Bastomsky, C. H., and P. V. N. Murthy. 1976. Enhanced *in vitro* hepatic glucuronidation of thyroxine in rats following cutaneous application on ingestion of polychlorinated biphenyls. Can. J. Physiol. Pharmacol. 54:23–26.

Bastomsky, C. H., and J. M. Wyse. 1975. Enhanced thyroxine metabolism following cutaneous application of microscope immersion oil. Res. Commun. Chem. Pathol. Pharmacol. 10:725–733.

Bastomsky, C. H., P. V. N. Murthy, and K. Banovac. 1976. Alterations in thyroxine metabolism produced by cutaneous application of microscope immersion oil: Effects due to polychlorinated biphenyls. Endocrinology 98:1309–1314.

Burstone, M S. 1958. Histochemical comparison of naphthol AS-phosphates for the demonstration of phosphatases. J. Natl. Cancer Inst. 20:601–614.

Hammond, A. L. 1972. Chemical pollution: Polychlorinated biphenyls. Science 175:155–156.

Hansen, L. G., C. S. Byerly, R. L. Metcalf, and R. F. Bevill. 1975. Effect of a polychlorinated biphenyl mixture on swine reproduction and tissue residues. Am. J. Vet. Res. 3:23–26.

Hansen, L. G., D. W. Wilson, and C. S. Byerly. 1976. Effects on growing swine and sheep of two polychlorinated biphenyls. Am. J. Vet. Res. 37:1021–1024.

Hurst, J. G., W. S. Newcomer, and J. A. Morrison. 1974. Some effects of DDT, toxaphene, and polychlorinated biphenyl on thyroid function in Bobwhite quail. Poult. Sci. 53:125–133.

Kasza, L., M. A. Weinberger, C. Carter, D. E. Hinton, B. F. Trump, and E. A. Brouwer. 1976. Acute, subacute, and residual effects of polychlorinated biphenyl (PCB) in rats. II. Pathology and electron microscopy of liver and serum enzyme study. J. Toxicol. Environ. Health 1:689–703.

Kimbrough, R. D. 1974. The toxicity of polychlorinated polycyclic compounds and related chemicals. CRC Crit. Rev. Toxicol. 2:445–498.

Kuratsune, M., T. Yoshimura, J. Matsuzaka, and A. Yamaguchi. 1972. Epidemiologic study on Yusho, a poisoning caused by ingestion of rice oil contaminated with a commercial brand of polychlorinated biphenyls. Environ. Health Perspect. 1:119–128.

Platonow, N. S., P. W. Saschenbrecker, and H. S. Funnell. 1971. Residues of polychlorinated biphenyls in cattle. Can. Vet. J. 12:115–118.

Vos, J. G., and R. B. Beems. 1971. Dermal toxicity studies of technical polychlorinated biphenyls and fractions thereof in rabbits. Toxicol. Appl. Pharmacol. 19:617–633.

Vos, J. G., and J. H. Koeman. 1970. Comparative toxicologic study with polychlorinated biphenyls in chickens with special reference to porphyria, edema formation, liver necrosis and tissue residues. Toxicol. Appl. Pharamcol. 17:656–668.

QUESTIONS AND ANSWERS

D. SCARPELLI: Has anyone demonstrated alterations in the basal metabolic rate in animals chronically exposed to PCB?

D. BARSOTTI: We have never seen any alteration in the microscopic structure of the thyroid gland in our work on the effects of PCB on rhesus monkeys.

W. COLLINS: Both histological and gross changes in thyroid glands are very subtle indeed. The major changes occur at the ultrastructural level.

D. SCARPELLI: What happens to the PCB? Is it recycled back into fat depots or is it excreted totally from the organism?

D. HINTON: We studied the livers of animals 1 year after cessation of a 1-month exposure to PCB. We found sustained high levels in body fat and approximately 50 to 100 ppm in liver. Furthermore, the liver did not show normal morphology even 1 year after exposure ended. The biological half-life of PCB in adipose tissue of rats in this study was 8 weeks in males and 12 weeks in females.[1] In ultrastructural studies of livers in some of these animals, altered

[1] Braunberg, R. C., R. E. Dailey, E. A. Brouwer, L. Kasza, and A. M. Blaschka. 1976. Acute, subacute, and residual effects of polychlorinated biphenyl (PCB) in rats. I. Biologic half-life in adipose tissue. J. Toxicol. Environ. Health 1:683–688.

smooth endoplasmic reticulum persisted 1 year after cessation of exposure.[2] This indicated to us the protracted nature of PCB effects.

G. KOLAJA: Your preservation of tissues for ultrastructural studies was excellent. Would you comment on your fixation techniques.

W. COLLINS: They were fixed by immersion in cold glutaraldehyde.

S. NIELSEN: Would you comment on the scalloping present in the colloid?

W. COLLINS: I have associated this with increased activity of the thyroid gland, but I cannot give you the precise reasons for it.

S. NIELSEN: In Aroclor dermatitis or chloracne there is considerable hyperkeratosis, but it has never been associated with hypothyroidism, as in your animals.

W. COLLINS: I am not familiar with the details of hypothyroidism in rats. But in dogs, hypothyroidism is associated with hyperkeratosis.

[2] Kasza, L., M. A. Weinberger, C. Carter, D. E. Hinton, B. F. Trump, and E. A. Brouwer. 1976. Acute, subacute, and residual effects of polychlorinated biphenyl (PCB) in rats. II. Pathology and electron microscopy of liver and serum enzyme study. J. Toxicol. Environ. Health 1:689–703.

Exposure of Infant Rhesus Macaques (*Macaca mulatta*) to Polychlorinated Biphenyl (PCB)

L. A. CARSTENS, D. A. BARSOTTI, and J. R. ALLEN

Recently, there has been considerable interest regarding possible injurious effects in the fetus and neonate following exposure to polychlorinated biphenyls (PCB's). Transplacental movement of PCB's has been found in a number of animal species (Platonow and Reinhart, 1973; Allen *et al.*, 1974; Török and Kriegel, 1975), as has continued transmammary exposure of infants to PCB's during nursing (Allen and Barsotti, 1976; Zabik *et al.*, 1976). Embryotoxic and teratogenic abnormalities have been observed in the offspring of lower animals exposed to PCB's (Villeneuve *et al.*, 1971; Cecil *et al.*, 1974). In addition, premature births and weakness of infants appear to be complications resulting from exposure to PCB's (Ringer *et al.*, 1972; DeLong *et al.*, 1973). Moreover, both intrauterine and neonatal exposure to PCB's may be responsible for abnormal psychological development (Shiota, 1976; Bowman *et al.*, 1978).

Possible effects of PCB's on reproduction in nonhuman primates were first observed in female rhesus monkeys (*Macaca mulatta*) that were fed a diet containing 25 ppm Aroclor 1248 for 2 months (Allen *et al.*, 1974). During this time, the monkeys consumed approximately 250 mg of PCB's. After removing the animals from the experimental diet for 3 months, investigators attempted to breed them. Three of the five animals conceived; the other two were bred repeatedly without conceiving. Two of the three animals that did become pregnant aborted during the second month of pregnancy. The third animal carried her infant to term. Although this infant was well developed, its body weight was considerably lower than the average rhesus infant (375 g versus

544 ± 101 g). This animal was sacrificed immediately following delivery in order to determine PCB concentrations in the tissue resulting from intrauterine exposure. Transplacental movement of PCB's was demonstrated by gas chromatographic analysis. Highest concentrations of PCB's in tissue occurred in the adrenals and adipose tissue—24 and 27 μg/g, respectively. These concentrations were approximately one-half of the PCB concentration in maternal fat at parturition (50 μg/g).

Thus, in this early study, there were suggestions that short-term exposure to PCB's affected the ability of the nonhuman primate to conceive and maintain pregnancy. There were also indications that the ingestion of PCB's by the mother affected the birth weight of the offspring and resulted in deposition of considerable amounts of PCB's in the tissues of the infant.

MATERIALS, METHODS, AND RESULTS

To clarify further the possible effects of PCB's on reproduction and on fetal and neonatal development, monkeys were fed diets containing PCB's at 2.5 and 5.0 ppm (Aroclor 1248) for approximately 18 months (Allen and Barsotti, 1976; Barsotti et al., 1976). After the animals had been on the experimental diets for 6 months, attempts were made to breed them. All of the eight animals on the 2.5-ppm diet conceived; however, only five were able to carry their infants to term. Six of the eight animals on the 5.0-ppm diet conceived. Four of these aborted early in gestation. A female that had received 5.0 ppm of PCB in the diet delivered a stillborn infant. Death was due to hypoxia that resulted from a difficult delivery. High PCB concentrations were found in its pancreas, adrenals, and lungs (63, 25, and 95 μg/g of tissue, respectively).

At birth all of the surviving infants were small (399 ± 22 versus 507 ± 59 g). Other than their small stature, low birth weights, and focal areas of hyperpigmentation of the skin, their general appearance was normal, as were radiographic evaluations and hemograms. Biopsies of the neonates' skin, including the epidermis, dermis, and attached underlying subcutaneous tissue, revealed PCB concentrations of 1.0 to 4.8 μg/g of tissue.

The infants were permitted to remain with their mothers. Their only source of nourishment was mothers' milk. During the nursing period the infants showed a constant weight gain, although below that of the controls. Within 2 months the infants began to show signs of PCB intoxication, characterized by development of facial acne, swelling of the eyelids, loss of eyelashes, and increased skin pigmentation. Following these clinical observations, four 5-cm³ milk samples were obtained from the females for PCB analysis. PCB's in three milk samples varied from 0.154 to 0.397 ppm/g of milk. In the fourth sample, the concentration of PCB's in the milk fat was 16.44 ppm. At this time, the PCB concentration in the infants' skin ranged between 86 and 136 μg/g tissue.

Three of the six nursing infants died from PCB intoxication: One male infant from the 5-ppm group expired on the 112th day, and a male and a female from the 2.5-ppm group expired on days 44 and 239 following birth. The infant that died on the 44th day of life had had the highest concentration of PCB in the skin at birth. The infant that survived for 239 days experienced intermittent periods of vomiting during the terminal 3 months of its life.

The infants were small for their age and exhibited focal hyperpigmented areas of the skin. The skin of the face was dry and crusty with isolated acneform lesions. The upper eyelids were swollen, the eyelashes were sparse, and the conjunctivae were moderately congested. The thymuses were rudimentary and the spleen and lymph nodes were small. Except for gastric mucosal thickening in the animal with periodic vomiting, other organs appeared normal under gross examination. Microscopically, cortical and medullary thymocytes were nonexistent. The lymph nodules of the spleen and lymph nodes were extremely small, the germinal centers inapparent, and the bone marrow was hypocellular. In all animals there was hypertrophy, squamous metaplasia, and keratinization of the meibomian glands of the eyelids and hair follicles. In the stomach of the animal that vomited, there was gastric mucosal hyperplasia with some glandular elements penetrating the muscularis mucosa. In addition, there were acute inflammatory cells in the edematous submucosa.

The three surviving infants were weaned at 4 months of age and placed on a milk substitute. These infants improved decidedly within 4 months after removal from their mothers. Their eyelashes became more plentiful, the swelling of the eyelids subsided, and the acneform lesions of the face disappeared. When the animals reached 8 months of age, the PCB concentrations were 20 μg/g of fat and 4 μg/g of skin.

Behavioral and learning tests were conducted on the three surviving PCB-exposed infants for more than 1 year following weaning (Bowman *et al.*, 1978). The investigators found that these animals were hyperactive when compared with controls in two locomotor activity tests and that they made more errors in the first two of three discrimination-reversal problems. These effects have persisted despite improvement in the physical status of the animals.

Following the exposure to 2.5 and 5.0 ppm PCB for 18 months, the adult female rhesus monkeys were removed from the experimental diets for 1 year prior to rebreeding. During this time their physical status improved markedly, and a normal menstrual cycle was reestablished (Allen and Barsotti, 1977). When bred, all of them conceived (Table 1). Infants born to these females were small, and PCB's were detected in their tissues at birth. Three infants, who were born to and nursed by the mothers that had received PCB's in their diets at 5 ppm, died. One infant died from the 2.5-ppm group. Lymph nodes and thymus glands were rudimentary in all of these infants, as well as in one stillborn from the 5.0-ppm group. The eight surviving infants from both groups were separated from their mothers after 4 months of nursing and are under psychological evaluation.

DISCUSSION

The need for additional information on the potential injurious effects of PCB exposure on the human fetus and neonate has been emphasized in reports of studies on nonhuman primates. This need is particularly apparent from the results of a national survey of human breast milk, which showed the average PCB concentration in milk fat to be 1.7 ppm (Savage, 1976). Since data indicate transplacental movement of the PCB's and high PCB concentrations in the milk fat of lactating mammals, it is likely that a large percentage of the infant population has been

TABLE 1   Breeding Performance of Rhesus Macaques Exposed to PCB's (Aroclor 1248) in Their Diet for 18 Months

| Effect | Breeding Performance, Number of Animals Affected/Number of Animals in-Group | | | |
| | During PCB Exposure | | 1 Year After PCB Exposure | |
	2.5 ppm	5.0 ppm	2.5 ppm	5.0 ppm
Total impregnated	8/8	6/8	8/8	7/7
Absorption and resorption	3/8	4/8	1/8	1/7
Stillbirths	0/8	1/8	0/8	1/7
Normal births	5/8	1/8	7/8	5/7

exposed to PCB's during critical development stages. Although the initial group of rhesus infants were exposed to concentrations far greater than human infants would receive, the effects in their siblings suggest that much lower exposures could produce physiological and psychological alterations.

As a result of the increasing PCB concentrations to which the human population is exposed (Kutz and Strassman, 1976) and the apparent increased susceptibility of the fetus and neonate to PCB intoxication, it will be necessary to conduct detailed studies on humans and animals to establish the magnitude of the problem. In future research, animals should be exposed to PCB concentrations that approximate human exposure. Investigators should collect data concerning immunologic incompetence and learning and behavioral deficits in infants that are exposed to low PCB concentrations. Moreover, effects related to intrauterine and transmammary exposure should be studied independently.

Reproductive abnormalities in the adult female may also be attributable to PCB's. The PCB's may alter the menstrual cycle, inhibit embryonic implantation in the uterus, and cause early abortions. All of the above-mentioned facets are subject to evaluation in laboratory animals and could be correlated with clinical studies in exposed humans.

SUMMARY

Morbidity and mortality resulted from intrauterine and transmammary exposure of infant rhesus monkeys to PCB's in tests where the mothers had consumed 2.5 to 5.0 ppm of Aroclor 1248 (Monsanto, St. Louis, Mo.) in their diets throughout gestation and lactation. In addition to acne, periorbital edema, and loss of eyelashes, atrophic changes occurred in the thymus and other lymphoid tissues. There was also depletion of bone marrow. Surviving infants showed behavioral and learning deficits. Similar changes have been recorded in the infants born to the PCB-exposed adults following 1 year on a control diet. These data indicate that low-level PCB exposure in infant nonhuman primates, which may occur *in utero* or from nursing, is capable of producing serious injurious effects.

ACKNOWLEDGMENTS

This investigation was supported in part by U.S. Public Health Service grants ES00958, ES00472, GM00130, and RR00167 from the National Institutes of Health and the University of Wisconsin Sea Grant Program. Part of this research was conducted in the University of Wisconsin–Madison Biotron, a controlled environmental research facility supported by the National Science Foundation and the University of Wisconsin.

REFERENCES

Allen, J. R., and D. A. Barsotti. 1976. The effects of transplacental and mammary movement of PCB's in infant rhesus monkeys. Toxicology 6:331–340.

Allen, J. R., and D. A. Barsotti. 1977. Response of infant primates to intrauterine and neonatal exposure to polychlorinated biphenyls. Proceedings of the Great Lakes Regional Meeting of the American Chemical Society, Stevens Point, Wisconsin, June 6–8, 1977. Unpublished.

Allen, J. R., L. A. Carstens, and D. A. Barsotti. 1974. Residual effects of short-term, low level exposure of nonhuman primates to polycholorinated biphenyls. Toxicol. Appl. Pharmacol. 30:440–451.

Barsotti, D. A., R. J. Marlar, and J. R. Allen. 1976. Reproductive dysfunction in rhesus monkeys exposed to low levels of polychlorinated biphenyls (Aroclor 1248). Food Cosmet. Toxicol. 14:99–103.

Bowman, R. E., M. P. Heironimus, and J. R. Allen. 1978. Correlation of PCB body burden with behavioral toxicology in monkeys. Pharmacol. Biochem. Behav. 9(1):49–56.

Cecil, H. C., J. Bitman, R. J. Lillie, G. F. Fries, and J. Verrett. 1974. Embryotoxic and teratogenic effects in unhatched fertile eggs from hens fed polychlorinated biphenyls (PCB's). Bull. Environ. Contam. Toxicol. 11:489–495.

DeLong, R. L., W. G. Gilmartin, and J. G. Simpson. 1973. Premature births in California sea lions: Association with high organochlorine pollutant residue levels. Science 181:1168–1169.

Kutz, F. W., and S. C. Strassman. 1976. Residues of polychlorinated biphenyls in the general population of the United States. Pp. 139–143 in Proceedings of the National Conference on PCB's, EPA-560/6-75-004. Environmental Protection Agency Office of Toxic Substances, Washington, D.C.

Platonow, N. S., and B. S. Reinhart. 1973. The effects of polychlorinated biphenyls (Aroclor 1254) on chicken egg production, fertility, and hatchability. Can. J. Comp. Med. 37:341–346.

Ringer, R. K., R. J. Aulerich, and M. Zabik. 1972. Effects of dietary polychlorinated biphenyls on growth and reproduction of mink. Proc. Am. Chem. Soc. 12:149–154.

Savage, E. P. 1976. Preliminary Report on Polychlorinated Biphenyls (PCB's) in Mothers' milk. Paper presented at Conference on PCBs in Mothers' Milk, August 1976. National Institutes of Health, Bethesda, Md.

Shiota, K. 1976. Postnatal behavioral effects of prenatal treatment with PCB's (polychlorinated biphenyls) in rats. Okajimas Folia Anat. Jpn. 53:105–114.

Török, P., and H. Kriegel. 1975. Distribution of ^{14}C PCB during fetal period. Arch. Toxicol. 33:199–207.

Villeneuve, D. C., D. L. Grant, K. Khera, D. L. Clegg, H. Baer, and W. E. J. Phillips. 1971. The fetotoxicity of a polychlorinated biphenyl mixture (Aroclor 1254) in the rabbit and in the rat. Environ. Physiol. 1:67–71.

Zabik, M. E., C. Beeble, and R. Schemmel. 1976. Effects of dam's diet on PCB accumulation in nursing Osborne Mendel pups. Arch. Environ. Contam. Toxicol. 4:246–256.

QUESTIONS AND ANSWERS

R. GARMAN: Is chloracne a direct effect of PCB's or is it secondary to liver injury?

D. BARSOTTI: We think it is a direct effect. The greatest hair loss is first noticed in the areas where sebaceous glands are localized, presumably because PCB's are lipophilic and concentrate to a greater extent in these glands.

D. KELLY: The changes that you described in the thymus are reminiscent of those seen in young animals infected with viruses such as canine distemper, feline leukemia, and panleukopenia. Is there any suggestion that the PCB's interfere with the immune status and may possibly predispose animals to viral infection?

D. BARSOTTI: I do not have any direct evidence. However, we have seen *Shigella* infections in our colonies only after they were intoxicated by PCB's. We do not know whether this is due to immune suppression or some other mechanism.

S. NIELSEN: Were there any changes in mucus-secreting cells, either in the trachea, bronchi, pancreatic duct, bile duct, or colon?

D. BARSOTTI: No, not in these animals, other than the changes I have described in the gastrointestinal tract.

S. NIELSEN: Some of the effects you have shown closely resemble those obtained when absorption of vitamin A is impaired.

D. BARSOTTI: That's true. However, we have not studied that particular aspect.

Z. RUBEN: The mucosal invasion into the submucosa of the stomach is interesting. Do you classify this as a neoplasm or merely as a developmental abnormality?

D. BARSOTTI: We see hypertrophy and hyperplasia in gastric epithelium of PCB-treated animals. There is limited invasion accompanied by ulceration. We have not completely classified this lesion at present.

L. KOLLER: Did you see lesions in the lymph node that were similar to those you described in spleen, thymus, and bone marrow? Also, since you're using Aroclor 1248, did you encounter any lesions in the liver?

D. BARSOTTI: Yes, we did see atrophy of lymph nodes. With reference to the use of Aroclor 1248, we found liver lesions were very common. In fatal cases, there was liver necrosis.

B. ZOOK: The severity of the bone marrow lesions suggests that there might be an anemia, and perhaps also hypogammaglobulinemia. Were these present?

D. BARSOTTI: We found a shift in the albumin:globulin ratio, as well as anemia and thrombocytopenia.

Pathology of Rhesus Macaques (*Macaca mulatta*) Exposed to Tetrachlorodibenzo-ρ-Dioxin (TCDD)

J. R. ALLEN, D. A. BARSOTTI, and J. P. VAN MILLER

There is growing concern about the potential effects on public health of widely produced chemicals that persist indefinitely in the environment. One of the most toxic of these industrially synthesized chemicals is 2, 3,7,8-tetrachlorodibenzo-ρ-dioxin (TCDD), which is produced in minute amounts as a side product during the industrial synthesis of 2,4, 5-trichlorophenol. Whenever reaction chambers are inadvertently overheated, large quantities of TCDD are produced. Overheating, in combination with other exothermic reactions, has led to a number of explosions. Following accidental industrial exposure of humans to TCDD, chloracne, liver and kidney lesions, emphysema, myocardial degeneration, reflex irregularities, asthenia, and porphyria cutanea tarda have been observed (May, 1973). The recent industrial accident in Italy, at a site where trichlorophenol was being produced, led to extensive exposure of humans and animals to TCDD and brought worldwide attention to the toxicological hazards of this chemical (Rawls and O'Sullivan, 1976).

While the industrial accidents have been the most publicized, there are numerous other incidents in which humans and animals have been exposed to TCDD. The herbicide Agent Orange, a combination of 2,4, 5-trichlorophenoxyacetic acid (2,4,5-T) and 2,4-dichlorophenoxyacetic acid (2,4-D), was used extensively in Viet Nam as a defoliant. Skin disorders, convulsions, abortions, and birth defects were found in domestic animals from heavily sprayed areas (Rose and Rose, 1972). In humans, chloracne, fatigue, spontaneous abortions, a higher infant mortality, and increased incidence of liver cancer were observed (Tung, 1973).

346

Waste oil containing products of trichlorophenol production including TCDD have been used to control dust in horse arenas. The concentration of TCDD in the soil of these areas has been recorded in excess of 30 ppm (Carter *et al.*, 1975). Deaths in the horse population were preceded by severe acne, alopecia, weight loss, conjunctivitis, joint stiffness, laminitis, hematuria, and abortions (Case and Coffman, 1973). Chloracne, cystitis, pyelonephritis, and arthralgia were experienced by exposed humans.

Tolerable levels of TCDD in the environment and in the food chain are not known. Particularly significant are the recent reports of TCDD levels as high as 60 parts per trillion (ppt) in fat from cattle that have grazed on lands treated with 2,4,5-T (R. T. Ross, Environmental Protection Agency, personal communication, 1977). Detectable levels of TCDD have also been reported in human milk fat (10 to 40 ppt) (M. Meselson, Harvard University, personal communication, 1977).

In the recent experiments described below, nonhuman primates were fed low levels of TCDD for an extended period to clarify the pathological effects produced by this compound. These data should be helpful in postulating the effects of low-level TCDD exposure in humans.

METHODS AND RESULTS

Sixteen adult rhesus monkeys, each weighing approximately 5.6 kg, were housed in a controlled environment whose facility simulated that of the breeding season in their natural habitat. Data on their general health, hematology, and serum chemistry were obtained during a 6-month observation period. After establishing these control parameters, the animals were divided into two groups of eight animals each. One group was placed on a diet that contained 500 ppt of TCDD. The other group received the same diet without TCDD. Food intake for each animal in the experimental group was quantitated over a 9-month period. After 6 months on the experimental diet and at the appropriate time of their menstrual cycles, the females were mated with males fed the control diet. After 9 months, the experimental animals were removed from the TCDD diet, placed on a control diet, and observed for an additional 3 months. Total TCDD intake for these animals was estimated at (mean ± standard deviation) 5.8 ± 0.9 μg after 3 months, 11.3 ± 1.7 μg after 6 months, and 17.9 ± 2.0 μg following 9 months on the diet.

By the third month of TCDD exposure, the monkeys had developed periorbital edema, loss of facial hair and eyelashes, acne, accentuated

TABLE 1 Hemograms and Body Weights of Rhesus Macaques Fed 500 ppt of TCDD for 9 Months

Animal Number	Initial Body Wt., kg	Terminal Experimental Values				Terminal Body Wt., kg
		White Blood Cells, x10³/mm³	Platelets, x10³/mm³	Hemoglobin, g%	Hematocrit, %	
7[a]	6.19	2.3	23	4.0	12.0	5.86
9[a]	6.74	2.4	44	6.9	21.5	5.74
23[b]	6.62	3.8	28	12.6	40.0	5.03
31[b]	6.02	10.4	480	10.7	35.5	5.35
32[a]	5.00	4.1	50	6.0	19.5	4.22
38[a]	5.63	3.8	34	6.6	22.0	5.22
41[b]	5.73	8.0	340	11.5	44.5	5.44
49[a]	5.39	8.7	54	8.5	29.5	4.46
X̄ pre-experiment (N = 8)		9.3 ± 1.2[c]	327 ± 57[c]	13.9 ± 0.5[c]	42.9 ± 1.8[c]	

[a] Values at necropsy.
[b] Values at 12 months on trial.
[c] ± 1 S.D.

hair follicles, and dry scaly skin. These changes were particularly prominent in six of the eight animals following 6 months of TCDD exposure. Hemograms of five animals from the experimental group showed a decrease in the hemoglobin and hematocrit by the sixth month. The hematological disorders became more extensive in six of the seven animals that survived following 9 months of TCDD exposure (Table 1). However, blood urea nitrogen, total serum lipid, serum cholesterol, serum glutamic–pyruvic transaminase (SGPT), total serum protein, and albumin:globulin ratios were not altered appreciably during the experiment except in moribund animals where a slight decrease in serum albumin and increases in SGPT existed. Animals in the experimental group lost weight throughout the trial even though food intake remained constant.

All of the control animals conceived after three matings. Of the three animals in the experimental group that conceived (animal nos. 9, 38, 41), two aborted (nos. 9 and 38) during the second month of pregnancy. The third animal experienced an uncomplicated pregnancy. Three of the remaining animals (nos. 7, 23, 49) failed to conceive even after repeated breedings. One of the animals (no. 32) died before mating. This animal had a previous history of abortions and was not evaluated in the reproduction study.

By the seventh month of exposure, one animal (no. 32) became severely pancytopenic. Prior to death the peripheral blood smear of this animal was practically devoid of platelets and immature red and white blood cells. During the subsequent 4 months, four additional animals experienced similar clinical changes and died (nos. 7, 9, 38, 49). The three surviving animals experienced considerable hair loss and moderate periorbital edema. One animal (no. 23) developed a severe leukopenia and thrombocytopenia during the twelfth month of the experiment.

In addition to skin lesions, the major findings at necropsy included focal areas of hemorrhage in the skin, heart, lung, stomach, intestine, adrenals, skeletal musculature, brain, and bone marrow. There were irregularities in the growth patterns of the fingernails and toenails, some with gangrenous necrosis (Figure 1). Ascites and subcutaneous edema were also apparent. Hypertrophy, dilatation, edema, and hydropic degeneration of the myocardium occurred in all of the animals. There was marked dilatation of the biliary ducts. The common bile duct frequently measured over 1-cm across (Figure 2). However, the dilatation of the bile ducts was not associated with obstruction or jaundice. Moderate hyperkeratosis of the skin and cystic keratosis of the hair follicles and sebaceous glands were conspicuous (Figure 3). A marked

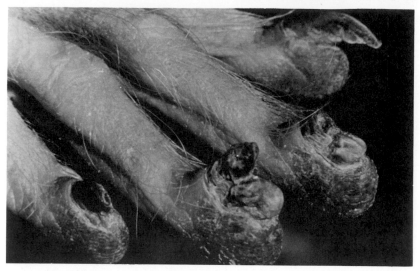

FIGURE 1 Toes of an animal that died after a 9-month exposure to a diet containing 500 ppt of TCDD. Marked thickening and irregular growth of the toenails are depicted, as well as swelling of the terminal phalanges. In the more severely affected animals there was gangrenous necrosis of the terminal phalanges.

decrease of hematopoiesis in the bone marrow (Figure 4) was present in all of the animals. In addition, the lymphoid tissue was hypocellular.

There were hypertrophy and hyperplasia of the mucous-secreting cells and associated paucity of parietal cells in the gastric epithelium. Invasion of the submucosa by these cells was also observed. Ulceration and mucinous cysts were common in the modified gastric mucosa. A generalized hypertrophy and hyperplasia of the lining epithelium occurred throughout the biliary system. The transitional epithelium of the urinary tract and goblet cells of the palpebral conjunctivae were also hyperplastic. Metaplastic changes were observed in the bronchial epithelium and the epithelium of the salivary gland ducts, bile ducts, and pancreatic ducts. Portions of the epithelium were replaced by mucus-secreting cells. Squamous metaplasia was also recorded in the sebaceous glands of the eyelids.

DISCUSSION

Indications are that profound cellular alterations develop in many tissues following the ingestion of minute concentrations of TCDD by primates for 9 months (Allen *et al.*, 1977). The most severe effects of TCDD involve the hematopoietic system. As exposures lengthen, cellular degeneration in bone marrow and lymphoid tissue becomes more severe. These observations agree with previous studies where mixtures of dioxins including TCDD were fed to nonhuman primates (Allen and Carstens, 1967). In addition to the depletion of circulating lymphocytes, atrophy of the thymus, spleen, and lymph nodes has also been observed in several species. These tissue alterations may be related to immune suppression, which has been recorded in animals exposed to TCDD (Vos *et al.*, 1973).

Widespread epithelial hypertrophy, hyperplasia, and metaplasia are common features of TCDD intoxication in mammals (Allen and Carstens, 1967; Allen *et al.*, 1977). These cellular changes appear to be

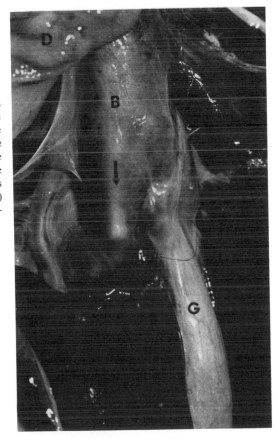

FIGURE 2 A probe (→) has been introduced through the ampulla of Vater in the duodenum (D) to demonstrate edema and dilatation in the common bile duct (B). The hepatic and cystic ducts as well as the gallbladder (G) were also dilated in this animal.

of particular significance in light of recent reports (Van Miller and Allen, 1977) of an increased incidence of neoplasms in rats exposed to low concentrations of TCDD. In most instances, the tumors in the rats developed in tissues analogous to those showing cellular alterations in the monkeys. These observations, along with the reports that tumors in humans are attributable to exposure to TCDD following industrial explosions (May, 1973) and to defoliants containing TCDD (Tung, 1973), suggest a carcinogenic action of the compound.

There seems to be considerable difference in the level of TCDD exposure necessary to produce intoxication. After a single exposure of nonhuman primates to TCDD, concentrations of 50 to 70 μg/kg body weight were required to produce death in 50% of the animals within 45 days (McConnell *et al.*, 1977). However, in the experiment described above, death occurred in over half of the monkeys when approximately 3 μg of TCDD/kg body weight were administered over 9 months. The pathological alterations caused by these two dose levels were quite similar.

These observations indicate that special attention should be given to humans exposed to TCDD. Perhaps of foremost importance are the

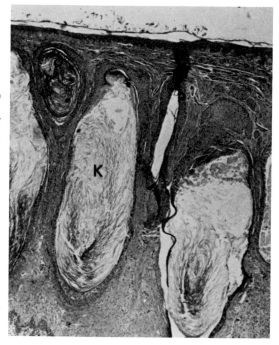

FIGURE 3 Keratinization (K) of the meibomian glands of the eyelids was present in all of the monkeys that died of TCDD intoxication. (Hematoxylin and eosin stain. ×17.)

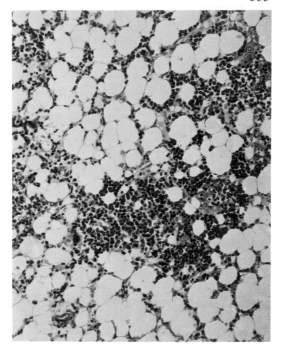

FIGURE 4 Bone marrow taken from an animal that died with severe pancytopenia after ingesting 11.2 μg of TCDD. Note small foci of darker cells that are dispersed among the lucent fat cells. (Hematoxylin and eosin stain. ×166.)

alterations that occur in the hematopoietic system. Anemia, thrombocytopenia, and leukopenia invariably developed in nonhuman primates after exposure to TCDD. The altered lymphopoiesis may be associated with immune suppression. The possibility of reproductive abnormalities also exists. Altered levels of serum progesterone and estradiol associated with difficulty in conception and early abortion have been observed in female monkeys exposed to low concentrations of TCDD (J. R. Allen, unpublished observations). Testicular atrophy has also developed in male monkeys given small amounts of dioxin (Allen and Carstens, 1967). In addition, the widespread hypertrophy, hyperplasia, and metaplasia that occurred in the epithelium of monkeys exposed to TCDD (Allen *et al.*, 1977) and the increased incidence of neoplasia in rats exposed to low concentrations of this compound (Van Miller and Allen, 1977) suggest a carcinogenic action of TCDD.

SUMMARY

Female rhesus monkeys fed a diet containing 500 ppt of TCDD developed acne, periorbital edema, loss of hair and eyelashes, and anemia within 6 months. Five of the eight animals conceived after 6 months of exposure to TCDD; however, four of these aborted early in gestation. Following 9 months of exposure to TCDD, two animals died due to aplastic anemia, leukopenia, and thrombocytopenia. Three additional animals expired during the tenth and eleventh month of the experiment due to severe pancytopenia, even though the TCDD diets were discontinued after the ninth month. These data indicate that exposure to extremely low concentrations of TCDD can produce widespread deleterious effects in primates.

ACKNOWLEDGMENTS

This investigation was supported in part by U.S. Public Health Service grants ES00472, GM00130, and RR00167 from the National Institutes of Health and the University of Wisconsin Sea Grant Program. Part of this research was conducted in the University of Wisconsin–Madison Biotron, a controlled environmental research facility supported by the National Science Foundation and the University of Wisconsin. Primate Center Publication No. 16-051.

REFERENCES

Allen, J. R., and L. A. Carstens. 1967. Light and electron microscopic observations in *Macaca mulatta* monkeys fed toxic fat. Am. J. Vet. Res. 28:1513–1526.
Allen, J. R., D. A. Barsotti, J. P. Van Miller, L. J. Abrahamson, and J. J. Lalich. 1977. Morphological changes in monkeys consuming a diet containing low levels of 2,3,7,8-tetrachlorodibenzo-*p*-dioxin. Food Cosmet. Toxicol. 15:401–410.
Carter, C. D., R. D. Kimbrough, J. A. Liddle, R. E. Cline, M. M. Zacks, and W. F. Barthel. 1975. Tetrachlorodibenzodioxin: An accidental poisoning episode in horse arenas. Science 188:738–740.
Case, A. A., and J. R. Coffman. 1973. Waste oil: Toxic for horses. Vet. Clin. North Am. 3:273–277.
May, G. 1973. Chloracne from the accidental production of tetrachlorodibenzodioxin. Br. J. Ind. Med. 30:276–283.
McConnell, E. E., J. A. Moore, and D. W. Dalgard. 1977. Toxicity of 2,3,7,8-tetrachlorodibenzo-*p*-dioxin (TCDD) in rhesus monkeys following a single oral dose. Toxicol. Appl. Pharmacol. 43:175–187.
Rawls, R. L., and D. A. O'Sullivan. 1976. Italy seeks answers following toxic release. Chem. Eng. News 54:27–35.
Rose, H. R., and S. P. R. Rose. 1972. Chemical spraying as reported by refugees from South Vietnam. Science 177:710–712.
Tung, T. T. 1973. Le cancer primaire du foie au Vietnam. Chirurgie 99:427–436.
Van Miller, J. P., and J. R. Allen. 1977. Chronic toxicity of 2,3,7,8-tetrachlorodibenzo-*p*-dioxin in rats. Fed. Proc. 36:396.
Vos, J. G., J. A. Moore, and J. Zinkl. 1973. Effects of TCDD on the immune system of laboratory animals. Environ. Health Perspect. 5:149–162.

QUESTIONS AND ANSWERS

G. CHOULES: What is the LD_{50} for TCDD?

D. BARSOTTI: The LD_{50} for monkeys exposed to a single oral dose of this compound ranges between 50 and 70 $\mu g/kg$ body weight. In our study we used only 2 to 3 $\mu g/kg$ throughout the entire 9-month period of feeding.

G. CHOULES: Were there any striking changes in blood chemistry with respect to blood urea nitrogen concentration and albumin:globulin ratios?

D. BARSOTTI: We monitored these throughout the entire experiment, and found changes terminally.

J. BOGAN: Do you know whether dioxins are known to be contaminants of PCB's?

D. BARSOTTI: Commercial mixtures of PCB's contain dioxin was well as other substances including dibenzofurans.

J. ZINKL: I don't believe that dioxins have ever been detected in PCB's, although the chlorinated dibenzofurans have been detected in European PCB's. Chlorinated naphthalenes might also be considered as part of this whole syndrome, since these have been found in European PCB's.

D. BARSOTTI: Right, I stand corrected.

Brain Cholinesterase Activities of Passerine Birds in Forests Sprayed with Cholinesterase Inhibiting Insecticides

JOSEPH G. ZINKL, CHARLES J. HENNY, and
PATRICK J. SHEA

Cholinesterase (ChE) inhibitors (organophosphates and carbamates) have been increasingly used as insecticides since the initiation of the bans or restrictions on chlorinated hydrocarbons. They block synaptic transmissions in the cholinergic parts of the nervous system by binding to the active site of acetylcholinesterase (AChE), which normally hydrolyzes the neurotransmitter acetylcholine. Thus, ChE inhibitors permit excessive acetylcholine accumulations at synapses and disrupt nerve impulse transfer. AChE inhibition probably causes death in air-breathing vertebrates by blocking neurotransmissions in the respiratory center of the brain or at myoneural junctions of respiratory muscles (O'Brien, 1967; Corbett, 1974).

When the U.S. Forest Service sprayed insecticides in 1975 and 1976 on small forest plots in Montana and Oregon, we had the opportunity to study the effect of ChE inhibitors on the wild fauna. We determined brain ChE activities of passerines, because depression of this enzyme has been used as an indicator of organophosphate and carbamate poisoning (Coppage and Matthews, 1974; Stickel, 1974).

MATERIALS AND METHODS

During the summer of 1975, the U.S. Forest Service conducted a pilot test for control of western spruce budworm (*Choristoneura occidentalis*) in the Beaverhead National Forest of southwestern Montana (Dewey, 1975). Three 400- to 800-ha (1,000- to 2,000-acre) plots were sprayed from a helicopter with 1.13 kg/ha (1 lb/acre) of the AChE

356

inhibitor 1-naphthyl *N*-methylcarbamate (carbaryl, Sevin-4-oil). In the same study, three other plots were sprayed with 1.13 kg/ha of another AChE inhibitor, 0,0-dimethyl-2,2,2-trichloro-1-hydroxy-ethylphosphonate (tricholorfon, Dylox®). Partial results of this study have been reported by Zinkl *et al.* in 1977. In June 1976 in the Wallowa–Whitman National Forest of northeastern Oregon, acephate (Orthene®, 0-methyl *S*-methyl *N*-acetylphosphoramidothioate) was applied at 1.13 kg/ha and at 2.26 kg/ha (2 lb/acre) on two separate 128-ha plots during a safety test of selected chemicals on nontarget species (Shea, 1975). Carbaryl was applied at 2.26 kg/ha on two other 128-ha (320-acre) plots. In September, acephate was applied at 0.56 kg/ha (0.5 lb/acre) on nine 24-ha plots and at 1.13 kg/ha on three other 24-ha plots of the Wallowa–Whitman National Forest during an efficacy study against larch casebearer (*Coleophora laricella*) (Hard *et al.*, 1976).

Birds required for brain ChE analysis were collected by mist netting or by shooting with shotguns. The netted birds were killed by carbon dioxide asphyxiation. The carcasses were frozen on dry ice immediately after collection to prevent deterioration of the ChE enzymes (J. G. Zinkl and R. H. Hudson, U.S. Fish and Wildlife Service, 1975, unpublished data). Brain ChE activity was usually determined within 12 h after the birds were collected. In the fall 1976 study, a few birds were stored up to 5 days.

After removal from the calvarium, brains were homogenized in cold 0.1 *M* phosphate buffer (pH 7.4) at a 1:5 dilution. They were then diluted to either 1:50 or 1:100 with the phosphate buffer just prior to analysis. The ChE activity was determined by the method of Ellman *et al.* (1961) as adapted for brain tissue by Dieter and Ludke (1975). The reagents for this technique were obtained in kit form from BMC Corporation, Dallas, Texas, and Irvine, California. A Spectronic 88 spectrophotometer (Bausch & Lomb) fitted with a water-jacketed micro flow-through cell was used to determine the activity. Optical density readings were taken every 30 s for 3 min to assure that the reaction was linear. All analyses were carried out at 25°C.

Activities were calculated in mU/mg of brain. One unit is defined as the conversion of one mole of substrate to product(s) in 1 min. The mean brain ChE activity and its standard deviation were calculated for each species. Depressed activities for birds from treatment areas were evaluated individually. It was necessary to establish criteria by which activities could be judged normal or depressed. To do this two values were calculated, the mean less 2 standard deviations and the mean less 20% of the mean (Ludke *et al.*, 1975). Since these calculations give different values, the one giving the lowest value was used when determining whether an activity was depressed.

RESULTS

Birds from unsprayed forests with habitats similar to those of the sprayed areas were used as controls (Table 1). The brain ChE activity in the exposed birds from the two summer studies was similar, but in birds from the fall exposure was less. Therefore, only the brain ChE activities of the corresponding control birds were measured to determine inhibition in the exposed birds' brains.

The effect of trichlorfon spraying on brain ChE activity was evaluated in 10 passerine species. Of 28 birds collected on the day of spray (day 0), 17 on day 1, 17 on day 2, 21 on day 3, and 28 on day 5, only two western tanagers (*Piranga ludoviciana*) collected on day 0 had depressed brain ChE activities. Their activities were decreased 21% and

TABLE 1   Brain Cholinesterase Activities of Birds from Unsprayed Forests

Species	Brain Cholinesterase Activities, mU/mg Brain, $\bar{X} \pm 1$ SD. Number of Birds in Parentheses.		
	Montana—Summer	Oregon—Summer	Oregon—Fall
Dark-eyed junco (*Junco hyemalis*)	33.2 ± 1.1 (6)	33.3 ± 3.3 (20)	27.6 ± 2.4 (10)
Mountain chickadee (*Parus gambeli*)	33.8 ± 1.3 (3)	32.6 ± 1.7 (12)	25.3 ± 2.5 (7)
American robin (*Turdus migratorius*)	26.6 ± 3.9 (10)	29.7 ± 2.4 (16)	24.3 ± 2.6 (7)
Chipping sparrow (*Spizella passerina*)	23.5 ± 1.8 (7)	23.0 ± 2.0 (20)	20.5 ± 2.1 (3)
Golden-crowned kinglet (*Regulus satrapa*)	—[a]	26.0 ± 2.9 (4)	22.8 ± 2.3 (3)
MacGillivray's warbler (*Oporornis tolmiei*)	34.5 ± 1.9 (5)	30.3 ± 1.2 (5)	—[a]
Flycatchers (*Empidonax* spp.)	22.2 ± 2.6 (8)	22.8 ± 2.3 (13)	—[a]

[a] Not measured.

27% below the mean. An evening grosbeak (*Hesperiphona vespertina*) from day 3 had 19.7% inhibition (Zinkl *et al.*, 1977).

Of the 10 species represented by 48 birds collected on day 0, 8 on day 1, 7 on day 2, and 23 on day 5 in the 1975 1.13 kg/ha carbaryl sprays, only a mountain chickadee (*Parus gambeli*) from day 0 had brain ChE inhibition (21%) (Zinkl *et al.*, 1977). In the 1976 2.26 kg/ha carbaryl spray, 9 species were collected but only 2 Cassin's finches (*Carpodacus cassinii*) collected on day 1 had 23% and 26% inhibition. None of the 16 birds from day 0, the other 13 from day 1, 12 from day 2, or 12 from day 4 had brain ChE inhibition.

In marked contrast to the slight effect of trichlorfon and carbaryl, acephate caused marked and widespread brain ChE depression (Table 2). Only an occasional bird with brain ChE inhibition was collected on the day of spray, but by day 1 or day 2 brain ChE depression was evident. All birds from days 1, 2, and 6 from the areas sprayed with 2.26 kg/ha had depressed activities. (Birds were not collected from the area after day 6.) Brain ChE inhibition was present on the 1.13 kg/ha areas 33 days after spray but not after 89 days. In the fall study, depression was found on days 2 and 5 but not on days 10 and 15. The percent of birds with inhibition and the degree of inhibition was less in the fall 1.13 kg/ha areas than in the summer 1.13 kg/ha area.

On all the acephate sprays, the most marked brain ChE depression was in dark-eyed juncos (*Junco hyemalis*). The greatest ChE inhibition (65%) occurred in a dark-eyed junco collected 15 days after spray from the summer 1.13 kg/ha area. On the 2.26 kg/ha area a dark-eyed junco and a golden-crowned kinglet (*Regulus satrapa*) collected on day 6 were inhibited 54%, the most marked inhibition found in this area. Of the 14 species collected from the acephate areas, only pine siskins (*Spinus pinus*) did not have brain ChE depression. In all other species collected in the summer, depression of 30% to 50% was common; in the fall the inhibition was less, 25% to 40%.

DISCUSSION

Differences in brain ChE depression occur in birds dying of organophosphate poisoning. Japanese quail (*Coturnix coturnix japonica*), dying after being fed 1,400 ppm parathion for up to 5 days, had 50% inhibition (Ludke *et al.*, 1975). Ninety percent inhibition occurred in ring-necked pheasants (*Phasianus colchicus*) that died after a single oral dose of various organophosphate insecticides (Bunyan *et al.*, 1968). Brain ChE was 95% inhibited in ring doves (domestic *Strep-*

TABLE 2 Brain Cholinesterase Depression in Birds Collected from Forests Sprayed with Acephate

Number of Birds with Depressed Brain Cholinesterase Activity/Number of Birds Collected, and the Percent Depression of the Activity in Each Bird. Data Presented by Number of Days after Spray. (0 = day of spray)

Species	2.26 kg/ha Spraying, Summer 1976				1.13 kg/ha Spraying, Summer 1976								1.13 kg/ha Spraying, Fall 1976					0.56 kg/ha Spraying, Fall 1976				
	Day No.:				Day No.:								Day No.:					Day No.:				
	0	1	2	6	0	1	2	4	7	15	33	89	0	2	5	10	15	0	2	5	10	15
Dark-eyed junco (Junco hyemalis)	0/1[a]	2/2 44[b] 43	2/2 46 42	2/2 54 51	0/1	—	2/2 33 26	1/1 29	—	1/1 65	0/2	0/5	—	2/4 40 28	0/4	0/3	—	—	2/6 36 26	4/6 43 35 26 18	0/4	—
Mountain chickadee (Parus gambeli)	0/1	2/2 36 26	—	1/1 36	0/1	1/1 26	—	1/1 47	—	1/1 38		0/3	0/7	1/4 29	0/4	0/3	0/2	0/7	3/5 35 23 22	0/2	0/3	0/3
American robin (Turdus migratorius)	1/2 23	—	2/2 37 36	1/1 21	0/2	1/2 30	1/1 31	1/1 40	1/1 44	2/2 46 37	—	—	—	1/4 32	3/3 37 26 25	0/1	—	0/1	0/2	0/5	0/1	—

Species																				
Chipping sparrow (*Spizella passerina*)	—	—	—	0/1 33	1/1 31	2/2 49 36	1/1 30	0/1	—	—	—	0/3	1/4 29	1/3 24	0/2	—					
Golden-crowned kinglet (*Regulus satrapa*)	0/1 35 34	2/2 28	1/1 54	—	—	—	—	—	0/1	2/4 24 21	4/6 35 32 30 21	1/4 31	—	0/4	—	0/1	1/3 24	1/3 33	0/1	0/3	
MacGillivray's warbler (*Oporornis tolmiei*)	—	—	—	1/3 23	1/2 37	2/2 38 32	2/2 51 35	1/1 45	2/2 50 40	2/2 43 31	—	—	—	—	—	—	—	—	—	—	
Flycatchers (*Empidonax* spp.)	—	—	—	0/1	1/2 47	3/3 43 45 37	1/2 36	3/3 49 43 37	2/2 56 43	1/1 25	—	—	—	—	—	—	—	—	—	—	
Other species[c]	0/2 38 43	1/1	1/1	0/5 33 38	2/5 33 38	2/4 39 36	0/4	0/3	1/3 32	1/8 28	0/1 31	1/1 31	—	0/1	0/6	0/1	2/2 31 23	1/3 24	0/1	—	
TOTAL	1/7	7/7	6/6	1/14	7/13	11/13	8/13	6/9	8/12	4/8	1/16	0/9	11/23	4/19	0/10	0/6	0/13	9/22	7/22	0/12	0/6

[a] Numerator = number of birds with depressed brain ChE activities; denominator = number of birds collected.

[b] Percent depression of brain ChE activity.

[c] Summer 2.26 kg/ha = yellow-rumped warbler (*Dendroica coronata*) and western tanager (*Piranga ludoviciana*); summer 1.13 kg/ha = pine siskin (*Spinus pinus*), yellow-rumped warbler (*Dendroica coronata*), warbling vireo (*Vireo gilvus*). Swainson's thrush (*Catharus ustulata*); fall, both sprayings = red-breasted nuthatch (*Sitta canadensis*).

topelia risoria) that had been killed with 42.4 mg/kg trichlorfon. Ring doves given 21.2 mg/kg were inhibited 83% when killed with carbon dioxide 2 h after being dosed. Other doves given this same dose survived (J. G. Zinkl and R. H. Hudson, 1975, unpublished data). Brain ChE activity was inhibited 83% in homing pigeons (domestic rock doves, *Columba livia*) that had been killed with a single oral dose of 195 mg/kg trichlorfon (J. G. Zinkl and R. H. Hudson, 1975, unpublished data). These limited data suggest that brain ChE inhibition of at least 80% is required to kill birds with a single oral dose of an organophosphate insecticide, but they may die with only 50% inhibition when continuously exposed. Brain ChE inhibition of killed birds may differ with the various insecticides and with species sensitivity.

No birds collected from either the trichlorfon or the carbaryl areas were in danger of dying when 50% inhibition is used as the criterion. However, using 20% depression as an indicator of significant exposure (Ludke *et al.*, 1975), both insecticides affected some canopy dwellers, which were likely exposed more than ground-dwelling species during aerial application. The rather minimal effect of these insecticides is similar to that reported by Kurtz and Studholme (1974), who found only low residues in birds of eastern U.S. forests that had been sprayed with trichlorfon and carbaryl.

Brain ChE activity of many birds was depressed nearly 50% after the summer acephate spray. The prolonged depression suggested that these birds may have been in danger from acephate poisoning. Signs compatible with ChE inhibition were seen on three occasions on the 1.13 kg/ha sprayed area. A warbling vireo (*Vireo gilvus*) caught in a mist net on day 1 was salivating profusely. Its brain ChE activity had decreased 38%. An American robin (*Turdus migratorius*) had difficulty maintaining a perching position, and a mountain chickadee had tremors. Neither of these birds was collected.

In the fall acephate study, the inhibition on the 1.13 kg/ha study was not as marked, widespread, or prolonged as in the summer study. Probably the smaller plot size and the more transient nature of the fall birds resulted in less exposure and less brain ChE depression. Perhaps meteorologic and environmental conditions in the fall resulted in more rapid degradation of acephate.

Marked brain ChE inhibition was not present on day 0, even though collections started 2 h after spraying and continued for about 6 h. Thus, there was either an accumulative effect, which became detectable on the next day, or acephate was converted to a more potent ChE inhibitor. Loblolly pine (*Pinus taeda*) seedlings deacetylate acephate to another organophosphate methamidophos (0-methyl *S*-methyl phosphoramidothioate) (Werner, 1974). In rats, the acute oral LD_{50} of methamidophos is 7.5 mg/kg, while that of acephate is 400 mg/kg (Christensen and Luginbyhl, 1975). To mallards (*Anas platyrhynchos*), the respective LD_{50}'s are 8.5 mg/kg and 234 mg/kg (R. H. Hudson, U.S. Fish and Wildlife Service, 1975, personal communication). Perhaps the day between spraying and the finding of depressed ChE activity was required to produce methamidophos concentrations adequate to cause inhibition. Residue analysis of forest biota and a few bird brains indicated the presence of both methamidophos and acephate (R. Roberts, U.S. Forest Service, 1977, personal communication). Thus, the brain ChE depression may have been caused in part by methamidophos derived from deacetylation of acephate.

Spraying forest with acephate at 0.56, 1.13, or 2.26 kg/ha caused marked, widespread, and prolonged brain ChE depression in passerine birds. Methamidophos, derived from the acephate, may have been partly responsible for the depression. It is not known whether the acephate exposure was life-threatening or caused detrimental subacute effects. In contrast to the effects of acephate, 1.13 kg/ha trichlorfon or 1.13 or 2.26 kg/ha carbaryl cause minimal brain ChE depression in passerines.

SUMMARY

Brain cholinesterase activities were determined in passerines collected from northwestern forests that had been sprayed with trichlorfon, acephate, and carbaryl at 0.56, 1.13 and 2.26 kg/ha. Trichlorfon and carbaryl inhibited cholinesterase activity slightly in only a few birds, primarily canopy dwellers. In contrast, acephate caused marked inhibition of cholinesterase activity in nearly all birds collected. The inhibition was present even 33 days after spraying. Some birds from the acephate-sprayed forests exhibited clinical signs compatible with acute acetylcholinesterase inhibition.

ACKNOWLEDGMENT

The work leading to this paper was funded in whole or in part by the U.S. Department of Agriculture (USDA) Douglas-fir Tussock Moth Expanded Research and Development Program, USDA Forest Service Cooperative Agreements 21-199 and PSW-14.

REFERENCES

Bunyan, P. T., D. M. Jennings, and A. Taylor. 1968. Organophosphorous poisoning. Diagnosis of poisoning in pheasants owing to a number of common pesticides. J. Agric. Food Chem. 16:332–339.

Christensen, H. E., and T. T. Luginbyhl, eds. 1975. Registry of Toxic Effects of Chemical Substances. National Institute for Occupational Safety and Health. U.S. Department of Health, Education, and Welfare, Rockville, Md. 1,296 pp.

Coppage, D. L., and E. Matthews. 1974. Short-term effects of organophosphorous pesticides on cholinesterase of estuarine fish and pink shrimp. Bull. Environ. Contam. Toxicol. 11:483–488.

Corbett, J. R. 1974. The Biochemical Mode of Action of Pesticides. Academic Press, Inc., New York. 330 pp.

Dewey, A. E. 1975. Pilot control project of carbaryl, trichlorfon, and *Bacillus thuringiensis* against western spruce budworm. Beaverhead and Gallatin National Forests, Montana. State and Private Forestry, Forest Service, Northern Region, U.S. Department of Agriculture, Missoula, Mont. 19 pp.

Dieter, M. P., and J. L. Ludke. 1975. Studies on combined effects of organophosphates and heavy metals in birds. 1. Plasma and brain cholinesterase in coturnix quail fed methyl mercury and orally dosed with parathion. Bull. Environ. Contam. Toxicol. 13:257–262.

Ellman, G. L., K. D. Courtney, V. Andres, Jr., and R. M. Featherstone. 1961. A new and rapid colorimetric determination of acetylcholinesterase activity. Biochem. Pharmacol. 7:88–95.

Hard, J., S. Meso, and G. Markin. 1976. Field experiment of the insecticide orthene for control of larch casebearer in Oregon. Pacific Southwest Forest and Range Experiment Station, Forest Service, U.S. Department of Agriculture, Berkeley, Calif. 21 pp.

Kurtz, D. A., and C. R. Studholme. 1974. Recovery of trichlorfon (Dylox) and carbaryl (Sevin) in songbirds following spraying of forest for gypsy moth. Bull. Environ. Contam. Toxicol. 11:78–84.

Ludke, J. L., E. F. Hill, and M. P. Dieter. 1975. Cholinesterase (ChE) response and related mortality among birds fed ChE inhibitors. Arch. Environ. Contam. Toxicol. 3:1–21.

O'Brien, R. D. 1967. Insecticides: Action and Metabolism. Academic Press, Inc., New York. 332 pp.

Shea, P. J. 1975. Safety tests of selected chemicals on nontarget organisms. Pacific Southwest Forest and Range Experiment Station, Forest Service, U.S. Department of Agriculture, Berkeley, Calif. 24 pp.

Stickel, W. H. 1974. Effects on wildlife of newer pesticides and other pollutants. Proc. West. Assoc. State Game Fish Comm. 53:484–491.

Werner, R. A. 1974. Distribution and toxicity of root absorbed ^{14}C-Orthene and its metabolites in loblolly pine seedlings. J. Econ. Entomol. 67:588–591.

Zinkl, J. G., C. J. Henny, and L. R. DeWeese. 1977. Brain cholinesterase activities of birds from forests sprayed with trichlorfon (Dylox) and carbaryl (Sevin-4-oil). Bull. Environ. Contam. Toxicol. 17:379–386.

QUESTIONS AND ANSWERS

G. FOX: It seems strange that acephate would affect juncos, since they forage on the ground rather than in the tree tops.

J. ZINKL: We saw a depression in brain acetylcholinesterase activity with trichlorfon and carbaryl in a few canopy birds, since they would naturally be exposed to higher levels. With acephate spray, the effects were not noticed until the day after spraying. This would give time for insects to fall out of the trees and be eaten from the forest floor. Most of us think of juncos as seedeaters, but in truth they are quite opportunistic and will consume insects when they are available. We are conducting laboratory studies with the dark-eyed junco (*Junco hyemalis*), trying to determine the LD_{50}, the amount of acephate needed to cause 40% to 60% depression of acetylcholinesterase activity in brain, and the amount that is lethal.

G. FOX: Is there any cessation of song or other behavioral changes after spraying with these compounds?

J. ZINKL: Dr. Henny's group is analyzing the data from its study on these populations after spraying and has observed a decrease in activity and singing on the day after spraying; but this is just a single observation. In another preliminary study, the Fish and Wildlife Service found that there was an increase in the use of the sprayed area by Cooper's hawks (*Accipiter cooperii*), suggesting that sprayed birds may become more susceptible to predation.

G. FOX: Fenitrothion spraying in eastern Canada is accompanied by almost total cessation of song in some species of birds. This is the monitor method that the Canadian Wildlife Service is using as an estimate of mortality. At this point we do not know whether the birds just quit singing or whether they die.

J. ZINKL: We have searched for dead birds and have not been able to find them. However, finding dead birds in thick forest is almost impossible.

G. CHOULES: Ellin and Vicario (1975)[1] have demonstrated that cholinesterase activity increases an average of 5.5% per degree Celsius above the prescribed 25°C, and red blood cell activity changes 3.0% for the same temperature increase. Therefore, all assays should be done at 25°C to make results from different laboratories comparable to one another.

[1] Ellin, R. I., and P. Vicario. 1975. Delta pH method for measuring blood cholinesterase: A study on the effect of temperature. Arch. Environ. Health 30:263.

Polybrominated Biphenyls:
A Recent Environmental Pollutant

STUART D. SLEIGHT

The polybrominated biphenyls (PBB's) are a relatively recent addition to the seemingly endless list of environmental contaminants. Production of a flame retardant (Firemaster BP-6) containing PBB began in 1970. By 1974 annual production was nearly 2.27 million kg. Before a contamination incident in Michigan (described below), PBB had received relatively little attention from research workers. Fries *et al.* (1973) reported a similarity between PBB and polychlorinated biphenyl (PCB) retention in fat and excretion in feces, eggs, and milk. Fecal excretion of approximately 50% of an oral dose given to a cow occurred within 168 h (Willett and Irving, 1976). The initial half-life for PBB in milk after cessation of dosing was 10.5 days (Gutenmann and Lisk, 1975), but the decline was slower thereafter with a half-life of 58 days (Fries and Marrow, 1975). Placental transfer was also observed (Willett and Irving, 1976).

The marked ability of PBB to induce hepatic microsomal enzymes in rats has been amply demonstrated (Dent *et al.*, 1976a,b; Garthoff *et al.*, 1977; Moore *et al.*, 1976). Moore's group recorded this effect in rat pups nursing dams fed 1 ppb of PBB. Garthoff *et al.* (1977) compared the biochemical changes caused by PCB to those caused by PBB in rats. They found that PBB was about five times more active in inducing microsomal mixed function oxidase activity and also that it caused greater impairment of mitochondrial function than PCB. All liver lipid fractions were increased more by PBB than by PCB.

Hepatic enlargement is a consistent feature of PBB toxicosis. The

366

degree of enlargement in rats was greater with PBB than with PCB (Garthoff *et al.*, 1977). Enlargement was seen with octobromobiphenyl (Norris *et al.*, 1974; Lee *et al.*, 1975), as well as with the product responsible for the contamination in Michigan (Moore *et al.*, 1976; Sleight and Sanger, 1976). Decabromodiphenyl oxide did not cause hepatic enlargement or other discernible toxicologic effects (Kociba *et al.*, 1975).

In our studies, microscopic lesions consisted of extensive swelling and vacuolation of hepatic cells in rats fed 100 ppm or 500 ppm of PBB for 30 or 60 days. Slight swelling and vacuolation were seen in rats fed 10 ppm of PBB. Rat pups nursing dams fed a diet containing 10 ppm of PBB also had slight hepatic lesions. Electron micrographs revealed increases in mitochondrial size after feeding low levels of PBB. Myelin bodies, increased smooth endoplasmic reticulum, and prominent vacuoles were observed after feedings of higher concentrations (Sleight and Sanger, 1976).

The addition of 1,000 ppm of iodine to the diets of rats fed as much as 100 ppm of PBB did not have an appreciable additive effect. However, metaplasia of the epithelium of terminal bronchioles was seen only in rats fed diets containing 1,000 ppm of iodine and 100 ppm of PBB. In rats fed 10 or 100 ppm of PBB, results of serum protein electrophoresis were interpreted as indicating hepatic injury. Further confirmation of hepatic injury was evidenced by altered electrophoretic patterns of serum lipids and lactic dehydrogenase (LDH) isozymes. The effects were minimal in rats fed an iodine-deficient diet and 10 ppm of PBB. Livers of rats fed an iodine-deficient diet and 1 ppm of PBB were somewhat smaller than livers of rats fed an iodine-deficient diet without PBB. This was surprising, since there has been a slight and often significant increase in liver size in rats fed diets containing 1 ppm of PBB (Sleight *et al.*, 1978).

Farm animals contaminated with PBB have been described as having widely varied manifestations of illness. Jackson and Halbert (1974) recorded clinical signs in a 400-cow herd that had been fed a heavily contaminated feed (4,000 ppm of PBB) for less than 3 weeks. They described anorexia, loss of milk production, infertility, hematocysts, elongated hooves, delayed parturition, lameness, and thickening of the skin. Five of 12 young cattle died when fed the suspect feed for 6 weeks. Thirty-three cows were removed from the herd 9 months after exposure to PBB and were observed for the next 6 months. Even though fat concentrations averaged 800 ppm, the cows were healthy during this time and at necropsy had no lesions attributable to PBB (Michigan State University, 1976).

Researchers concluded that low-level contamination of feed with PBB could not be specifically related to syndromes reported in Michigan cattle (Mercer et al., 1975; Moorhead et al., 1977). Moorhead et al. (1977) studied the pathology of experimentally induced PBB toxicosis and observed signs and lesions only in those cattle fed as much as 25 g/day. Nonspecific signs included anorexia, emaciation, dehydration, lacrimation, diarrhea, depression, and abortion. Specific lesions included cystic dilatation of the renal tubules, hyperkeratosis, and cystic hyperplasia of the gallbladder. Feeding 0.25 mg/day or 250 mg/day caused no signs or lesions.

Mink were adversely affected by relatively low doses of PBB (Michigan State University, 1976). Reproduction and kit survival significantly decreased at concentrations as low as 1 ppm in the diet. Mortality reached 90% when adult mink were fed 6.25 ppm of PBB. Effects were most likely related to lowered feed consumption. Depression of food consumption by PBB also occurred in chickens (Ringer and Polin, 1977). Egg production, hatchability, and chick survival were also decreased. Nearly 60% of the daily intake of PBB was excreted in the egg yolk. Similar results were reported in Japanese quail (Babish et al., 1975). Investigators also observed higher residues in the tissues from males than from females. Guinea pigs died when fed diets containing 100 or 500 ppm of PBB mainly because of feed refusal (Sleight and Sanger, 1976). Ku et al. (1978) fed 20 and 200 ppm of PBB to young pigs. Liver size was increased and growth rate was depressed by PBB. Other effects of PBB have included porphyria in Japanese quail (Strik, 1973) and menstrual irregularities in monkeys fed as little as 0.3 ppm of PBB for 9 months (Lambrecht et al., 1978).

PBB's do not appear to be potent teratogens in rats and mice. Preache et al. (1976) found that 200 ppm of PBB in the diet of mice increased the percentage of dead and resorbed fetuses when fed from gestation day 4 through 18. External terata were not significantly increased. Corbett et al. (1975) found no terata in the offspring of rats fed as much as 1,000 ppm of PBB. Some terata were recorded in mice, but the incidence was no higher than in the controls. Ficsor and Wertz (1976) found no PBB-induced terata in rats, but did observe that PBB increased the effects of colchicine on bone marrow metaphase and mitotic indices. To my knowledge, there have been no reports to date of mutagenesis caused by PBB.

Bioaccumulation of PBB by fish apparently is comparable to PCB. Waterfowl such as ducks also have considerable capacity for accumulating PBB (Hesse, 1975).

The Michigan Contamination

In 1973, approximately 455 kg of Firemaster BP-6 was accidentally substituted for magnesium oxide during the formulation of dairy feed at a facility in Michigan. This PBB contamination was not confirmed until May 1974, nearly 9 months after it occurred. In the meantime, there was widespread distribution of contaminated feed to hundreds, perhaps thousands, of dairy livestock and poultry farms. The public was not aware that it was consuming PBB-contaminated milk, meat, and eggs. As soon as the PBB contamination was confirmed, the Food and Drug Administration (FDA) adopted action levels for PBB of 1 ppm in meat and milk (fat basis), 0.3 ppm in feed, and 0.1 ppm in eggs. Six months later the tolerance levels were set at 0.3 ppm for meat and milk, and 0.05 ppm for feed and eggs.

The PBB contamination has had drastic economic, social, and political effects. Thousands of cattle, swine, and poultry and tons of eggs, milk, butter, cheese, and feed have been destroyed and buried. Thousands, perhaps millions, of people consumed meat and farm products contaminated by unknown concentrations of PBB during the many months before the cause of the contamination was discovered. Millions of dollars in payments have been made to farmers by the companies involved, and there are perhaps half a billion dollars in unsettled claims and in pending lawsuits (Carter, 1976; Dunckel, 1975; Welborn *et al.*, 1975).

PBB has been a feature story in the nation's news media for the past 4 years. Dozens of stories have appeared describing illness in people allegedly sick from the effects of PBB, or depicting the plight of farmers whose animals were dying as a result of "PBB poisoning." "PBB action committees" have been formed and have initiated a recall petition against Michigan Governor William G. Milliken. A major labor union has endorsed the recall and called for the resignation of the Director of the Michigan Department of Agriculture (MDA).

Groups of citizens have boycotted Michigan farm products and pressured food stores into posting signs proclaiming whether the meat and milk sold in the store originated within or beyond the borders of Michigan.

Politicians have responded in various ways. A Special Investigating Committee was appointed by the Michigan State Senate to investigate the PBB problem (Welborn *et al.*, 1975). Various bills were introduced. One proposal, which dealt with regulation of the manufacture and distribution of feed, has become law. Appropriations were made to pay for testing and destroying animals and food products, but no money was included to indemnify farmers for their losses. The Michigan Department of Health (MDH), in cooperation with the FDA and the Center for Disease Control (CDC), assessed the short-term effects of PBB on human health. The data did not show that PBB caused any identifiable human ailments. A long-term, epidemiologic evaluation is under way (Welborn *et al.*, 1975).

Despite assurances from the FDA and MDA that all dairy products, meat, and eggs meeting FDA guidelines were safe for human consumption, despite assurances from the U.S. Department of Agriculture, FDA, and MDA that food entering commercial channels was in compliance with the guidelines, and despite a lack of confirmation of human disease related to PBB or of animal disease related to low-level contamination, the PBB issue became more political and emotional as time went on. By late 1975, the contamination crisis was over, but the issue was perpetuated by the news media, citizens' and farmers' groups, a few scientists, and some politicians. In March 1976, Governor Milliken appointed a Scientific Advisory Panel on PBB to advise him on whether the present action guidelines for meat, milk, poultry, and eggs were safe, whether they should be lowered, and, if so, to what level. Based mainly on the structural similarity of PBB to PCB and on the published information on precancerous and cancerous lesions of the livers of cancer-susceptible rats and mice fed high concentrations of PCB, the panel recommended, essentially, a zero tolerance for PBB (Bernstein, 1976). A year after the panel's recommendation, a bill to lower the action levels to 20 ppb was passed by the Michigan House of Representatives and Senate in October 1977. This belated action, nearly 4 years after the contamination occurred, can have little or no impact on human health since virtually all of the PBB has been removed from the food chain. Thousands of cattle may be killed to remove less than 0.02% of the original 455 kg or so of PBB that contaminated animal feed.

An additional factor was introduced when some political leaders arranged to have Dr. Irving J. Selikoff, of the Environmental Sciences Laboratory at Mount Sinai School of Medicine, evaluate possible human health problems related to PBB. A preliminary, incomplete report stated that "adverse health effects may occur in some people as the result of PBB exposure" (Selikoff, 1977). The survey's preliminary results were quickly endorsed by those who commissioned the study but were criticized by many, including the chairman of the Governor's Scientific Advisory Panel, who concluded that there was room for legitimate disagreement "based upon the results presented, the lack of a control, the absence of body burden data, and the composition of the group of people studied'' (Bernstein, 1977).

Ironically, the results of the Selikoff study and of the long-term study by the MDH, FDA, and CDC have little bearing on the legislation to lower the action levels to 20 ppb in fat. In the early stages of the contamination, estimates are that some people were consuming PBB at the rate of 400 to 3,200 μg/kg/day (Kolbye and Cordle, 1976). A 75-kg person consuming 0.25 kg of meat contaminated at 0.3 ppm in the fat would receive a dosage of PBB of approximately 0.1 μg/kg/day. Even this degree of exposure would be highly unlikely at present, and any exposure would be sporadic. It is extremely improbable that such minute dosages of PBB have any toxicologic significance. Yet, the controversy continues.

SUMMARY

In 1973 and 1974, livestock feed was contaminated by PBB's at a facility in Michigan. To protect humans against exposure to PBB, millions of dollars worth of farm animals and products were destroyed. There was widespread publicity and considerable controversy concerning the public health implications of the contamination. Relatively little experimental data were available as a basis for recommendations. Work at Michigan State and elsewhere indicates that PBB's act similarly to PCB's but are more potent microsomal enzyme inducers and produce more extensive hepatic lesions in rats. Guinea pigs, mink, and monkeys are apparently more sensitive to PBB than rats.

REFERENCES

Babish, J. G., W. H. Gutenmann, and G. S. Stoewsand. 1975. Polybrominated biphenyls: Tissue distribution and effect on hepatic microsomal enzymes in Japanese quail. J. Agric. Food Chem. 23(5):879–882.

Bernstein, I. A. 1976. PBB Scientific Advisory Panel Report to William G. Milliken, Governor, State of Michigan on Polybrominated Biphenyls, May 24. Pp. 30–34 in Hearings before the Subcommittee on Science, Technology, and Space of the Committee on Commerce, Science, and Transportation, U.S. Senate, March 28, 1977.

Bernstein, I. A. 1977. Perspectives on the polybrominated biphenyl (PBB) contamination incident in the State of Michigan. Pp. 6–30 in Hearings before the Subcommittee on Science, Technology and Space of the Committee on Commerce, Science, and Transportation, U.S. Senate, March 28, 1977.

Carter, L. J. 1976. Michigan's PBB incident: Chemical mix-up leads to disaster. Science 192:240–243.

Corbett, T. H., A. R. Beaudoin, R. G. Cornell, M. D. Anver, R. Schumacher, J. Endres, and M. Szwabowska. 1975. Toxicity of polybrominated biphenyls (Firemaster BP-6) in rodents. Environ. Res. 10:390–396.

Dent, J. G., K. J. Netter, and J. E. Gibson. 1976a. Effects of chronic administration of polybrominated biphenyls on parameters associated with hepatic drug metabolism. Res. Commun. Chem. Pathol. Pharmacol. 13:75–82.

Dent, J. G., K. J. Netter, and J. E. Gibson. 1976b. The induction of hepatic microsomal metabolism in rats following acute administration of a mixture of polybrominated biphenyls. Toxicol. Appl. Pharmacol. 38:237–249.

Dunckel, A. E. 1975. An updating on the polybrominated biphenyl disaster in Michigan. J. Am. Vet. Med. Assoc. 167:838–841.

Ficsor, G., and G. F. Wertz. 1976. Polybrominated biphenyl non-teratogenic C-mitosis synergist in rat. Abstr. 29. Seventh Annual Meeting, Environmental Mutagen Society. Mutat. Res. 38:388.

Fries, G. F., and G. S. Marrow. 1975. Excretion of polybrominated biphenyls into the milk of cows. J. Dairy Sci. 58:947–951.

Fries, G. F., L. W. Smith, H. C. Cecil, J. Bitman, and J. R. Lillie. 1973. Retention and excretion of polybrominated biphenyls by hens and cows. Proc. 165th Natl. Meet., Am. Chem. Soc., Pestic. Chem. Div., Dallas, Tex. American Chemical Society, Washington, D.C.

Garthoff, L. H., L. Friedman, T. Farber, K. Locke, T. Sobotka, S. Green, N. E. Hurlen, E. Peters, G. Story, F. M. Moreland, C. Graham, J. Keys, M. J. Taylor, J. Rothlein, and E. Sporn. 1977. Biochemical changes caused by ingestion of Aroclor 1254 (a commercial polychlorinated biphenyl mixture) or Firemaster BP-6 (a commercial polybrominated biphenyl mixture). Supplement in *Toxic Substances Part 2*, Serial No. 95-28, U.S. Government Printing Office, Washington, D.C., pp. 1298–1331.

Gutenmann, W. H., and D. J. Lisk. 1975. Tissue storage and excretion in milk of polybrominated biphenyls in ruminants. J. Agric. Food Chem. 23:1005–1007.

Hesse, J. L. 1975. Water pollution aspects of polybrominated biphenyl production: Results of surveys in the Pine River in the vicinity of St. Louis, Michigan. Second National Conference on Complete WateReuse: Water's Interface with Energy, Air and Solids, Palmer House, Chicago, May 4–8. Sponsored by the American Institute of Chemical Engineers and the U.S. Environmental Protection Agency. Tech. Trans. 20 pp.

Jackson, T. F., and F. L. Halbert. 1974. A toxic syndrome associated with the feeding of polybrominated biphenyls-contaminated concentrate to dairy cattle. J. Am. Vet. Med. Assoc. 165:437–439.

Kociba, R. J., L. O. Frauson, C. G. Humiston, J. M. Norris, C. E. Wade, R. W. Lisowe, J. F. Quast, G. C. Jersey, and G. L. Jewett. 1975. Results of a two-year dietary feeding study with decabromodiphenyl oxide (DBDPO) in rats. J. Combust. Toxicol. 2:267–285.

Kolbye, A. C., Jr., and F. Cordle. 1976. Statement on polybrominated biphenyls (PBB's) read before the Michigan State Department of Agriculture, Lansing, Mich., June 10. Pp. 1028–1031 in Hearings before the Subcommittee on Science, Technology, and Space of the Committee on Commerce, Science, and Transportation. U.S. Senate, March 31, 1977.

Ku, P., M. G. Hogberg, A. L. Trapp, P. S. Brady, and E. R. Miller. 1978. Polybrominated biphenyl (PBB) in the growing pig diet. Environ. Health Perspect. 24:13–18.

Lee, K. P., R. R. Hebert, H. Sherman, J. G. Aftosmis, and R. S. Waritz. 1975. Bromine tissue residue and hepatotoxic effects of octabromobiphenyl in rats. Toxicol. Appl. Pharmacol. 34:115–127.

Lambrecht, L. K., D. A. Barsotti, and J. R. Allen. 1978. Response of nonhuman primates to a polybrominated biphenyl mixture. Environ. Health Perspect. 24:139–146.

Mercer, H. D., A. Furr, G. Meerdink, R. J. Condon, W. Buck, R. H. Teske, and G. Fries. 1975. 1975 Michigan State dairy herd survey. A report on herd health status of animals exposed to polybrominated biphenyls (PBB). Food and Drug Administration, Beltsville, Md. 12 pp.

Michigan State University Agricultural Experiment Station. 1976. MSU research on PBB's. Michigan Science in Action. Information Services, Michigan State University, E. Lansing, Mich. 11 pp.

Moore, R. W., G. Dannan, and S. D. Aust. 1976. Induction of drug metabolizing enzymes in rats nursing from mothers fed polybrominated biphenyls. Fed. Proc. 35:708.

Moorhead, P. D., L. B. Willett, C. J. Brumm, and H. D. Mercer. 1977. Pathology of experimentally induced polybrominated biphenyl toxicosis in pregnant heifers. J. Am. Vet. Med. Assoc. 170:307–313.

Norris, J. M., J. W. Ehrmantraut, C. L. Gibbons, R. J. Kociba, B. A. Schwetz, J. Q. Rose, C. G. Humiston, G. L. Jewett, W. B. Crummett, P. J. Gehring, J. B. Tirsell, and J. S. Brosier. 1974. Toxicological and environmental factors involved in the selection of decabromodiphenyl oxide as a fire retardant chemical. J. Fire Flam./Comb. Toxicol. 1:52–77.

Preache, M. M., S. Z. Cagen, and J. L. Gibson. 1976. Perinatal toxicity in mice following maternal dietary exposure to polybrominated biphenyls. Abstr. 192, 15th Annual Meeting, Society of Toxicology, Atlanta, March 14–18.

Ringer, R. K., and D. Polin. 1977. The biological effects of polybrominated biphenyls in avian species. Fed. Proc. 36:1894–1898.

Selikoff, I. J. 1977. PBB health survey of Michigan residents, November 4–10, 1976. Initial report of findings to Governor Milliken. Mt. Sinai School of Medicine, City University of New York, N.Y. 18 pp.

Sleight, S. D., and V. L. Sanger. 1976. Pathologic effects of polybrominated biphenyls in the rat and guinea pig. J. Am. Vet. Med. Assoc. 169:1231–1235.

Sleight, S. D., S. Mangkoewidjojo, B. T. Akoso, and V. L. Sanger. 1978. Polybrominated biphenyl toxicosis in rats fed an iodine-deficient, iodine-adequate, or iodine-excess diet. Environ. Health Perspect. 24:341–346.

Strik, J. J. T. W. A. 1973. Chemical porphyria in Japanese quail (*Coturnix c. japonica*). Enzyme 16:211–223.

Welborn, J. A., R. Allen, G. Byker, A. DeGrow, and J. Hertel. 1975. The contamination crisis in Michigan: Polybrominated biphenyls, a report from the Senate Special Investigating Committee, July. State of Michigan, Lansing. 152 pp.

Willett, L. B., and H. A. Irving. 1976. Distribution and clearance of polybrominated biphenyls in cows and calves. J. Dairy Sci. 59(8):1429–1439.

QUESTIONS AND ANSWERS

L. KOLLER: What are the permissible levels of PBB's in food?

S. SLEIGHT: The levels are 0.3 ppm in meat and milk, and 0.05 ppm in eggs and animal feed.

J. WEIS: What did they do with the thousands of PBB-poisoned cattle that they slaughtered?

S. SLEIGHT: They buried the carcasses in a site in northern Michigan.

J. WEIS: Do you expect PBB's to get into soil and possibly be cycled into plants?

S. SLEIGHT: It is possible that the buried carcasses will contaminate the soil. However, they picked a relatively isolated site where it would be highly unlikely that the PBB's would get into groundwater.

D. JONES: We just completed a pilot study in cats fed brominated biphenyls. Both adults and kittens were subjected to a behavioral test measuring their ability to traverse a 91-cm-long, 1.9-cm-wide balance beam. All animals were able to accomplish this prior to the administration of PBB's. After the kittens were fed 100 mg/kg/day for 30 days, they were unable to traverse the beam.

ABSTRACTS
AND
POSTERS

Response to the call for papers by the organizing group preparing the programs for the symposium was considerable. More than 100 abstracts were received for consideration. In an effort to accommodate the maximum number of presentations, some of the abstracts not selected for oral presentation were scheduled for the poster session. The remaining abstracts are included in the proceedings because they clearly reflect the fact that more relevant information exists than presented in the papers and posters, and they provide additional reference material.

Retardation of Bone Formation in Dogs Exposed to Lead

COLIN ANDERSON and KENNETH D. DANYLCHUK

When animals or humans are exposed to lead in the environment, a large proportion of this metal is stored in bone. Although the clinicopathological aspects of lead intoxication and the results of many experimental observations are well known, little attention has been focused on the "first order" effect of lead on bone activity.

Dogs subjected to 1.3 mg of lead per kilogram of body weight per day for 201 days were found to have a decrease in bone formation rate at the cell (10%), tissue (15%), and organ (22%) levels when compared to litter-mate controls. The changes observed in our exposed animals occurred over more than two periods of bone remodeling and were not accompanied by any clinical or laboratory evidence suggestive of either localized or systemic effects of lead intoxication. The biochemical parameters used to assess abnormalities in gastrointestinal and renal treatment of calcium and phosphate absorption and excretion showed no alteration, confirming that any observed effect on bone must be direct.

Our findings indicate that bone turnover rates in exposed animals possessing a similar remodeling pattern to that for humans are influenced by chronic low-level lead exposure before the previously described clinical manifestations of chronic lead toxicity may draw attention to this condition. Since lead has a half-life of 20 years or more in bone, from which lead is difficult to remove, prolonged exposures to low doses of lead in the environment may result in a significant decrease in the total skeletal mass of an individual in old age.

The Effect of Atmospheric Sulfur Dioxide on Plants

LOUISE E. ANDERSON and JEFFREY X. DUGGAN

Sulfur dioxide is a noxious industrial, and occasionally natural, atmospheric pollutant that causes extensive damage to higher plants. Several plant enzymes, oxidative phosphorylation in plant mitochondria, and oxygen evolution in spinach chloroplasts can be affected by millimolar levels of sulfite. However, the atmospheric concentration of sulfur dioxide, even in heavily polluted areas, is usually considerably lower than that necessary to generate the sulfite concentrations that would inhibit these enzymes (or systems) *in vivo*. In 1976, Anderson and Avron[1] reported that light modulation of enzyme activity in

[1] Anderson, L. E., and M. Avron. 1976. Light modulation of enzyme activity in chloroplasts. Generation of membrane-bound vicinal-dithiol groups by photosynthetic electron transport. Plant Physiol. 57:209–213.

higher plants involves generation of membrane-bound vicinal dithiol groups within chloroplasts, apparently by reduction of disulfide bonds. We have now found that short-term fumigation of intact pea (*Pisum sativum*) plants with sulfur dioxide inactivates the light effect modulation (LEM) system probably by sulfitolysis of disulfide bonds. Therefore, the major harmful effect of atmospheric sulfur dioxide on higher plants may be due to disruption of light regulation of photosynthetic carbon metabolism.

These studies were supported by grants from the U.S. National Science Foundation and the U.S.–Israel Binational Science Foundation.

Studies of Environmental Pollutants on Air-Breathing Fish (*Clarias batrachus,* L.) and Non-Air-Breathing Fish (*Tilapia mossambica* Peters)

SUDIP K. BANERJEE, PRATAP K. MUKHOPADHYAY, and PADMAKAR V. DEHADRAI

The culture of air-breathing walking catfish (*Clarias batrachus*) in Indian swamps is threatened by contamination from heavy metals and pesticides. The studies of the nutrition, biochemistry, and pathobiology of these fish exposed under various environmental conditions have been undertaken. Studies of both the air-breathing catfish (*C. batrachus*) and the non-air-breathing Mozambique mouthbrooder (*Tilapia mossambica*) show that the growth, survival, and lysosomal- and membrane-bound microsomal enzyme activities of their gills are affected by exposure to cadmium. Both species lost weight over the 4-week period of the experiment, but the mouthbrooder was less resistant than the catfish to cadmium toxicity, showing 100% mortality at 250 ppm of cadmium compared to none for the catfish. Exposure to cadmium markedly increased the activity of lysosomal- and membrane-bound microsomal enzyme in the gills of the catfish, but did not change the activities in the gills of the mouthbrooder. Cadmium inhibited the activity of alkaline phosphatase in the gills of the catfish but stimulated it in the gills of the mouthbrooder. A significant alteration in the isozyme pattern of gill alkaline phosphatase was not observed. Such interspecific differences in the responses of fish may be helpful in monitoring the extent of cadmium toxicity in fish.

Malathion, an organophosphorus insecticide used widely in India, alters the activities of drug metabolizing, lytic, and digestive enzymes in fish. In a malathion-contaminated environment, degeneration of different tissues of fish and marked changes in the levels of serum indices, such as serum glucose, ascorbic acid, inorganic phosphate, and protein, have been observed in histopathological and gel electrophoretic studies.

Decontamination of Cattle Exposed to Parathion Dip

G. LEW CHOULES, WILLIAM C. RUSSELL, and F. JAMES SCHOENFELD

A Utah livestock owner mistakenly dipped 185 cattle in a 1,600-ppm emulsion of ethyl parathion. Five days after the incident occurred, the Utah State Veterinarian requested that a joint investigational team from Dugway Proving Ground and the Utah State Veterinarian's office be sent to the livestock farm. The animals were sprayed with alkaline detergent, which removed only 50% to 75% of the parathion. Excitement caused by the spraying was detrimental to the cattle and 100 deaths occurred shortly after spraying. The remainder were dipped in detergent and 2% hypochlorite several days later. The dip removed 87% of the remaining parathion, but more deaths occurred leaving only 25 survivors. Treatments with atropine (0.5 to 1.0 mg/kg of body weight) and Protopam chloride (1 g per animal) were given from time to time when veterinarians were available, but brought only temporary relief to the animals.

Laboratory experiments with guinea pigs were undertaken to find a more effective method for decontaminating animals exposed to parathion. We developed an emulsifiable solvent composed of kerosene, laundry detergent, and butanol. The mixture consistently removed over 97% of the parathion from the skin and hair of exposed guinea pigs when it was sprayed on, then rinsed off with water. Detergent treatment with or without alkali yielded variable results leaving as much as a 16% residue of the parathion. Sodium hypochlorite proved dangerous because it converted up to 75% of the parathion to paraoxon, which is 3 to 5 times more toxic. Permanganate and hydrosulfite were ineffective.

Effect of Cadmium on Reproduction and Fetal Health in *Peromyscus* (White-Footed Mice)

MARY FLEMING FINLAY, ANDREW PHILLIPS, and NARVIE KENNEDY

The effect of cadmium exposure on reproduction and fetal health in the native American rodent genus *Peromyscus* was investigated under laboratory conditions. Laboratory-bred deer mice (*P. maniculatus*) or old field mice (*P. polionotus*) were randomly mated in monogamous pairs. Sibling matings were avoided. For 1 year each mated pair was exposed to 1 ppm of cadmium, as cadmium acetate dissolved in drinking water. The results and comparisons with controls maintained on distilled water under similar conditions are given in Table 1. For both species, similar control data were obtained from Dawson.[2]

[2] Dawson, W. D. 1964. Fertility and size inheritance in a *Peromyscus* species. Evolution 19:44–55.

TABLE 1   Effects on First Generation of *Peromyscus maniculatus* and *Peromyscus polionotus*

| | Species | | | |
| | *P. maniculatus* | | *P. polionotus* | |
	Control (N = 6)[a]	Exposed to 1 ppm (N = 6)	Control (N = 10)	Exposed to 1 ppm (N = 8)
Fertile, %[b]	100	66	60	37
Fertility[c]	0.56	0.27	0.30	0.16
Productivity[d]	2.16	1.03	0.94	0.53
Birthweight, g	1.70 ± 0.02	1.60 ± 0.04	1.62 ± 0.02	1.60 ± 0.05
Survival to weaning (21 days), %	93	80	86	52

[a] The number of mated pairs.
[b] The number of matings producing one or more live offspring.
[c] The number of pregnancies to term/mating month.
[d] The number of newborn mice/mating month.

Mice born from the cadmium-exposed matings were maintained continually on a cadmium solution after weaning. When mature, these mice were mated. Sibling mating was avoided. They were maintained on cadmium and observed for the effects as described in Table 1. The results for the second generation of cadmium-exposed mice are shown in Tables 2 and 3.

For each group, control and experimental matings were started at the same time to minimize differences due to seasonal breeding.

TABLE 2   Effects on Second Generation of *Peromyscus maniculatus*

Effect	Control (N = 10)	Exposed to 1 ppm (N = 10)
Fertile, %	100	80
Fertility	0.55	0.40
Productivity	2.27	1.52
Birthweight, g	1.80 ± 0.03	1.68 ± 0.03
Survival to weaning (21 days), %	80	69

TABLE 3   Effects on Second Generation of
Peromyscus polionotus

Effect	Control (N = 10)	Exposed to 1 ppm (N = 5)
Fertile, %	60	66
Fertility	0.30	0.13
Productivity	0.94	0.38
Birthweight, g	1.62 ± 0.02	NA
Survival to weaning (21 days), %	86	64

These results suggest a reduction in fertility of mice exposed to cadmium at 1 ppm, particularly in old field mice. Birth weights and fetal survival seem to be affected little. Additional exposures are being conducted.

This study was supported by RR08117 (Minority Biomedical Support), National Institutes of Health.

Morphological Effects of Petroleum and Chlorobiphenyls on Fish Tissues

JOYCE W. HAWKES

Both pelagic and benthic marine fish, as well as one species of freshwater fish, were exposed either to Prudhoe Bay crude oil incorporated in their food or to the seawater-soluble fraction (SWSF) of crude oil administered via a flow-through system. In all experiments, exposures were sublethal. They ranged from high doses (1 part oil to 1,000 parts food) to low doses (100 ppb SWSF).

Cellular damage was observed in skin, gill, liver, eye lens, and intestine. Scanning electron micrographs revealed empty mucous cells in the skin of English sole (*Parophrys vetulus*) exposed to 13% SWSF for 5 days. Gills from the starry flounder (*Platichthys stellatus*) exhibited areas of epithelial sloughing after a 5-day exposure to 100 ppb SWSF.

Morphological changes in the liver of trout (*Salmo gairdneri*) fed 1 part Prudhoe Bay crude oil per 1,000 parts food for 14 and 75 days included depletion of lipid and glycogen and a pronounced increase in rough endoplasmic reticulum. After 8 months on the same feeding regime, trout had fibrotic replacement around the sinusoids and other vascular elements of the liver, as well as alterations of the eye lens. The volumes of the eye lenses increased twofold and hydration appeared to be the primary factor contributing to the size increase. Recently, coho salmon (*Oncorhynchus kisutch*) held in net pens in Puget Sound were accidentally exposed to repeated diesel fuel spills for 3 months. These fish were subsequently found to have hydrated and cloudy lenses. Their inability to avoid objects and feed normally indicated they were blind. No bacteriological, parasitic, or other pathological conditions were found, and trace levels of petroleum hydrocarbons were demonstrated in tissues from the exposed fish.

Sloughing of the intestinal mucosa was observed in marine chinook salmon (*O. tshawytscha*) exposed to a model mixture of 5 ppm petroleum hydrocarbons in food for 27 days. A total of 5 ppm chlorobiphenyls (biphenyl, 2-chlorobiphenyl, 2,2'- and 2,4'-chlorobiphenyl, 2,5,2'-trichlorobiphenyl, and 2,5,2',5'-tetrachlorobiphenyl) with or without 5 ppm petroleum was also incorporated in food for 27 days, and intestinal sloughing was most severe when the fish were simultaneously exposed to both petroleum and chlorobiphenyls.

This study was supported by the Environmental Protection Agency (Interagency Agreement EPA-IAG-EG93) and by the Bureau of Land Management through interagency agreement with the National Oceanic and Atmospheric Administration and the Outer Continental Shelf Environmental Assessment Program (#R7120819).

Comparative Toxicity of Polychlorinated Biphenyl (PCB) and Polybrominated Biphenyl (PBB) in the Livers of Holtzman Rats: Light and Electron Microscopic Alterations

DAVID E. HINTON, BENJAMIN F. TRUMP, LOUIS KASZA, MORRIS A. WEINBERGER, LEONARD FRIEDMAN, and LARRY H. GARTHOFF

The comparative toxicity of polychlorinated biphenyl (PCB) and polybrominated biphenyl (PBB) in the livers of male Holtzman rats (*Rattus norvegicus*) was studied. A total of 51 4-week-old animals were fed the compounds in their diets at levels of 0, 5, 50, and 500 ppm for 5 weeks and then sacrificed. Investigators observed that the mean liver:body weight ratios of the 50 and 500 ppm groups had increased. Histopathological examination of the livers of rats in those two groups revealed fatty degenerative change resulting from both PCB and PBB exposure. This change was more marked in the 500 ppm groups. Various sized lamellar cytoplasmic inclusions were detected in livers of animals fed 500 ppm of either compound. However, these inclusions were more numerous in the PBB-treated rats. Several of the animals that had been fed 50 ppm of PBB had developed a few inclusions. In rats that received 500 ppm of

PBB, hypertrophic and degenerative hepatocytes were present around central veins. At the periphery of such foci there were occasional multinucleated hepatocytes.

Electron microscopic examination of rats maintained on diets containing 5 ppm of either PCB or PBB showed a slight proliferation of smooth endoplasmic reticulum (SER), a moderate increase of lipid droplets and liposomes, a marked proliferation of Golgi condensing vesicles containing lipoprotein particles, and decreased numbers of mitochondria and lysosomes. In the 50 ppm group, similar but more marked ultrastructural alterations were observed. In addition, there was an increased number of branched and cupshaped profiles of mitochondria and a decreased number of Golgi condensing vesicles containing lipoprotein. Concentric membranous cytoplasmic whorls were encountered only in the rats treated with 50 ppm of PBB. In the 500 ppm groups, the number of mitochondria decreased in the rats fed both compounds. Moreover, the number of SER and liposomes increased markedly, concomitant with a decreased number of Golgi condensing vesicles containing lipoprotein granules. There were also membranous whorls in the 500 ppm groups.

These data suggest that a blockade in the transport of lipid occurs upon exposure to PCB and/or PBB. The blockade appears to be localized at the Golgi apparatus or between the Golgi and the endoplasmic reticulum. Since more changes at lower concentrations were encountered with PBB than with PCB, we concluded that the PBB is more toxic.

Toxicity of Isolated Water-Soluble C_6 Petroleum Hydrocarbons to Lobsters, *Homarus americanus*

BIRINENGI IDONIBOYE-OBU

Some of the equivocation concerning the effects of oil pollution may be removed through systematic studies of individual constituents of crude oils and oil products. Since crude oils are highly variable mixtures of hundreds of chemical compounds, each with its particular stereochemical configuration, reactivity with membranes, macromolecules, volatility, and solubility in body fluids, tissues, and organelles, their individual effects may be missed in studies using crude oil.

Three isolated C_6 petroleum hydrocarbons (*n*-hexane, cyclohexane, and benzene) have been used in laboratory investigations of toxicity to juvenile and adult lobsters obtained from Nova Scotia and Cape Cod, Massachusetts, respectively. Results showed that 9.5 ppm of *n*-hexane, 53.3 ppm of cyclohexane, and 111.8 ppm of benzene (6.25% saturation in water) were lethal to adult decapod crustaceans after 20 to 60 min and to juvenile lobsters (moult stages 4/5) after 2 to 5 min. Radioactive studies (using ^{14}C-benzene) indicate that the hepatopancreas, the nerve cord, and the heart, followed by the gills and the gut, are the principal target organs accumulating these toxicants. Ultrastructural and histological examination of organs excised from entire adult lobsters exposed for 15 min to 6.25% benzene in water showed severe mitochondrial damage, nuclear pyknosis, collapsed lamellae of endoplasmic reticulum, and other degenerative changes associated with acute toxicity. A rapid technique to rank isolated water-soluble hydrocarbon toxicity has been suggested elsewhere.[3,4] Data from such studies can form the basis for simple or elaborate experiments to study synergistic and antagonistic effects of various components in crude oil.

These studies indicate a greater importance than hitherto recognized of light hydrocarbons, which appear to be responsible for high mortality of decapod crustaceans. During oil spills, toxicity appears to be related to their increasing solubility, cyclic structure, and molecular weight. Furthermore, the role of n-hexane as a vector solvent for many organic compounds in toxicity studies needs a serious reappraisal in view of the data presented above.

A Model System Using Hamsters (*Mesocricetus auratus*) for Behavioral Toxicology of Noise

W. BEN ITURRIAN

Noise is probably the least understood pollutant, because mammals appear to adapt to it quickly. An animal model developed in this laboratory suggests that the price of this apparent adaptation is altered cellular, metabolic, reproductive, and behavioral processes that persist for several weeks. Immature animals are especially vulnerable.

Hamsters (*Mesocricetus auratus*) exposed to sound (95 dBA for 30 s) during a "critical period" (28–31 days of age) developed audiosensitivity. Fourteen days later, they displayed transient seizure susceptibility (ss) (Iturrian and Johnson, 1971).[5] Repeated exposure to the sound stimulus prior to the development of ss inhibited convulsions by desensitization. Behavioral adaptation, which was induced by desensitization, and audiosensitization were accompanied by profound aberrations in intracellular calcium distribution, par-

[3] Idoniboye-Obu, B. 1977. Recording bioelectric action potentials of marine decapod crustacea by remote electrodes: A bioassay procedure for monitoring hydrocarbon pollution. Environ. Pollut. 12:159–166.

[4] Idoniboye-Obu, B. 1977. Bioelectric action potentials of *Procambarus acutus acutus* (Girrard) in serially diluted solutions of selected C_6 hydrocarbons in water. Environ. Pollut. 14:5–24.

[5] Iturrian, W. B., and H. D. Johnson. 1971. Conditioned seizure susceptibility in the hamster induced by prior auditory exposure. Experientia 27:1193–1194.

ticularly in brain and heart mitochondria (Diebel *et al.*, 1975).[6] Desensitization decreased calcium binding capacity of the soluble fraction and increased microsomal protein concentration of neural and cardiac tissue. Additional alterations occurred in the cellular metabolism of phosphorus and certain drugs. Although the animals appeared to adapt behaviorally, desensitized hamsters had hyperactive locomotor activity scores and precocious sexual development.

Termination of ss resulted not from loss of audiosensitivity, but through development of more prominent inhibitory systems (Iturrian and Fink, 1969).[7] Desensitizing treatment apparently potentiates a physiological adaptative process that inhibits ss. Similar inhibitory mechanisms occur during adaptation following alcohol withdrawal (Comer and Iturrian, 1973).[8] Reserpine tranquilized the control rodents. In contrast, it blocked the effects of both desensitization and passive adaptation in audiosensitized hamsters and mice causing them to convulse (Iturrian and Johnson, 1975).[9] Other strains of mice fail to adapt to audiosensitization and remain susceptible to seizures for months.

The model is sensitive to aircraft noise (Etheredge *et al.*, 1970)[10] and detects an interaction between noise and low level X-irradiation (Iturrian and Peacock, 1972).[11] Audiosensitization and desensitization provide valuable biochemical and behavioral models with which to evaluate effects of acute auditory stress and the process(es) of adaptation.

Alteration of the Immune Response of Channel Catfish (*Ictalurus punctatus*) by Polychlorinated Biphenyls

DANIEL H. JONES, DONALD H. LEWIS, THOMAS E. EURELL, and MARVIN S. CANNON

Polychlorinated biphenyls (PCB's) are now recognized as being the most ubiquitous halogenated hydrocarbon pollutants in our environment. These compounds exist in the environment in low concentrations, but because of

[6] Diebel, D. R., C. M. Mokler, and W. B. Iturrian. 1975. Subcellular calcium distribution and seizure susceptibility in audiosensitized and desensitized hamster. Ga. J. Sci. 33:91–92.

[7] Iturrian, W. B., and G. B. Fink. 1969. Influence of age and brief auditory conditioning upon experimental seizure in mice. Dev. Psychobiol. 2:10–18.

[8] Comer, C. P., III, and W. B. Iturrian. 1973. Effects of sound on alcohol withdrawal and recovery mechanism. Pharmacologist 15:159.

[9] Iturrian, W. B., and H. D. Johnson. 1975. Infantile auditory exposure and unusual response to antipsychotic drugs. Proc. Soc. Exp. Biol. Med. 148:219–223.

[10] Etheredge, G. N., J. S. Gibson, L. R. Mills, and W. B. Iturrian. 1970. Extra-auditory effects of noise in immature mice. Bull. Ga. Acad. Sci. 28:43.

[11] Iturrian, W. B., and L. J. Peacock. 1972. Low-level X-radiation and audio-sensitization seizure. Neurosci. Abstr. II:250.

their chemical stability and lipid solubility, they undergo biomagnification at each trophic level of the food chain. This results in tissue residue levels in fish thousands of times greater than that found in the surrounding environment. Little is known, however, regarding the effects of chemical pollutants on the immune response of fish exposed to such chemicals.

Each of 12 channel catfish (*Ictalurus punctatus*) (six controls and six experimentals) was individually held in 120-l aquaria under controlled environmental conditions. A commercial PCB, Aroclor 1232, was dissolved in oil and administered intraperitoneally at a dosage of 70 mg/kg. Control fish received an equivalent amount of oil. Blood samples were taken from each fish and electrophoretic profiles and agglutinating antibody titers to the bacterium *Aeromonas hydrophila* were determined prior to administration of either Aroclor 1232 or oil, 1 week following administration, and prior to challenge. Following an immunization period of 3 weeks, the fish were challenged with 10^3 of the virulent organism *A. hydrophila*.

Although all immunized fish demonstrated agglutinating titers (240^{-1} or greater) to the bacteria, 100% mortality was observed in the PCB-treated fish, while immunized control fish were not affected. Examination of immunoelectrophoretic profiles of PCB-treated fish using a crossed immunoelectrophoresis technique revealed a significant decrease in beta globulin levels, which was not observed in control fish. Alpha globulin levels were also found to be slightly elevated in PCB-treated fish as compared to control fish.

All PCB-treated fish demonstrated sufficient agglutinating antibody titers to protect them from the virulent challenge. To ascertain whether the PCB's depressed cellular immunity, fish were treated with Aroclor 1232 at a dosage of 70 mg/kg intraperitoneally, while control fish received an equivalent amount of oil. Twenty-four hours later all fish were challenged with *A. hydrophila* (10^3 virulent organisms). Nine hours following the challenge, peritoneal washings were taken from fish in each of the two groups. Fifty macrophages were counted and the number of macrophages containing engulfed bacteria were recorded. PCB-treated fish had fewer macrophages containing engulfed bacteria than did the control fish (1.3 vs. 4.0), suggesting a depressed cellular immunity.

Vitamin A Effect on Lung Cell Cycle Kinetics after Exposures of Hamsters (*Mesocricetus auratus*) to Nitrogen Dioxide

JAMES C. S. KIM

Hamsters (*Mesocricetus auratus*) are widely used in experiments on inhalation and respiratory carcinogenesis because histological features of the hamster respiratory tract adequately resemble those found in humans and because spontaneous infection and tumors are rather rare. In addition, a vitamin A deficiency can be readily created in hamsters with distinct clinical symptoms by administering a special diet.

Among the irritant air pollutant gases, including ozone and sulfur dioxide, nitrogen dioxide is a biohazard of great importance because of its presence not only in air pollutant gases but also in cigarette smoke. Characteristic epithelial cell hyperplasia and hypertrophy can be induced in the lung with a single exposure of 10 ppm of nitrogen dioxide for 5 h. Although inhibitory effects of vitamin A in chemically induced colon and lung cancers have been demonstrated, the role of vitamin A and its possible modifying role in relation to pollutant gases have not been explored.

The effect of vitamin A on hamster lung cell kinetics after exposure to nitrogen dioxide has been measured by autoradiography, liquid scintillation counting, and electron microscopic and histologic techniques. We have demonstrated the definite cell kinetic effects of vitamin A in cell repair and injury by the use of more than 200 hamsters. When exposed to nitrogen dioxide gas, both acutely deficient and extremely high-dosed animals showed reduced cell regeneration. Alveolar epithelial cell regeneration was retarded. In contrast, more widespread epithelial regeneration occurred in animals maintained with high doses (500 μg) of dietary vitamin A. However, this regeneration pattern was dependent on animal age and individual variation. Other nutritional factors such as fatty acids and vitamin E, which were present in each animal, might also influence overall utilization of vitamin A *in vivo*.

The study was supported by a National Institute of Environmental Health Sciences Research Grant (ESO-1166-0A1) and by the Environmental Protection Agency.

Newborn Calf Losses Associated with Energy Conversion Facilities in North Dakota

DONALD H. HASTINGS

Selenium deficiency is unexpected in North Dakota, a state generally regarded as having adequate amounts of selenium in its soil and its livestock diets. A considerable number of stillborn and weak calves appeared in a herd of 400 beef cows that wintered at a ranch 1.6 km from a thermoelectric plant and oil refinery. A second ranch 9.6 km from another thermoelectric complex was similarly affected. The dead calves were found to have myopathy resembling white muscle disease, which is reported to be associated with a dietary deficiency of selenium, a trace element that is an important component of the cellular enzyme, glutathione peroxidase.

Calf losses encountered at each ranch were reversed 24 to 48 h following an injection of a selenium pharmaceutical and are now prevented by the feeding of wheat or wheat bran, an excellent source of selenium, during the last 60 days of pregnancy.

Thermoelectric plants, which burn lignite coal, and oil refineries produce large quantities of sulfur dioxide. The growing alfalfa plant is capable of absorbing sulfur dioxide through its leaf stomata.[12] This results in the presence of high levels of sulfate in the alfalfa, which is consumed by the cattle. Studies have shown that ingested sulfate depresses the level of selenium in ruminants.[13] It is noteworthy that both ranches were located in areas prone to forage fumigation by breakup atmospheric inversion. Analysis of alfalfa at these ranches showed high sulfate and normal selenium levels.

Investigators are continuing to find methods of preventing the problem by monitoring forages or blood of the dams. Further studies are being conducted to find better methods of diagnosing the marginally selenium-deficient calf. Also under investigation are the roles of other influences, such as stress and concurrent trace element deficiencies or excesses.

This study was partially funded by the North Dakota Beef Commission.

Alkylmercurial Encephalopathy in Nonhuman Primates

ROBERT H. GARMAN and HUGH L. EVANS

Twenty squirrel (*Saimiri sciureus*) and 12 macaque monkeys (*Macaca arctoides* and *M. nemestrina*) were examined clinically and histologically after being on acute, subacute, or chronic dosage regimens of methyl mercury. The primary site of brain damage in these nonhuman primates was the cerebral cortex, although neuron degeneration was occasionally observed in the corpus striatum, thalamus, hypothalamus, brainstem, and cerebellar nuclei. In the cerebral cortex, there was eosinophilic neuron degeneration characterized by nuclear pyknosis and retracted, brightly eosinophilic cytoplasm. An associated microgliocytosis was evident as was a prominent enlargement (both nuclear and cytoplasmic) of astrocytes in the vicinity of the eosinophilic neurons. Eventually, only large gemistocytic astrocytes remained in heavily damaged areas of cortex.

One half of each monkey brain was dissected into various anatomical regions—9 for the squirrel monkeys and 16 for the macaques. The concentration of mercury in these regions was determined by scintillation counting (^{203}Hg) or atomic absorption. Degree of neuron damage was not correlated with

---

[12] Thomas, M. D., R. H. Hendricks, and G. R. Hill. 1950. Sulfur metabolism of plants; effects of sulfur dioxide on vegetation. Ind. Eng. Chem. 42:2231–2235.

[13] Underwood, E. J. 1977. P. 324 in Trace Elements in Human and Animal Nutrition, 4th ed. Academic Press, New York, N.Y.

mercury concentration. As one example, two areas of highest mercury concentration were the calcarine cortex and the lateral geniculate body. While the former region was highly susceptible to neuron degeneration, the latter rarely contained evidence of change.

Mercury localization at the cellular level was determined by a photographic emulsion histochemical technique identical to that for standard autoradiography. This revealed the presence of large amounts of mercury in certain large neurons (e.g., within the gasserian and dorsal root ganglia), which were histologically normal. In areas of neuron damage, little mercury was found in the neurons, but large amounts were present within the cytoplasm of astrocytes. Heavy concentrations of mercury were also detected within the cytoplasmic granules of perivenular mast cells in many areas of the brain.

These studies suggest that the astrocyte may be the primary target cell for methyl mercury, that some neurons are apparently able to acquire large amounts of intracellular mercury without cytologic change, and that the function of the brain's mast cells may alter after intoxication by methyl mercury.

Chromate Inhibition of Mucus Glycoprotein Secretion by Rat Tracheal Explants

JEROLD LAST

Rat tracheal explants were incubated in tissue culture medium containing various concentrations of sodium chromate. There was a dose-related inhibition of mucus glycoprotein secretion by the explants at all concentrations of sodium chromate evaluated between 0.03 mM and 2.16 mM. Plots of the amount of glycoprotein secreted versus the log of chromate concentration suggested that two types of inhibition occurred, one of which was observed only at higher concentrations of sodium chromate in the medium (≥ 0.5 mM). At concentrations of 0.5 mM and above, we also observed inhibition of precursor (D-(^{3}H)glucosamine) uptake and concurrent morphological damage in the epithelium of the tracheal slices. The histopathology findings included damage to ciliated cells, severe epithelial desquamation, and severe epithelial cell nuclear pyknosis.

We conclude from these experiments that we can confirm other studies[14] indicating that the *in vitro* incubation of tracheal slices in medium containing chromate ions causes morphologically observable damage to the epithelial cells, especially the ciliated cells; that concentrations of chromate sufficient to

[14] Mass, M. J., and B. P. Lane. 1976. Effect of chromates on ciliated cells of rat tracheal epithelium. Arch. Environ. Health 31:96–100.

elicit morphological damage also inhibit uptake of low molecular weight precursors of mucus glycoproteins into tracheal slices and also inhibit secretion of glycoproteins by such slices; that we can detect a hitherto unrecognized mode of action of the chromate ion, i.e., inhibition of glycoprotein secretion at concentrations that do not elicit morphological damage; and that determination of glycoprotein secretion by tracheal explants offers a sensitive, quantitative technique for studying the effects of soluble air and water pollutants on airway metabolism.

Current efforts are directed toward correlating these results with those obtained after incubation of tracheal explants from rats exposed to sodium chromate aerosols *in vivo*.

Environmental Pollutants: Whole Animal vs. *In Vitro* Models

JURI LINASK

To evaluate toxicity of trace chemical pollutants by exposing laboratory animals to the highest tolerated dose is time-consuming, expensive, and may be theoretically unreliable. As pharmacological and toxicological reactions at the tissue-cell level are quantitatively related to access of the chemical species to its site of action, critical attention must be directed toward tissue transport kinetics. For example, cellular absorption of chemical species can have linear, rectangular, hyperbolic, sigmoidal, or multiphasic relationships to the exposure level. Depending on the kinetic pattern, the highest tolerated dose evaluation mode may mask the potential toxicity or provide a false manifestation of toxicity. Since tissue interaction kinetics are complicated by excretory and humoral factors and by potentiation and competition from unknown plasma components, pollutant toxicants are best defined by closely correlating whole animal models with kinetic studies on isolated cells and organ systems. To prolong *in vitro* survival of functioning rat hearts, which could serve as models for chronic studies, trace element impurities in analytical grade reagents usually present in the 2 to 5 ppm range were established as one of the factors limiting survival.

Chelex 100 chromatography to remove heavy metal contaminants from the major media components (sodium chloride, sodium bicarbonate, potassium chloride, or Dextrose) necessitated subsequent readdition of zinc, iron, manganese, and copper in nM to μM amounts to a complex protein-free solution. The resulting perfusate permitted consistent 6- to 9-day maintenance of hearts. Preliminary X-ray fluorescence analysis of hearts perfused 5 min and 24 h indicates that other trace elements may be affecting survival, as selenium and strontium accumulated and bromine, rubidium, chromium, and nickel were washed out. This system can be used to evaluate the transport of pollutants over a prolonged period at extremely low concentrations.

Effects of Acute PCB Exposure on the Hepatic Mixed-Function Oxidative System in Channel Catfish (*Ictalurus punctatus*)

MICHAEL M. LIPSKY, JR., JAMES E. KLAUNIG, and DAVID E. HINTON

Polychlorinated biphenyls (PCB's) are widespread environmental contaminants that leave residues in terrestrial and aquatic species. Upon exposure to PCB, mammalian livers incur a marked increase in enzyme activity in the mixed-function oxidative system (MFOS). Accompanying enzyme induction is a proliferation of smooth endoplasmic reticulum and an increase in intracellular lipid. Since little is known about the effects of PCB on teleosts, this study was undertaken to determine biochemical and morphologic alterations in the livers of channel catfish (*Ictalurus punctatus*) that have been acutely exposed to PCB. Following six daily intraperitoneal injections of Aroclor 1254 (Monsanto) in mineral oil (50 mg/kg body weight), increases in microsomal MFOS components were observed. The terminal oxidase cytochrome P-450 increased 39% in PCB-treated fish and cytochrome b_5 increased twofold. Another microsomal electron transport component, NADPH cytochrome c reductase, increased twofold over controls. Microsomal protein content and the rate of aminopyrine demethylation did not differ significantly from controls, although the protein values were increased 18%.

Hepatocytes of exposed fish were enlarged with rounded profiles and increased numbers of lipid droplets. Electron microscopy revealed a number of subcellular alterations. Most prominent was proliferation of rough endoplasmic reticulum (RER) seen as increases in parallel cisternae, meandering tracts, and the appearance of circular, vesicular profiles. The RER was often dilated, containing an amorphous material of low to medium electron density. Configurations resembling "typical liposomes" were often observed. Large lipid droplets were frequently encountered, some appearing to coalesce into larger forms. Increases in smooth endoplasmic reticulum were minimal. They were restricted to foci at the cell periphery of lateral outgrowths of RER tracts. There was occasional nuclear atypia, primarily large indentations.

This work was supported by EPA grant number EPA-R-804866-01-0.

Vinyl Chloride: Ubiquitous Carcinogen

CAROLYN H. LINGEMAN

For nearly four decades, vinyl chloride (VC) has been used in ever-increasing amounts, primarily in the manufacture of plastic products. In its gaseous state, VC has been used as a propellant for aerosol spray products such as insecticides and hairsprays. Polymers of VC, usually combined with related compounds such as vinylidene chloride, have become widely distributed in the environment. vc-containing plastics are considered to be stable, but the gas can leach into beverages from plastic containers. Large amounts can be released into the air during combustion of vinyl plastics.

Inhalation of high concentrations of vc causes narcosis and death. Occupational exposures of humans to lower concentrations over a long period, particularly during synthesis of plastics, can cause fibrosis and hemangiosarcomas of the liver as well as vascular changes. Mice, rats, and hamsters exposed to vc by inhalation develop a variety of neoplasms including hemangiosarcomas of the liver. The effects of vc on other animals, including wildlife, are unknown.

Air and water near synthesizing and fabricating plants can be contaminated with vc. Thus, there is opportunity for vc to enter the food chain. Although only sparingly soluble in water, vc can be stored for long periods in body fat. In common with an ever-increasing list of man-made carcinogens, opportunities for prolonged exposure to vc exist for humans and animals. However, the full impact of such exposure probably will not be known for many years.

Experimentally Induced Effects of Methyl Mercury Exposure on the Developing Mouse

FRED C. OLSON and EDWARD J. MASSARO

Methyl mercury at 5 mg of mercury per kilogram of body weight (in phosphate buffered saline [PBS]) was administered subcutaneously, in a volume equivalent to 0.1% of body weight, to Swiss Webster CFW mice (weighing 27 to 31 g) on day 12, hour 6 (12^6) of gestation. Controls received PBS only.

In the fetuses of control animals, palate closure was completed by 14^{10}. Exposure to methyl mercury resulted in delayed palate closure in 66% of the fetuses examined on 15^6, 45% examined on 16^6, and 40% examined on 17^6.

Total fetal protein and DNA content were examined at consecutive 24-h intervals after methyl mercury administration up to 17^6. By 14^6, total protein had decreased maximally to 72% of the control level. Thereafter, it increased toward the control level. Alterations in DNA content followed a similar pattern and decreased to a maximum of 65% of the control level at 15^6.

Maternal exposure to methyl mercury decreased the rate of fetal protein synthesis. The rate of 4,5 ^{3}H-isoleucine incorporation into protein was decreased 5% at 12^9, 20% at 12^{12}, and 26% at 13^6 (end of observation). The calculated average decrease in protein synthesis (19%) between 12^{12} and 13^6 was in agreement with the measured decrease in protein content (22%) at 13^6, suggesting that reduction in the rate of protein synthesis was responsible for the decreased fetal protein content.

Administration of tritiated water at 12^{18} indicated that methyl mercury administered at 12^6 had no effect on the rate of placental blood flow or fetal water space. However, observation at 12^{18} indicated that the administration of methyl mercury at 12^6 reduced the concentration of certain fetal free amino acids (alanine, 23.0%; valine, 9.7%; methionine, 22.6%; isoleucine, 12%; leucine, 18.2%). At the same time, uptake of the nonmetabolizable amino acid, ^{14}C-cycloleucine, was decreased (19%). This suggested that the growth-inhibiting effects of methyl mercury were related, at least in part, to impaired placental/fetal transfer of amino acids.

This study was supported by contract no. 68-0201768 from the U.S. Environmental Protection Agency.

Teratogenic Effects of Carbaryl, Malathion, and Parathion on Developing Eggs of Medaka (*Oryzias latipes*)

HOWARD M. SOLOMON

The carbarmate insecticide carbaryl and the organophosphate insecticides malathion and parathion were tested for teratogenic effects on developing eggs of the medaka (*Oryzias latipes*), a Japanese freshwater fish.

Groups of 10 newly fertilized eggs were placed into dishes containing 10 ml embryo-rearing solution. The insecticide to be tested was dissolved in acetone and added to the dish. The range of insecticide concentrations employed was 0.5–30.0 ppm carbaryl, 5.0–40.0 ppm malathion, and 1.0–18.0 ppm parathion. The eggs, which were observed daily, remained exposed to the insecticide until hatching or death. In all, 658 eggs were used.

Circulatory system anomalies, including rudimentary heart development, pericardial edema, irregular heart beat, oscillation of blood between the atrium and ventricle, and clots in the intra- and extraembryonic circulatory systems, were often observed. The ED_{50} (effective dose) for parathion was 2.0 ppm; carbaryl, 2.5 ppm; and malathion, 10.0 ppm. Histological examination of representative embryos affected by the insecticide showed abnormalities only in heart structure.

Pollutants and Regeneration in Estuarine Killifish (*Fundulus* spp.) and Fiddler Crabs (*Uca* spp.)

JUDITH S. WEIS and PEDDRICK WEIS

Regeneration of caudal fins in the killifish (*Fundulus heteroclitus*) was retarded by treatment with 10 ppb DDT, malathion, parathion, or carbaryl. Although DDT was quite toxic, it did not retard regeneration as much as the other insecticides. Cadmium chloride at 0.01 and 0.1 mg/l retarded wound healing and subsequent regeneration. Although mercuric chloride had no effect at sublethal concentrations, 0.01 mg/l of methylmercuric chloride retarded regeneration significantly in the killifish (*F. confluentus*). Combinations of methyl mercury and cadmium resulted in no significant retardation. This indicates an inhibitory effect of these two chemicals on each other. The effect of methyl mercury was reduced when the salinity of the water was decreased.

Regeneration in the fiddler crabs (*Uca pugilator*) and (*U. pugnax*) was studied after single or multiple autotomy of limbs and exposure to insecticides or heavy metals. Exposure to DDT at 10 and 25 ppb accelerated the rate of limb regeneration and, in animals with multiple autotomy, shortened the time to ecdysis. These responses may be the result of heightened excitation of the nervous system and the resulting secretion of neuroendocrine factors promoting limb regeneration and molting. Parathion, malathion, and carbaryl, while toxic at 100 ppb, showed no specific effect on regeneration at 10 or 100 ppb. Cadmium chloride at 0.1 and 1.0 mg/l retarded the regeneration rate, but mercuric chloride inhibited regeneration only at 1.0 mg/l (not 0.1 mg/l), which proved toxic to many crabs. Methylmercuric chloride, however, inhibited regeneration at 0.5 mg/l and inhibited melanogenesis in the regenerates at lower concentrations (0.1 mg/l).

Research on the Effects of Environmental Pollutants in Animals

EDWIN I. PILCHARD

For this overview, the Current Research Information System (CRIS)[15] and the Smithsonian Science Information Exchange (SSIE)[16] were selected as the two major sources of information on research in progress on the effects of pollution on animals. Individual researchers were also contacted directly.

[15] Current Research Information System (CRIS), U.S. Department of Agriculture, Washington, D.C. 20250.
[16] Smithsonian Science Information Exchange (SSIE), Inc., 1730 M St., N.W., Washington, D.C. 20036.

TABLE 4   Research on Effects on Animals of Pollution from
Power Plants, 1975

Region of United States	Plant	Number of Sites	Animals	Pollutants
North Central	Coal	7	Fish: invertebrates	Heat, chlorine
South	Hydroelectric	1	Largemouth bass	Heat
	Coal	2	Fish	Heat, chlorine
	Nuclear	1	Several mammalian species	Not specified
West	Fossil fuel	2	Rodents, primates, freshwater biota	Effluent
	Nuclear	6	Albacore, clams, crabs	Plutonium, anti-foulants; heat
North East	Nuclear	2	Dairy cattle	Effluent

Energy is one important category. Pollution from energy generation presents a surfeit of problems to society and the scientific community. Energy from automotive gasoline containing lead concerns environmentalists and has stimulated studies of roadside biospheres.

Electrical power generation is an even greater energy-related concern. A query of SSIE and CRIS produced a list of seven in-progress investigations on the effects on animals of pollution from power plants (Table 4). These studies, which are being conducted throughout the United States, primarily involve aquatic animal life. In one study, investigators are describing effects from exposure to low-level radioactive substances in thyroid glands of dairy cattle that are downwind from two nuclear power plants near Lake Ontario.

Efforts to obtain an accurate picture of the scope and intensity of current research in the effects of environmental pollutants on animals have shown an essential need for research project information that is readily accessible to those needing it. A summary of the active research that is described by CRIS and SSIE shows that a wide range of pollution-related problems are being investigated in animals. Few of the projects appear to be supported adequately. The scientific field of pathology accounts for only one of every six fulltime-equivalent scientist years in studies of the effects of pollution on animals.

Conclusion

DANTE SCARPELLI

On behalf of the Organizing Group, I thank you all for attending and participating in this symposium. The National Academy of Sciences, the Northeastern Research Center for Wildlife Diseases, and the Registry of Comparative Pathology, by their joint support and sponsorship of the symposium, acknowledge the importance and pervasiveness of environmental pollutants and their impact on the biosphere, generally, and on wildlife, specifically. The charge to the committee was to organize a symposium to highlight the potential of wildlife as models for the detection and study of the effects of environmental pollutants and, hopefully, to verify their pathological effects in the laboratory. The nature and quality of the papers presented and the ensuing comments and discussion in the 3 days of this meeting would suggest that the committee was successful.

The papers presented covered a broad spectrum of research approaches employing a variety of methodologies and techniques. A number of the contributors reviewed and analyzed both laboratory and field studies of the interactions of pollutants with flora and fauna in such detail that it became obvious that many of these interactions are multifactorial and extremely complex. The deleterious effects of crude oil and related compounds on aquatic birds, fish, and invertebrates continue to occupy the attention of environmental scientists. From what we have heard here, the effects of such substances are not only acute, often fatal lesions, but also latent and chronic. For example, in certain molluscs and flatfish there is evidence that prolonged exposure to crude oil, their derivatives, and certain carcinogenic polycyclic

397

hydrocarbons such as benzo[a]pyrene may be responsible for the development of a variety of proliferative lesions, some of which possess morphological and biological features indicative of neoplasia. In one laboratory study, exposure to the carcinogens was carried out under controlled conditions of dosage and frequency to verify field studies where such data are not obtainable.

Heavy metal poisoning, which is one of the earliest recognized examples of the impact pollutants have made on humans and other animals, still remains an important topic. The application of modern analytical methods and of clinical enzymology is particularly encouraging, since such approaches will undoubtedly lead to an earlier diagnosis of a problem in the field and a more rapid institution of measures to control the pollution.

Many environmental pollutants not only caused demonstrable degenerative changes, but also were identified as either carcinogens or teratogens. Examples of environmentally related cancers were described in various species living in close contact with products of industry. It appeared that aquatic animals such as oysters and other bivalves were particularly prone to develop tumors, either because of contact with higher concentrations of carcinogens or because of higher susceptibility. Hatchability and development of fish and amphibian eggs were economical and excellent bioindicators of teratogens in aquatic pollutants.

Participants and Coauthors

A. ABOLINS, College of Fisheries, Fisheries Research Institute, University of Washington, Seattle, Washington 98195

E. N. ALBERT, Department of Anatomy, George Washington University, 2300 Eye Street, N.W., Room 426 Ross Hall, Washington, D.C. 20037

JAMES R. ALLEN, Department of Pathology and Experimental Pathology Unit, Regional Primate Research Center, University of Wisconsin, Madison, Wisconsin 53706

COLIN ANDERSON, Department of Pathology, Health Sciences Centre, The University of Western Ontario, London, Ontario, Canada N6A 5C1

LOUISE ANDERSON, Department of Biological Sciences, University of Illinois at Chicago Circle, Chicago, Illinois 60680

C. HAROLD BAER, Fish and Wildlife Service, National Fish and Wildlife Laboratory, U.S. Department of the Interior, Madison, Wisconsin 53706

SUDIP K. BANERJEE, Department of Biochemistry, University College of Science, Calcutta 700 019 India

DEBORAH A. BARSOTTI, Department of Pathology and Experimental Pathology Unit, Regional Primate Research Center, University of Wisconsin, Madison, Wisconsin 53706

WESLEY J. BIRGE, School of Biological Sciences, University of Kentucky, Lexington, Kentucky 40506

JEFFREY A. BLACK, School of Biological Sciences, University of Kentucky, Lexington, Kentucky 40506

JAMES A. BOGAN, Department of Veterinary Pharmacology, University of Glasgow Veterinary School, Bearsden Road, Bearsden, Glasgow, Scotland G61 1QH

CHRIS W. BROWN, Departments of Animal Pathology and Chemistry and Oceanography, University of Rhode Island, Kingston, Rhode Island 02881

ROBERT S. BROWN, Departments of Animal Pathology and Chemistry and Oceanography, University of Rhode Island, Kingston, Rhode Island 02881

M. S. CANNON, College of Medicine, Texas A&M University, College Station, Texas 77843

CHARLES C. CAPEN, Department of Veterinary Pathobiology, The Ohio State University, Columbus, Ohio 43210

LAURINE A. CARSTENS, Department of Pathology and Experimental Pathology Unit, Regional Primate Research Center, University of Wisconsin, Madison, Wisconsin 53706

CELESTER CARTER, Division of Pathology, Food and Drug Administration, Bureau of Foods, Washington, D.C. 20204

HAROLD W. CASEY, Department of Pathology, Armed Forces Institute of Pathology, Washington, D.C. 20306

G. LEW CHOULES, Environment and Ecology Branch, Environmental and Life Sciences Division, Dugway Proving Ground, Dugway, Utah 84022

DANIEL COHEN, Section of Epidemiology and Public Health, Department of Clinical Studies, School of Veterinary Medicine, University of Pennsylvania, Philadelphia, Pennsylvania 19174

WILLIAM T. COLLINS, Department of Veterinary Pathobiology, The Ohio State University, Columbus, Ohio 43210

JOHN A. COUCH, U.S. Environmental Protection Agency, Gulf Breeze Environmental Research Laboratory, Gulf Breeze, Florida 32561

LEE A. COURTNEY, U.S. Environmental Protection Agency, Gulf Breeze Environmental Research Laboratory, Gulf Breeze, Florida 32561

K. D. DANYLCHUCK, Department of Pathology, University of Western Ontario, London, Ontario, Canada N6A 5C1

THOMAS S. DAVIES, Ayerst Research Laboratories, Animal Health Division, Chazy, New York 12921

A. S. W. DEFREITAS, Division of Biological Sciences, National Research Council of Canada, Ottawa, Canada K1A 0R6

P. V. DEHADRAI, Central Inland Fisheries Research Institute, All India Coord. Project on Airbreathing Fish Culture, Barrackpore, DT 24 Parganas, West Bengal 743101, India

MICHAEL P. DIETER, Building 31, Room C-31, National Institute on Aging, Bethesda, Maryland 20014

KENNETH R. DIXON, Appalachian Environmental Laboratory, Frostburg State College, Frostburg, Maryland 21532

JEFFREY X. DUGGAN, Department of Biological Sciences, University of Illinois at Chicago Circle, Chicago, Illinois 60680

GERALD N. ESRA, Los Angeles City Zoo, Los Angeles, California 90033

T. E. EURELL, College of Veterinary Medicine, Texas A&M University, College Station, Texas 77843

HUGH L. EVANS, University of Rochester Medical Center, Rochester, New York 14627

S. P. FELTON, College of Fisheries, Fisheries Research Institute, University of Washington, Seattle, Washington 98195

G. W. FERGUSON, President, University of Connecticut, Storrs, Connecticut 06268

MARY F. FINLAY, Benedict College, Harden and Blanding Streets, Columbia, South Carolina 29204

STEVEN A. FOSS, U.S. Environmental Protection Agency, Gulf Breeze Environmental Research Laboratory, Gulf Breeze, Florida 32561

GLEN A. FOX, National Wildlife Research Center, Department of Fisheries and Environment, Ottawa, Ontario, Canada K1A 0H3

LEONARD FRIEDMAN, Division of Toxicology, Food and Drug Administration, Bureau of Foods, Washington, D.C. 20204

MILTON FRIEND, Fish and Wildlife Service, National Fish and Wildlife Health Laboratory, U.S. Department of the Interior, 1655 Linden Drive, Madison, Wisconsin 53706

ROBERT H. GARMAN, University of Rochester Medical Center, Rochester, New York 14627

LARRY H. GARTHOFF, Division of Toxicology, Food and Drug Administration, Bureau of Foods, Washington, D.C. 20204

M. A. J. GIDNEY, Division of Biological Sciences, National Research Council of Canada, Ottawa, Ontario, Canada K1A 0H3

ANDREW P. GILMAN, National Wildlife Research Centre, Canadian Wildlife Service, Department of Fisheries and Environment, Ottawa, Ontario, Canada K1A 0H3

EARL WAYNE GROGAN, Institute of Laboratory Animal Resources, National Research Council, 2101 Constitution Avenue, N.W., Washington, D.C. 20418

MAX A. HAEGLE, Fish and Wildlife Service, U.S. Department of the Interior, Madison, Wisconsin 53706

D. J. HALLETT, National Wildlife Research Centre, Canadian Wildlife Service, Department of Fisheries and Environment, Ottawa, Ontario, Canada K1A 0H3

WANDA M. HASCHEK, Oak Ridge National Laboratory, Union Carbide Corporation, Nuclear Division, P.O. Box Y, Oak Ridge, Tennessee 37830

DONALD H. HASTINGS, Midway Veterinary Clinic, P.O. Box 911, Bismarck, North Dakota 58501

DR. JOYCE W. HAWKES, Northwest Alaska Fisheries Center, National Marine Fisheries Service, Seattle, Washington 98112

CHARLES J. HENNY, Patuxent Wildlife Research, Pacific Northwest Field Station, Corvallis, Oregon 97330

DAVID E. HINTON, Medical Center School of Medicine, West Virginia University, Morgantown, West Virginia 26506

GERALD L. HOFFMAN, Environmental Research Laboratory, U.S. Environmental Protection Agency, Narragansett, Rhode Island 02882

EDWIN B. HOWARD, Section of Comparative Pathology, Los Angeles County, University of Southern California, 12824 Erickson Avenue, Downey, California 90242

RICK HUDSON, Fish and Wildlife Service, National Fish and Wildlife Health Laboratory, U.S. Department of the Interior, Madison, Wisconsin 53706

BIRINENGI IDONIBOYE-OBU, College of Science and Technology, Port Harcourt, Nigeria

W. B. ITURRIAN, Pharmacology Department, University of Georgia, Athens, Georgia 30601

WAYNE IWAOKA, College of Fisheries, Fisheries Research Institute, University of Washington, Seattle, Washington 98195

D. H. JONES, College of Veterinary Medicine, Texas A&M University, College Station, Texas 77843

B. S. JORTNER, Department of Pathology, New Jersey Medical School, Rutgers University, 195 University Avenue, Newark, New Jersey 07102

LOUIS KASZA, Division of Pathology, Food and Drug Administration, Bureau of Foods, Washington, D.C. 20204

DONALD F. KELLY, Department of Pathology, University of Bristol, The Medical School, University Walk, Bristol, England BS8 1TD

NARVIE KENNEDY, Route 2, Box 477, Jefferson, South Carolina 29718

JAMES C. S. KIM, Delta Primate Research Center, Three Rivers Road, Covington, Louisiana 70433

KEITH I. KING, Department of General Science, Oregon State University, Corvallis, Oregon 97331

WILLIAM B. KINTER, Mount Desert Island Biological Laboratory, Salsbury Cove, Maine 04672

JAMES E. KLAUNIG, Department of Pathology, School of Medicine, University of Maryland, 31 S. Greene Street, Baltimore, Maryland 21201

GERALD KOLAJA, Veterinary Medicine Division, Biomedical Laboratory, Department of the Army, Edgewood Arsenal, Aberdeen Proving Ground, Maryland 21010

LOREN D. KOLLER, School of Veterinary Medicine, Oregon State University, Corvallis, Oregon 97331

MARSHA L. LANDOLT, College of Fisheries, Fisheries Research Institute, University of Washington, Seattle, Washington 98195

JEROLD A. LAST, California Primate Research Center, University of California, Davis, California 95616

DONALD H. LEWIS, College of Veterinary Medicine, Texas A&M University, College Station, Texas 77843

DR. JURI LINASK, Bockus Institute, Graduate Hospital, 19th and Lombard Streets, Philadelphia, Pennsylvania 19146

CAROLYN H. LINGEMAN, National Cancer Institute, National Institutes of Health, Bethesda, Maryland 20014

MICHAEL M. LIPSKY, JR., Department of Pathology, School of Medicine, University of Maryland, 31 S. Greene Street, Baltimore, Maryland 21201

DONALD J. LISK, Department of Pathology, New York State College of Veterinary Medicine, Cornell University, Ithaca, New York 14853

CLAYTON G. LOOSLI, Department of Pathology, School of Medicine, University of Maryland, Los Angeles, California 90033

V. M. LUCKE, Department of Pathology, University of Bristol, The Medical School, University Walk, Bristol, England BS8 1TD

EDWARD J. MASSARO, Department of Biochemistry, State University of New York, Buffalo, New York 14214

DENNIS L. MEEKER, Fish and Wildlife Service, National Fish and Wildlife Health Laboratory, U.S. Department of the Interior, Madison, Wisconsin 53706

GEORGE MIGAKI, Registry of Comparative Pathology, Armed Forces Institute of Pathology, Washington, D.C. 20306

DAVID S. MILLER, Mount Desert Island Biological Laboratory, Salsbury Cove, Maine 04672

MICHAEL C. MIX, Department of General Science, Oregon State University, Corvallis, Oregon 97331

D. G. MORGAN, Department of Pathology, University of Bristol, The Medical School, University Walk, Bristol, England BS8 1TD

P. K. MUKHOPADHYAY, Central Inland Fisheries Research Institute, All India Coord. Project on Airbreathing Fish Culture, Barrackpore, DT 24 Paraganas, West Bengal 743101, India

BRIAN D. MURPHY, Computer Sciences Division, Oak Ridge National Laboratory, Union Carbide Corporation, Nuclear Division, Oak Ridge, Tennessee 37830

JAMES R. NEWMAN, Environmental Science and Engineering, Inc., P.O. Box 13454, University Station, Gainesville, Florida 32604

I. NEWTON, Institute of Terrestrial Ecology, 12 Hope Terrace, Edinburgh EH9 2AS, Scotland

SVEND W. NIELSEN, Department of Pathology and Northeastern Research Center for Wildlife Diseases, University of Connecticut, Storrs, Connecticut 06268

R. J. NORSTROM, National Wildlife Research Center, Canadian Wildlife Service, Department of Fisheries and Environment, Ottawa, Ontario, Canada K1A 0H3

ARLAND E. OLSON, Department of Veterinary Science, Utah State University, Logan, Utah 84322

FRED C. OLSON, Department of Biochemistry, State University of New York, Buffalo, New York 14214

DAVID B. PEAKALL, National Wildlife Research Centre, Canadian Wildlife Service, Department of Fisheries and Environment, Ottawa, Ontario, Canada K1A 0H3

HOWARD B. PETERSON, Department of Agricultural and Irrigation Engineering/UMC41, Utah State University, Logan, Utah 84322

A. PHILLIPS, Benedict College, Harden and Blanding Streets, Columbia, South Carolina 29204

K. PIERSON, College of Fisheries, Fisheries Research Institute, University of Washington, Seattle, Washington 98195

EDWIN I. PILCHARD, Emergency Programs, U.S. Department of Agriculture, Federal Building, Room 759, 6505 Belcrest Road, Hyattsville, Maryland 20782

JOHN S. REIF, Section on Epidemiology and Public Health, Department of Clinical Studies, School of Veterinary Medicine, University of Pennsylvania, Philadelphia, Pennsylvania 19174

RICHARD V. RHEINBERGER, Environmental Research Laboratory, U.S. Environmental Protection Agency, Narragansett, Rhode Island 02882

J. RUSSELL ROBERTS, Division of Biological Sciences, National Research Council of Canada, Ottawa, Ontario, Canada K1A 0R6

DOUGLAS E. ROSCOE, Northeastern Research Center for Wildlife Diseases, University of Connecticut U-89, Storrs, Connecticut 06268

W. C. RUSSELL, Environment and Ecology Branch, Environmental and Life Sciences Division, Dugway Proving Ground, Dugway, Utah 84022

SAUL B. SAILA, Departments of Animal Pathology and Chemistry and Oceanography, University of Rhode Island, Kingston, Rhode Island 02881

PATRICK F. SCANLON, Department of Fisheries and Wildlife Sciences, Virginia Polytechnic Institute and State University, Blacksburg, Virginia 24061

DANTE G. SCARPELLI, Department of Pathology, School of Medicine, Northwestern University, 303 E. Chicago Avenue, Chicago, Illinois 60611

PATRICK J. SHEA, Pacific Southwest Forest and Range Experiment Station, U.S. Forest Service, Davis, California 95616

F. J. SCHOENFELD, Utah State Veterinarian, B-45 State Capital Building, Salt Lake City, Utah 84114

JAMES L. SHUPE, Veterinary Sciences Department UMC56, Utah State University, Logan, Utah 84322

STUART D. SLEIGHT, Department of Pathology, School of Veterinary Medicine, Michigan State University, East Lansing, Michigan 48824

HOWARD M. SOLOMON, Stein Research Center, Jefferson Medical College, 920 Chancellor Street, Philadelphia, Pennsylvania 19107

R. A. STEHN, New York Cooperative Wildlife Research Unit, Fernow Hall, Cornell University, Ithaca, New York 14853

SHERMAN F. STINSON, National Cancer Institute, National Institutes of Health, Bethesda, Maryland 20014

C. THOMAS, Benedict College, Harden and Blanding Streets, Columbia, South Carolina 29204

STEVE R. TRENHOLM, Department of General Science, Oregon State University, Corvallis, Oregon 97331

BENJAMIN F. TRUMP, School of Medicine, University of Maryland, 31 S. Greene Street, Baltimore, Maryland 21201

J. P. VAN MILLER, Department of Pathology and Experimental Pathology Unit, Regional Primate Research Center, University of Wisconsin, Madison, Wisconsin 53706

MORRIS A. WEINBERGER, Division of Pathology, Food and Drug Administration, Washington, D.C. 20204

JUDITH S. WEIS, Department of Zoology and Physiology, Rutgers University, 195 University Avenue, Newark, New Jersey 07102

PEDDRICK WEIS, Department of Anatomy, New Jersey Medical School, Newark, New Jersey 07103.

A. G. WESTERMAN, School of Biological Sciences, University of Kentucky, Lexington, Kentucky 40506

C. R. WILPIZESKI, Thomas Jefferson University Medical Center, Philadelphia, Pennsylvania 19107

JAMES T. WINSTEAD, U.S. Environmental Protection Agency, Gulf Breeze Environmental Research Laboratory, Gulf Breeze, Florida 32561

RICHARD E. WOLKE, Departments of Animal Pathology and Chemistry and Oceanography, University of Rhode Island, Kingston, Rhode Island 02882

PAUL P. YEVICH, Environmental Research Laboratory, U.S. Environmental Protection Agency, Narragansett, Rhode Island 02882

DAVID YOUNG, Southern California Coastal Water Research Project, 1500 East Imperial, El Segundo, California 90245

JOSEPH G. ZINKL, Department of Clinical Pathology, University of California, Davis, California 95616

BERNARD C. ZOOK, Animal Research Facility, George Washington University Medical Center, 2300 Eye Street, N.W., Washington, D.C. 20037

Attendees

RANDIE E. ABOUD, Registry of Comparative Pathology, Armed Forces Institute of Pathology, Washington, D.C. 20306

ROBERT S. ANDERSON, Sloan–Kettering Institute for Cancer Research, 145 Boston Post Road, Rye, New York 10580

RICHARD ARIMOTO, Biological Sciences Group, University of Connecticut U-42, Storrs, Connecticut 06268

JAMES BETHUNE, Department of Natural Resources Conservation, University of Connecticut U-87, Storrs, Connecticut 06268

JOEL E. BODAMMER, National Marine Fishing Service, Oxford Laboratory, Oxford, Maryland 21654

MALIN B. BONNETT, National Marine Fisheries Service, Auke Bay Laboratory, P.O. Box 155, Auke Bay, Alaska 99821

DANIEL BRANSTETTER, Aberdeen Proving Ground, Maryland 21010

EVERETT BRYANT, Department of Pathobiology, University of Connecticut U-89, Storrs, Connecticut 06268

CARROLL BURKE, Department of Pathobiology, University of Connecticut, Storrs, Connecticut 06268

GISELA CLEMONS, Lawrence Berkeley Laboratory, Building 74, Berkeley, California 94720

STEVEN COHEN, School of Pharmacy, University of Connecticut U-92, Storrs, Connecticut 06268

RICHARD DIETERS, Department of Pathobiology, University of Connecticut U-89, Storrs, Connecticut 06268

THEODORE GIRSHICK, Department of Pathobiology, University of Connecticut U-89, Storrs, Connecticut 06268

RICHARD GRILLO, Biological Sciences Group, University of Connecticut U-42, Storrs, Connecticut 06268

DAVID GROMAN, Department of Pathobiology, University of Connecticut U-89, Storrs, Connecticut 06268

MICHAEL HANNON, N.E. Center for Zoonosis, 43 Colonial Road, Beacon, New York 12508

MARGARET HARRIS, San Mateo Animal Hospital, 2320 Palm Avenue, San Mateo, California 94403

TOM HARRIS, San Mateo Animal Hospital, 2320 Palm Avenue, San Mateo, California 94403

WILLIAM HAWKINS, Department of Anatomy, University of South Alabama, Mobile, Alabama 36688

RICHARD HERBST, ERT, 696 Virginia Road, Concord, Massachusetts 01742

WILLIAM G. HUBER, College of Veterinary Medicine, Washington State University, Pullman, Washington 99164

RICHARD JAKOWSKI, Department of Pathobiology, University of Connecticut, Storrs, Connecticut 06268

JOHN JACKANGELO, New Jersey Division of Fish, Game and Shell Fisheries, P.O. Box 1809, Trenton, New Jersey 08625

DON KINSMANN, Department of Animal Industries, University of Connecticut U-40, Storrs, Connecticut 06268

E. J. KUZIA, New York State Department of Environmental Conservation, Field Pesticide Research Unit, 8314 Fish Hatchery Road, Rome, New York 13440

LINDA LOWENSTINE, Harvard Medical School, N.E. Regional Primate Research Center, 1 Pine Hill Drive, Southboro, Massachusetts 01772

D. A. MAIO, Air Force Office of Scientific Research, Bolling Air Force Base, Washington, D.C. 20332

J. MARKOFSKY, Orentreich Foundation for the Advancement of Science, Inc., 910 5th Avenue, New York, New York 10021

KATHRYN H. MARTIN, SUNY at Oswego, 326 Piez Hall, Oswego, New York 13126

R. MASAKE, Department of Veterinary Pathology, Cornell University, Ithaca, New York 14853

NANCY A. MUCKENHIRN, Institute of Laboratory Animal Resources, National Academy of Sciences, 2101 Constitution Avenue, N.W., Washington, D.C. 20418

DENNIS O'CONNOR, Department of Pathobiology, University of Connecticut U-89, Storrs, Connecticut 06268

CLARK OLSON, North Carolina State University, Raleigh, North Carolina 27607

JETTY F. PAYNE, Fisheries and Marine, Newfoundland Biology Station, 3 Water Street East, St. John's, Newfoundland

LOUIS PIERRO, Department of Animal Genetics, University of Connecticut U-92, Storrs, Connecticut 06268

ROBERT PIROZOK, Department of Pathobiology, University of Connecticut U-89, Storrs, Connecticut 06268

SUSAN PORTEUS, Humane Society of the United States, 2100 L Street, N.W., Washington, D.C. 20037

WILLIAM A. PRIESTER, Clinical Epidemiology Branch, National Cancer Institute, National Institutes of Health, Bethesda, Maryland 20014

B. RAU, Department of Pathobiology, University of Connecticut U-89, Storrs, Connecticut 06268

LARRY RENFRO, Biological Sciences, University of Connecticut U-42, Storrs, Connecticut 06268

TERRY RETTIG, Veterinary Services Section, Bureau of Wildlife, New York State Department of Environmental Conservation, Wildlife Resources Center, Delmar, New York 12054

WAYNE H. RISER, School of Veterinary Medicine, University of Pennsylvania, 3800 Spruce Street, Philadelphia, Pennsylvania 19174

GEORGE W. ROBINSON, 969 Thayer Avenue, No. 2, Silver Spring, Maryland 20910

SOREN ROSENDAL, Department of Pathobiology, University of Connecticut U-89, Storrs, Connecticut 06268

ZEKE RUBEN, Department of Pathobiology, University of Connecticut U-89, Storrs, Connecticut 06268

JACK SCHULTZ, Biological Sciences, University of Connecticut U-42, Storrs, Connecticut 06268

JOSEPH SIMON, College of Veterinary Medicine, University of Illinois, Urbana, Illinois 61801

LETA STROLLE, Department of Pathobiology, University of Connecticut U-89, Storrs, Connecticut 06268

THANE THURMOND, Department of Pathobiology, University of Connecticut U-89, Storrs, Connecticut 06268

HELEN TRYPHONAS, Health and Welfare Canada, Health Protection Branch, Tunney's Pasture, Ottawa, Ontario, Canada K1A 0L2

LEANDER TRYPHONAS, Division of Toxicology, Bureau of Chemical Safety, Food Directorate, National Health and Welfare, Ottawa, Ontario, Canada K1A 0L2

JOHN WATSON, Biological Sciences Group, University of Connecticut U-42, Storrs, Connecticut 06268

CAMMY WATTS, ERT, 696 Virginia Road, Concord, Massachusetts 01742

LEE WILLIAMS, Department of Pathobiology, University of Connecticut U-89, Storrs, Connecticut 06268

Author Index

409

Subject Index

411